Neural Bases of Motor Behaviour

NATO ASI Series

Advanced Science Institutes Series

A Series presenting the results of activities sponsored by the NATO Science Committee, which aims at the dissemination of advanced scientific and technological knowledge, with a view to strengthening links between scientific communities.

The Series is published by an international board of publishers in conjunction with the NATO Scientific Affairs Division

A Life Sciences **B Physics**	Plenum Publishing Corporation London and New York
C Mathematical and Physical Sciences **D Behavioural and Social Sciences** **E Applied Sciences**	Kluwer Academic Publishers Dordrecht, Boston and London
F Computer and Systems Sciences **G Ecological Sciences** **H Cell Biology** **I Global Environmental Change**	Springer-Verlag Berlin, Heidelberg, New York, London, Paris and Tokyo

PARTNERSHIP SUB-SERIES

1. **Disarmament Technologies**	Kluwer Academic Publishers
2. **Environment**	Springer-Verlag / Kluwer Academic Publishers
3. **High Technology**	Kluwer Academic Publishers
4. **Science and Technology Policy**	Kluwer Academic Publishers
5. **Computer Networking**	Kluwer Academic Publishers

The Partnership Sub-Series incorporates activities undertaken in collaboration with NATO's Cooperation Partners, the countries of the CIS and Central and Eastern Europe, in Priority Areas of concern to those countries.

NATO-PCO-DATA BASE

The electronic index to the NATO ASI Series provides full bibliographical references (with keywords and/or abstracts) to more than 50000 contributions from international scientists published in all sections of the NATO ASI Series.
Access to the NATO-PCO-DATA BASE is possible in two ways:

– via online FILE 128 (NATO-PCO-DATA BASE) hosted by ESRIN,
Via Galileo Galilei, I-00044 Frascati, Italy.

– via CD-ROM "NATO-PCO-DATA BASE" with user-friendly retrieval software in English, French and German (© WTV GmbH and DATAWARE Technologies Inc. 1989).

The CD-ROM can be ordered through any member of the Board of Publishers or through NATO-PCO, Overijse, Belgium.

Series D: Behavioural and Social Sciences – Vol. 85

Neural Bases of Motor Behaviour

edited by

Francesco Lacquaniti

Istituto di Fisiologia umana,
Università di Cagliari,
Cagliari, Italy
and
Istituto Scientifico S. Lucia,
Rome, Italy

and

Paolo Viviani

Faculté de Psychologie et des Sciences de l'Education,
Université de Genève,
Carouge, Switzerland
and
Dipartimento di Scienze Cognitive,
Fondazione San Raffaele,
Milan, Italy

Kluwer Academic Publishers

Dordrecht / Boston / London

Published in cooperation with NATO Scientific Affairs Division

Proceedings of the NATO Advanced Study Institute on
Multi-Sensory Control of Movement
Trieste, Italy
3–12 July 1994

A C.I.P. Catalogue record for this book is available from the Library of Congress.

ISBN 0-7923-4009-4

Published by Kluwer Academic Publishers,
P.O. Box 17, 3300 AA Dordrecht, The Netherlands.

Kluwer Academic Publishers incorporates the publishing programmes of
D. Reidel, Martinus Nijhoff, Dr W. Junk and MTP Press.

Sold and distributed in the U.S.A. and Canada
by Kluwer Academic Publishers,
101 Philip Drive, Norwell, MA 02061, U.S.A.

In all other countries, sold and distributed
by Kluwer Academic Publishers Group,
P.O. Box 322, 3300 AH Dordrecht, The Netherlands.

Printed on acid-free paper

Printed in the Netherlands

TABLE OF CONTENTS

FOREWORD

This volume contains the proceedings of the NATO Advanced Study Institute (ASI) "Multisensory Control of Movement" held at SISSA, Trieste (Italy), July 3-12, 1994. The ASI took the format of a Summer School that we organized with the collaboration of a scientific committee formed by A. Berthoz (Paris), H. Collewijn (Rotterdam), H. Heuer (Dortmund), M. Jeannerod (Lyon), and J.F. Stein (Oxford).

The School was attended by sixty students selected from twice as many applicants from fourtheen countries. Their contribution to the discussions, and the presentation of their own ongoing research organized in informal sessions proved very stimulating.

The main focus of the School was on brain mechanisms involved in multisensory control of movement. This broad topic was addressed with a multidisciplinary approach, integrating contributions from neuroanatomy, neurophysiology, psychophysics, neurology, and cognitive science. This choice is reflected in the content of the present volume.

Lacquaniti reviews recent advances in the field of motor control, with a special emphasis on the role of reference frames and coordinate transformations. Visuomotor channels are discussed by Jeannerod within the context of the control of object-oriented actions. Johnson, Ferraina and Caminiti describe the functional architecture of cortical networks involved in reaching under visual guidance. Fogassi, Gallese, Fadiga and Rizzolatti present new data on space coding in inferior premotor cortex (area F4). The import of neurological findings for understanding human motor control is illustrated by Freund. Stein reviews the role of posterior parietal cortex and cerebellum in the visual guidance of movement, also in relation with dyslexia. Petit, Orssaud, Tzourio, Mazoyer and Berthoz illustrate the neural circuits involved in the control of saccadic eye movements, as revealed by neuroimaging techniques. Roucoux, Missal, Crommelinck and Lefevre review the role of the superior colliculus in fast and slow gaze orientation. The role of cortico-collicular paths in the control of arm movements is discussed by Stuphorn, Bauswein and Hoffmann. Johansson covers extensively the role of afferent and memory information in the control of manipulation. Heuer illustrates the behavioral approach to the study of timing control in skilled movements. Viviani illustrates some of the conceptual difficulties involved in modelling the representation of motor competences.

In addition to the institutional support by the NATO Scientific Affairs Division, the School was also funded by a EEC Human Capital and Mobility grant awarded to TECS-SISSA, Trieste. Moreover, travel grants to some U.S. student participants were awarded by the National Science Foundation (USA). We gratefully acknowledge all these sources of support.

The smooth running of the School would not have been possible without the valuable assistance of Lidia Bogo, Rosanna De Jurco and Rosanna Sain who managed to solve both expected, and unexpected problems arising in the course of the event.

Francesco Lacquaniti *Paolo Viviani*

LECTURERS AND CONTRIBUTORS

(The contributors who did not attend the ASI are indicated by an asterisk)

*E. BAUSWEIN, Allgemeine Zoologie und Neurobiologie, Ruhr-Universitaet Bochum, Universitaetstrasse 150, D-44780 Bochum *(Germany)*

A. BERTHOZ, Laboratoire de Physiologie de la Perception et de l'Action, CNRS-Collège de France, 15, rue de l'Ecole de Médécine, 75270 Paris *(France)*

R. CAMINITI, Istituto di Fisiologia umana, Università La Sapienza, piazzale Aldo Moro 5, 00185 Roma *(Italy)*

*M. CROMMELINCK, Laboratoire de Neurophysiologie, Université Catholique de Louvain, avenue Hippocrate 54, 1200 Bruxelles *(Belgium)*

*L. FADIGA, Istituto di Fisiologia umana, Università di Parma, via Gramsci 14, 43100 Parma *(Italy)*

*S. FERRAINA, Istituto di Fisiologia umana, Università La Sapienza, piazzale Aldo Moro 5, 00185 Roma *(Italy)*

*L. FOGASSI, Istituto di Fisiologia umana, Università di Parma, via Gramsci 14, 43100 Parma (ITALY)

H-J. FREUND, Neurologische Klinik, Universitaet Duesseldorf, Moorenstrasse 5, DW-4000 Duesseldorf *(Germany)*

*V. GALLESE, Istituto di Fisiologia umana, Università di Parma, via Gramsci 14, 43100 Parma *(Italy)*

H. HEUER, Institut fur Arbeitsphysiologie, Ardeystrasse 67, W-4600 Dortmund *(Germany)*

K-P. HOFFMANN, Allgemeine Zoologie und Neurobiologie, Ruhr-Universitaet Bochum, Universitaetstrasse 150, D-44780 Bochum *(Germany)*

M. JEANNEROD, Vision et Motricité, INSERM U94, 16, avenue du Doyen Lépine, F-69500 Bron *(France)*

R. JOHANSSON, Fysiologiska Institutionen Umea Universitet, S-901 87 Umea *(Sweden)*

*P.B. JOHNSON, Department of Anatomy and Cell Biology, University of North Carolina, Chapel Hill, NC 27599 *(USA)*

F. LACQUANITI, Istituto Scientifico S. Lucia, via Ardeatina 306, 00179 Roma *(Italy)*

*P. LEFEVRE, Centre for systems engineering and applied mechanics, CESAME, Université Catholique de Louvain, 4, avenue G. Lemaitre, 1348 Louvain-la-Neuve *(Belgium)*

*B. MAZOYER, Groupe d'Imagerie Neurofonctionelle, Service Hospitalier F. Joliot, DRM-CEA, Orsay *(France)*

M. MISSAL, Laboratoire de Neurophysiologie, Université Catholique de Louvain, avenue Hippocrate 54, 1200 Bruxelles *(Belgium)*

*C. ORSSAUD, Laboratoire de Physiologie de la Perception et de l'Action, CNRS-Collège de France, 15, rue de l'Ecole de Médécine, 75270 Paris *(France)*

*L. PETIT, Laboratoire de Physiologie de la Perception et de l'Action, CNRS-Collège de France, 15, rue de l'Ecole de Médécine, 75270 Paris *(France)*

G. RIZZOLATTI, Istituto di Fisiologia umana, Università di Parma, via Gramsci 14, 43100 Parma *(Italy)*

A. ROUCOUX, Laboratoire de Neurophysiologie, Université Catholique de Louvain, avenue Hippocrate 54, 1200 Bruxelles *(Belgium)*

J. F. STEIN, Physiology Department, Oxford University, Parks Road, Oxford OX1 3PT *(United Kingdom)*

*V. STUPHORN, Allgemeine Zoologie und Neurobiologie, Ruhr-Universitaet Bochum, Universitaetstrasse 150, D-44780 Bochum *(Germany)*

*N. TZOURIO, Groupe d'Imagerie Neurofonctionelle, Service Hospitalier F. Joliot, DRM-CEA, Orsay *(France)*

P. VIVIANI, University of Geneva, FAPSE, 9, route de Drize, 1227 Carouge *(Switzerland)*

STUDENT PARTICIPANTS

G. BAUD-BOVY, University of Geneva, FAPSE, Carouge *(Switzerland)*

H. BEKKERING, Department of Movement Sciences, University of Limburg *(Netherlands)*

J.K. BOLINE, Department of Physiology/Brain Science Center, University of Minnesota, Minneapolis, MN *(USA)*

B. BOLHUIS VAN, Department of Medical Physics & Biophysics, Nijmegen *(Netherlands)*

E. BRENNER, Department of Physiology, Erasmus University, Rotterdam *(Netherlands)*

P. BRUNO, University of Trieste, DEEI, Trieste *(Italy)*

C. BUNEO, Department of Physiology, University of Minnesota, Minneapolis, MN *(USA)*

M. BURSTEDT, Department of Physiology, University of Umea, Umea *(Sweden)*

A. CASABONA, Institute of Human Physiology, Catania *(Italy)*

V. CHATURVEDI, Department of Medical Physics & Biophysics, Nijmegen *(Netherlands)*

Y. COELLO, Department of Psychology, Edinburgh, Scotland *(U.K.)*

S. COLLINS, MRC Human Movement & Balance Unit, London WCI 3BG *(U.K.)*

M. COLEMAN, Department of Psychology, Edimburgh, Scotland *(U.K.)*

P. CONTRERAS-VIDAL, Arizona State University, Tempe, AZ *(USA)*

E. DAPRATI, Istituto di Fisiologia umana, Parma *(Italy)*

C. DE' SPERATI, Department of Psychology, Università of Torino, Torino *(Italy)*

M. DESMURGET, INSERM U94 Vision et Motricité - 69500 Bron *(France)*

C. DOORENBOSCH, Faculty of Human Movement Sciences, Free University of Amsterdam *(Netherlands)*

S. DUBOIS, Laboratory of Neurophysiology, Faculty of Medecine UCL, 1200 Brussels *(Belgium)*

N. GANTCHEV, CNRS-NBM, Marseille *(France)*

A. GEMMERT VAN, Department of Experimental Psychology, Nijmegen *(Netherlands)*

L. GOFFART, INSERM U94 Vision et Motricité, 69500 Bron *(France)*

S. GOODBODY, MRC Human Movement & Balance Unit, London We l 3BG *(U.K.)*

J. GOOSSENS, Department of Medical Physics and Biophysics, Nijmegen *(Netherlands)*

J. GRAAF DE, INSERM U94 Vision et Motricité, 69500 Bron *(France)*

P. GRANDGUILLAUME, INSERM CREARE, Université Pierre et Marie Curie, Paris *(France)*

P. GRIBBLE, Department of Psychology, McGill University, Montreal, Quebec *(Canada)*

C. GROENINGEN VAN, Human Physics, University of Utrecht, Utrecht *(Netherlands)*

A. GRUART, Laboratory of Neuroscience, University of Seville, Seville *(Spain)*

E. GUIGON, INB-CNR, 20131 Milano *(Italy)*

S. HANNETON, LPPA-CNRS, Collège de France, Paris *(France)*

D. HECK, Max-Planck Institute for Biological Cybernetics, 72076 Tubingen *(Germany)*

P. KALUZNY, Institute of Physiology, University of Fribourg, 1700 Fribourg *(Switzerland)*

O. KAZENNIKOV, Institute of Physiology, University of Fribourg, 1700 Fribourg *(Switzerland)*

K. KESSLER, Neurology Department, 40225 Duesseldorf *(Germany)*

A. LEDEBT, Ecole des Hautes Etudes en Sciences Sociales PRI Apprentissage, Cognition et Culture, 75006 Paris *(France)*

C. MALAPANI, Fédération de Neurologie et INSERM U1289, Hôpital de la Salpétrière, Paris (*France*)

J. MCINTYRE, LPPA-CNRS, Collège de France, Paris *(France)*

M. MISSAL, Laboratoire de Neurophysiologie, Université Catholique de Louvain, 1200 Bruxelles *(Belgium)*

T.G. NIELSEN, Department of Medical Informatics and Image Analysis, Aalborg University *(Denmark)*

D. OPITZ, MPI Biological Cybernetics, 72076 Tubingen *(Germany)*

E. OTTAVIANO, Laboratory of Bioengineering, University of Pavia *(Italy)*

L. OUT, Faculty of Human Movement Sciences, Free University, Amsterdam *(Netherlands)*

N. PORT, University of Minnesota, Minneapolis, MN *(USA)*

X.H. QI, Brain Research Institute, Zurich University, Zurich *(Switzerland)*

Y. ROSSETTI, Vision et Motricité, INSERM U94, 69500 Bron *(France)*

J. SANGALS, IfADo, Dortmund *(Germany)*

M. SANTELLO, Applied Physiology Research Unit, School of Sport and Exercise Sciences, University of Birmingham *(U.K.)*

S. SHANBHAG, Neurosurgery Lab./Prog. in Neuroscience, University of Minnesota, Minneapolis, MN *(USA)*

J.P. SPENCER, Infant Motor Dev. Lab., Indiana University, Bloomington, IN *(USA)*

F. SULTAN, MPI of Biological Cybernetics, 72076 Tubingen *(Germany)*

S. SWINNEN, Motor Control Laboratory, K.U. 3001 Heverlee *(Belgium)*

I. TONI, Istituto di Fisiologia umana, Parma *(Italy)*

M. TRULSSON, Umea University, Department of Physiology, Umea *(Sweden)*

J. VAAL, Human Movements Sciences, Free University, Amsterdam *(Netherlands)*

P. VINDRAS, FAPSE, University of Geneva, Carouge *(Switzerland)*

J. VOLKMANN, Department of Neurology, University of Duesseldorf, 40255 Duesseldorf *(Germany)*

P. WEISS, Department of Neurology, University of Duesseldorf, 40255 Duesseldorf *(Germany)*

X.H. YU, Institute of Physiology, Fribourg University *(Switzerland)*

F. ZAAL, Faculty of Human Movement Sciences, Free University, Amsterdam *(Netherlands)*

CONTROL OF MOVEMENT IN THREE-DIMENSIONAL SPACE

F. LACQUANITI
Istituto di Fisiologia umana, Università di Cagliari, 09124 Cagliari; and Istituto Scientifico S. Lucia I.N.B.-C.N.R., 00179 Rome

1. Introduction

Goal-directed movements require accurate perception of the spatial location of complex stimuli and the translation of this spatial information into motor commands adequate for specific effectors. Several sensory systems (e.g., visual, auditory, vestibular, tactile, proprioceptive systems) contribute spatial information, and most efficient localization of the stimuli is generally accomplished by integrating multimodal sensory information (Knudsen and Brainard 1995). On the other hand, movements often involve the coordination of several body segments, as in orienting the eyes, head and arm together toward a target (Jeannerod 1988).

Although multisystemic sensorimotor coordination is advantageous for best achievement of the behavioral goal, it creates a potential babel of spatial codes at the level of the Nervous System. At the periphery, each sensory and motor apparatus is endowed with its own anatomical geometry and functional specialization. This is reflected in a correspondingly large diversity of modes of neural processing along the pathways that originate from each sensor or that are directed to each effector. Thus, a general problem of sensorimotor coordination is that multiple and diverse forms of encoding of spatial information co-exist at the input and output stages. Ultimately, however, this heterogeneity of spatial representations needs to be reconciled at the level of the Central Nervous System (CNS), in order to set a congruent correspondence of information across all sensory and motor channels, correspondence that is required for most goal-directed movements.

How is then spatial information for movement represented centrally ? What is computed and where in the brain ? These issues have received considerable attention in recent years (cf. Andersen et al. 1993; Arbib 1991; Bizzi et al. 1991; Georgopoulos 1991; Jeannerod 1991; Kalaska and Crammond 1992; Lacquaniti 1989; Paillard 1991; Pellionisz and Llinás 1980; Robinson 1992; Simpson and Graf 1985; Soechting and Flanders 1992; Stein 1992). One view is that at several brain sites spatial information is encoded at the neuronal level in terms of changes of activity along specific coordinate axes. According to this view, there exist multiple representations of the movement in different allocentric (i.e., world-based) or egocentric (i.e.,

F. Lacquaniti and P. Viviani (eds.), Neural Bases of Motor Behaviour, 1–40.

body-based) reference frames; distinct processing stages transform spatial information between such frames to create unitary representations in supramodal frames. According to an alternative view, central neurons do not explicitly code spatial information in any specific reference frame; instead, space codes are distributed over a whole neuronal population.

Irrespective of the specific neural mechanisms involved, however, coordinate transformations are always implicit in sensorimotor coordination. Representations of space and movement in well defined reference frames are an emergent property at the systems level, and their existence is reflected at the level of overt behaviour. In this chapter we discuss these issues by selecting cases in point from different areas of neurophysiology.

2. Representing spatial information for movements: statement of the logical problem

2.1 APPLICATION OF GEOMETRIC CONCEPTS TO THE DESCRIPTION OF MOVEMENT GENERATION

2.1.1 Where to move and how to move. The sequence of events taking place in movement generation can be described in simple geometric terms. Consider the case of pressing a switch with a finger. The location of the switch must be specified relative to the external world and to our body. This external location, that could be initially derived by sensory cues (say, visual or auditory) or cognitive cues (memory or imagery), represents the desired final position of the finger when it lands on target. Location (and force) at the contact target represents the final result of the motor action; its specification answers the question of "*where* do we want to move". However, this does not suffice to describe the movement. *How* do we get at the target ? Movement of the finger in space generally requires coordinated motion of several joints of the arm and hand. Only after the desired angular motion of each joint is specified, is it possible to compute the torque necessary to produce the corresponding joint motion. Finally, muscle contractions must be apportioned to produce the computed joint torques.

2.1.2 Frames of reference and coordinate transformations. Each step in this sequence of computations can be described in a reference frame. We know from geometry that a point in space is defined by the set of coordinates relative to an assigned frame of reference. In principle, many such frames of reference can be used. Thus, finger location can be identified by the corresponding coordinates (for instance, cartesian or polar coordinates) relative to a given origin, placed on the body or in the external world. Contact forces are similarly defined by using force values instead of position values along the coordinate axes. Rotations and torques at the joints are defined instead in the intrinsic space of the corresponding angular coordinates (e.g. flexion-extension, abduction-adduction etc). Finally, the pulling directions of the individual muscles identify the coordinate axes for describing muscle forces. Each of these sets of geometrical variables defines a possible space of representation of the movement. Because movements unfold simultaneously in the extrinsic space and in the intrinsic space, a process of coordinate transformations converts desired finger movement into joint rotations, and the latter into joint torques and muscle forces.

2.2 DOES THE BRAIN REPRESENT MOVEMENTS IN SPECIFIC REFERENCE FRAMES?

The fact that sensorimotor coordination can be formally described in reference frames does not imply that the brain encodes movements necessarily in such a way. In general we may ask in which cases it could be convenient *a priori* for a biological system to encode spatial information in reference frames, and what evidence we have that this type of code is used.

2.2.1 Local representations. With respect to the first question, one can note that reference frames are useful to encode information pertaining to a homogeneous space. Although physical space is normally homogeneous, there are many cases in which phenomenologically distinguishable points exist that make the space practically inhomogeneous. Spatial inhomogeneity due to physical texture or differential cognitive significance can create a variety of landmarks which allow local references to be established in the environment. Moreover, discrete categorization, instead of continuous geometrical coordinates, could be used to classify items in spatial terms. In typing, for instance, the location of a keybutton to be pressed could be defined in "keyboard space" in terms of proximity relations with other keys (for instance, the letter H is to the left and above the letter N). In a different example, a stimulus applied to a point of the body surface could be localized relative to anatomical landmarks (such as joints, nails, lips etc) in a topological map of the body scheme. Note that, because the skin is deformable and textured, the geometrical structure of this map may differ drastically from that of a rigid homogeneous space endowed with a fixed reference frame. In both these examples, addresses or labels could be conveniently used instead of the geometrical coordinates of the target.

2.2.2 Global representations. Local representations can be used to define intrinsic spatial properties of an object or body part. However, they do not easily generalize in the construction of a more global representation of peripersonal space, as when separate objects must be located relative to each other and to our body. Furthermore, local topologies might be conveniently used for encoding spatial information within one sensory (or motor) domain, but they do not easily transfer to other sensory channels or motor effectors. Thus, tactile texture is often quite different from visual texture in the previous example of a stimulus applied to the skin, and the local representations within the somatosensory and visual domains would be correspondingly different.

Global representations endowed with a fixed reference frame, on the other hand, allow this transfer. For instance, orienting toward a visible, audible, touchable target involves at the input stage a sensory code in a retinotopic map for the visual cues, in a head-centered map for the auditory cues, and in a somatotopic map for the tactile cues. The possibility of transforming these diverse maps in a common frame (say, referenced to the head) would set multisensory information in register, ensuring sensor-fusion and efficient sensorimotor coordination.

2.2.3 Coordinate-free representations. Intermediate representations in supramodal reference frames might then be useful to solve the problem of coordinate transformations. However, they are not the only solution to the problem; there exist alternative computational solutions that are coordinate-free. For instance, neural networks have been recently introduced in neurosciences to simulate coordinate transformations involved in movement generation (Burnod et al 1992; Zipser and Andersen 1988; Anastasio and Robinson 1990). These networks learn to associate the required movement to a given sensory stimulus by changing progressively the synaptic weights of the inner units (*hidden layer*) until the desired goal is achieved. After learning, the tuning functions of the hidden layer can approximate closely those described electrophysiologically at specific brain sites, but they may not reflect any specific coordinate system (Robinson 1992). This is because the initial connectivity pattern is random, and the final tuning will be different from one implementation to the next, and will be idiosyncratic to each neuron. This example has not been chosen to suggest that the brain works as these artificial neural networks (in fact, some of the rules implemented in neural networks are hardly plausible from a biological standpoint), but only to indicate the existence of alternative computational schemes. Thus, even in the case in which representations in geometric terms exist at the sensory input and at the motor output, they may not exist at central neural stages involved in sensorimotor integration.

2.2.4 Experimental strategies for the study of reference frames. What criteria need be satisfied to propose that specific forms of neural representations or neural processing reflect a given reference frame ? It could be suggested that, from the observed behaviour, a specific geometrical structure should be recognizable and differentiable from that associated with alternative reference frames. Thus external measurements of the variable under investigation (the psychophysical determination of perceptual localization of a stimulus, or movement analysis of motor performance, or recording of neural activity at a given brain site) should be skewed toward the coordinate axes of this hypothetical frame of reference.

Let us consider a few examples related to the task of pointing to a target with the hand in three-dimensional (3D) space. If one considers repeated measurements and finds that the variances of the spatial distribution of the final positions are uncorrelated along the coordinate axes centred at the head, then one is entitled to propose that a head-centred frame is used for this form of sensorimotor coordination, as opposed to hand-centred or shoulder-centred frames, for instance. One might find also that the pattern of cell activity recorded from a given brain area varies systematically along one or more of the three coordinate axes centred at the head, the optimal directions of activation of individual neurons tending to cluster around such axes.

In the following we consider a number of different sensory and motor systems. In each case we address the problem of spatial representations at two different levels of analysis: 1) at the level of the global behaviour of the system, and 2) at the level of neural coding at specific sites of the CNS. It will become readily apparent that in some cases neural operations at single sites may not conform with any specific frame, but spatial representations and coordinate transformations are an emergent property at the systems level. Different approaches need then be considered, encompassing psychophysics in behavioral studies, electrophysiological recording in neurophysiology, and modelling in computational studies.

3. Arm movements

Goal-directed movements of the arm rely on multisensory information that must be transformed into motor commands adequate to attain the aim. Multimodal integration is necessary to take into account the position of the target relative to the eyes, head, and arm. These interactions may result in intermediate neural representations that are coded neither in pure sensory coordinates nor in pure motor coordinates (Carrozzo and Lacquaniti 1994; Flanders et al. 1992). In the following we review first behavioral studies in man that suggest the existence of multiple representations in different reference frames, and we then consider some of the neural substrates involved in sensorimotor coordination of arm movements.

3.1 BEHAVIORAL STUDIES

Three main categories of goal-directed actions will be considered to illustrate the role of reference frames in the control of kinematics (motion) and kinetics (forces): reaching, drawing, and catching movements.

3.1.1 REACHING MOVEMENTS

Object prehension and manipulation involve a complex sequence of kinematic events, with a transport phase of the arm toward the object, a phase of hand preshape to match object size and shape, and a grasping phase. Only the transport phase (reaching) will be considered here, as hand preshape and grasping are discussed in the chapter by Jeannerod (see also Jeannerod 1988; Jeannerod et al. 1995). Point-to-point reaching consists of one or more discrete events in time, with an initial acceleratory phase and a final deceleratory phase. These movements normally describe relatively simple (often straight) paths in space, and are performed by means of coordinated angular motion at the shoulder, elbow and wrist joints.

We consider first the problems of how the desired final position of the hand is computed and which reference frames are putatively used for this computation, and then we consider the modular organization of the motor commands used for specifying final position. Finally, we examine briefly some problems related to the specification of the entire trajectory of the arm and its kinematics from movement inception to end.

3.1.1.1 REFERENCE FRAMES FOR COMPUTING MOTOR ERROR

To reach for a target in space, the brain must establish a correspondence between target position and final hand position. When the target is presented visually, its location is encoded in eye-centred coordinates, initially. Horizontal position (azimuth) and vertical position (elevation) of the target are mapped topographically on the retina, whereas distance is estimated by means of binocular cues (retinal stereodisparity and oculomotor vergence) and monocular cues (shading, texture etc). The vector of desired hand movement is defined by the difference between target location and initial hand location. We call motor error of the hand this vector difference.

This motor error cannot be easily derived in retinotopic coordinates because of the following reasons. First, orienting gaze (eye and head) toward the target at variable times prior to and during reaching changes the retinotopic map of the motor error in complex ways, but contributes important retinal and extraretinal information. (Extraretinal information depends on both proprioception and efference copy of gaze movement.) Moreover, hand position can be encoded visually also, but outside the field of view it is defined by proprioception and by efferent copy of the motor commands to arm muscles. How is then motor error for reaching computed ?

3.1.1.1.1 Shoulder-centred frame. The analysis of human pointing to visually presented, remembered targets in 3D space has led to the following hypothesis concerning visuomotor transformations for reaching (Flanders et al. 1992). Target location is defined visually in a spherical coordinate system (distance, azimuth and elevation) centered at the eyes. The origin of the spherical representation is subsequently translated, resulting in a head-centred and then a shoulder-centred representation of the target (Soechting and Flanders 1989a). This shift takes into account progressively the position of the eyes in the orbits and of the head relative to the trunk, in order to define target location with respect to the pivot of the arm (the shoulder) even during gaze changes. In the subsequent stage of the scheme, the intended, final arm position is computed from a linear transformation of the target (shoulder-centered) coordinates into the angular coordinates of the arm used for arm proprioception. Errors in pointing would arise because this linear transformation is only approximately correct, the exact transformation being nonlinear (Soechting and Flanders 1989b). (These errors would be normally corrected by visual feedback.) Finally, the desired movement vector is generated by subtracting the values of the initial angular coordinates of the arm, as derived by proprioception, from the values of the intended, final angular coordinates, as derived by the visuomotor transformation.

3.1.1.1.2 Body scheme in a geotropic reference. Limb proprioceptors (muscle spindles, skin and joint receptors) monitor variables, such as changes in muscle length, changes in strain of the skin and joint capsule, that are related to the anatomical angles of the joints (Gandevia and Burke 1992). Psychophysical evidence indicates that this proprioceptive information is integrated within the CNS to generate a perception of the angles of orientation of the limb segments in space (Soechting and Ross 1984). In particular, the orientation of the upper arm or forearm is identified by the respective angles of horizontal rotation (yaw) and vertical elevation, defined with respect to the vertical axis and the sagittal plane (Soechting and Ross 1984; Worringham et al. 1987). The existence of an absolute reference for limb proprioception is consistent with the hypothesis put forth by Paillard (1971) that egocentric reference frames are geotropic, in other words they are based on the earth invariant within which animals have been moulded, the gravitational vertical.

Thus, arm reaching is planned within the more general scheme of the spatial relations of the body segments relative to each other and to the external environment. In this body scheme, the vertical is normally estimated using a combination of labyrinthine gravitoinertial, visual, and somatosensory cues (see section 5.1.2). Because the head is effectively stabilized in space under several conditions (the orbito-meatal line is maintained within ±10° from the horizontal), it provides an inertial platform for monitoring both gravity direction and visual reference lines (Pozzo et al. 1990). Somatosensory information contributes to defining an internal estimate of the idiotropic vector aligned with the long axis of the body (Jeannerod and Biguer 1987; Mittelstaedt 1983). The orientation of the body can then bias the perceived vertical and result in misreaching in darkness (Jeannerod and Biguer 1987).

3.1.1.1.3 Hand-centred frame. The shoulder-centred frame of reference proposed by Flanders et al. (1992) does not represent the only possible frame used to reaching for a visual target. It has been argued that motor error can be also computed in a hand-centered frame, for instance when the hand enters in the field of view (Flanders et al. 1992). Gordon et al. (1994) have suggested that pointing performance in two-dimensional (2D) space is best explained in a hand-centered frame even when vision of the hand is prevented, because the distribution of final positions of the hand changes as a function of the starting position.

3.1.1.1.4 Head-centred frame. Although reaching movements can be planned in shoulder-centred or hand-centred coordinates by compensating for gaze position, this compensation may not be complete. First, it is known that pointing with the eyes immobile at variable azimuth angles (the subjects are not allowed to foveate the peripheral target) results in systematic deviations in the same direction as gaze deviation, suggesting a retinocentric contribution to the estimate of motor error (Vanden Abeele et al. 1993). Second, pointing with the head rotated by variable azimuth angle errs in a direction opposite to that of head rotation (Fookson et al. 1994;

Gentilucci et al.). One possible source of the head position effects may depend on proprioceptive signals derived from the neck muscles. Indeed, vibration of neck muscles, which activates muscle spindle afferents similarly to head rotation, can cause an illusory sense of head rotation and of motion relative to the head of a stationary visual target (Biguer et al. 1988). When asked to point to the stimulus after neck vibration, subjects mispoint in the perceived direction of motion (Biguer et al. 1988; Roll et al. 1991).The same stimuli may reduce the neglect of visual stimuli in patients with hemiattentional neglect (Karnath et al. 1993).

All these results pertain to reaching for a visual target. Altogether, they suggest that eye and head position can affect the internal representation of target position relative to the body midline. A surprising instantiation of the role of the head-centred frame comes from an experiment involving the manual exploration of the 3D-orientation of an elongated bar with the eyes closed (Carrozzo and Lacquaniti 1995). Even though vision plays no role in the perception of target orientation in this task, head rotation or vibration of neck muscles affect significantly the ability of subjects to match the perceived orientation of the object with their arm.

3.1.1.1.5 Object-centred frames. Context-related factors may influence the choice of the coordinate system used to plan a goal-directed movement. Thus, the sets of coordinates may correspond to the degrees of freedom of target position, when only a subset of the spatial coordinates of the target are unknown a priori. Quite often, targets are not simple points in neutral space, but objects on a textured background. Local, object-centred frames may then correspond to figure-ground ensembles (see section 2.2.1). For instance, the spatial location of a target is misperceived in the presence of a surrounding rectangular frame presented asymmetrically in the visual field (Bridgeman 1991).

3.1.1.1.6 A hierarchical organization of multiple reference frames. From the evidence reviewed above, one can conclude that the reference frame in which the motor error is computed is not unique, but may change as a function of the task and of the prevailing sensory or cognitive conditions (cf. Jeannerod 1991; Paillard 1982). When a sequence of movements is performed, different frames could be used in different movement segments (Berthoz 1991). The initial, ballistic phase of reaching could be planned in body-centred (head-centred or shoulder-centred) coordinates; the final phase of reaching involves accurate homing on the target under direct visual guidance and could rely on a motor error computed in hand-centered or object-centred coordinates.

These multiple frames are not entirely independent of each other, but there exists an ordered structure within which different movement representations are presumably reconciled. In this context, it has been hypothesized that the ensemble of different reference frames is hierarchically organized (Carrozzo and Lacquaniti 1995; McIntyre et al. in preparation; Berthoz 1991; Jeannerod 1991; Paillard 1991). According to this hypothesis, the head-centred frame would receive a higher priority in the hierarchy, and would keep the alignment among different frames (e.g., for the left- and right-side of the body) by adjusting the relative gain or offset of visuo-motor and kinestho-motor transformations, for instance by biasing the estimate of the idiotropic midsagittal plane. Alignment in register of different frames is made possible also by the use of a geotropic reference, that is shared by many subsystems (such as the shoulder-centred and head-centred frames).

3.1.1.2 MODULAR ORGANIZATION

There is growing evidence that, irrespective of the specific frame of reference used to plan the movement, the different spatial parameters of reaching are not treated in a unitary manner, but are processed in parallel and largely independent of each other (see Georgopoulos 1991). Thus, misreaching in direction is distinct from misreaching in distance, the latter being generally much greater than the former (Gordon et al. 1994; Soechting and Flanders 1989a), and the information transmitted by the movement being accordingly higher for direction than for distance (Soechting and Flanders 1989a). In reaching for a visually presented, remembered target, the rate of decay of directional information is faster than that of distance information at increasing delays from target extinction (Rossetti and Lacquaniti). This suggests that direction and distance are stored in separate channels in the buffer of visuospatial working memory. Also, chronometric studies have indicated that the central processing time involved in programming direction is longer than that involved in programming distance (Rosenbaum 1980). In the scheme of coordinate transformations proposed by Flanders et al. (1992) there exist two separate channels: arm elevation is computed from target distance and elevation, whereas arm yaw is computed from target azimuth.

Modular organization represents probably an especially parsimonious way to handle multidimensional information in a distributed, parallel system such as the brain. Several examples of modular organization will be encountered in subsequent sections when reviewing other forms of sensorimotor coordination.

3.1.1.3 REACHING KINEMATICS

3.1.1.3.1 Kinematic invariances. Behavioral data support the notion that there may exist neural representations not only of static motor error but also of the spatial path and kinematics of the limb (cf. Georgopoulos 1986; Hogan 1988; Jeannerod 1988; Soechting and Flanders 1991). Thus a number of lawful relationships have been described for the kinematic trajectories of the hand in external space, and of the individual limb segments in the angular coordinates of the joints. The spatial trajectories of both the hand and the joints are essentially unaffected by wide changes in speed and load (Lacquaniti et al. 1982; Soechting and Lacquaniti 1981). The velocity profile of the hand tends to be bell-shaped (Flash and Hogan 1985), while the velocity profiles of shoulder and elbow angular motions are temporally correlated (Hollerbach and Atkeson 1987; Lacquaniti and Soechting 1982).

3.1.1.3.2 Optimum principles. These invariances suggest that limb kinematics could be explicitly planned at high levels of organization and could be subsequently converted into the specification of the appropriate forces and torques. The same invariances, however, can be also accounted for without assuming such explicit kinematic plan. Optimization principles are able to predict the time course of reaching by specifying only the initial position, an intermediate via-point, and the final position of the hand (Bullock and Grossberg 1988; Feldman and Levin 1995; Flash and Hogan 1985; Uno et al. 1989). Note that such optimization principles may pertain to either kinematics (minimum jerk of hand kinematics, cf. Flash and Hogan 1985) or dynamics (minimum rate of change of joint torques or muscle forces, cf. Hirayama et al. 1993; Uno et al. 1989). Minimum jerk, for instance, constrains movements to follow a maximally smooth time course (Flash and Hogan 1985). The ability of this model to reproduce the experimental kinematics is contingent upon using hand coordinates to define limb position, rather than the angular coordinates of the joints (Hogan 1988). On the other hand, minimum torque change predicts the data by using the angular coordinates of the joints (Uno et al. 1989).

3.1.2 DRAWING

3.1.2.1 Hand kinematics. In contrast to point-to-point reaching movements, drawing (and handwriting) movements are more or less continuous in time, and describe complex, generally curved paths. The geometrical shape of pen trajectory is a global property that is often preserved through wide changes in size and speed of movement (Lacquaniti 1989; Viviani and Terzuolo 1980). Although kinematic and figural aspects of drawing are a priori completely independent, they become strictly coupled as a result of neural control. The instantaneous tangential velocity of the hand is inversely related (by a power law) to the local curvature of the path as it is traced by the hand (Lacquaniti et al. 1983; Viviani and Terzuolo 1982). The gain factor of this law is piecewise constant, depending on the linear extent of each curve segment and on the average speed of movement (Lacquaniti et al. 1984). Thus these apparently continuous movements are in fact segmented. Superposition of harmonic oscillations in two orthogonal directions of the plane of motion predicts the power law. In this model, the shape of the figure is determined by modulating amplitude and phase of the oscillations in a segmented fashion. The power law can be also accounted for by the minimum-jerk theory (Viviani and Flash 1995). Thus, as in the case of reaching, only the boundary conditions at the starting point, via-points and final point may need to be specified, without a detailed kinematic plan.
3.1.2.2 Inter-joint coordination. How is the motion of all limb segments that participate in drawing co-ordinated ? The algorithm that has been proposed, on the basis of a detailed analysis of movement patterns, involves oscillatory changes in the orientation angles of elevation and yaw at the shoulder and elbow (Soechting et al. 1986). The parameters describing the spatial location of the figure drawn in external space are related linearly to the phase differences between the angular coordinates of the shoulder and elbow. Because the algorithm involves linear relations between joint angles and hand coordinates whereas the exact relations are nonlinear, drawing in specific regions of the workspace (such as the sagittal plane) is affected by characteristic distortions. This algorithm is closely reminiscent of that described in section 3.1.1.1.1 for the visuomotor transformations of reaching. Thus, there exist considerable similarities in some general organizational principles underlying the generation of reaching and drawing, despite the apparent differences of these 2 types of movement.

3.1.3 CATCHING

3.1.3.1 Internal models. In interceptive tasks, such as catching, information about the approaching object must be preprocessed in order to plan the dynamic interaction well in advance of its occurrence. In particular, the time, location and momentum of the impact of the object on the hand need be accurately estimated, and limb kinematics and kinetics accordingly controlled. Recent work has shown that internal models of the dynamical properties of objects and limbs are constructed in the brain in a geotropic frame, and are used to predict the dynamic characteristics of a forthcoming interaction and to adjust adaptively the parameters of the neural controller (Lacquaniti 1996). Thus, a priori knowledge of the most likely path and law of motion of the object (linear motion uniformly accelerated by gravity in free-fall, parabolic motion with constant horizontal velocity during ballistic throws etc.) is used in conjunction with visual on-line information to pre-stiffen arm muscles in catching tasks. Muscle activation is time-locked to the time-to-contact, computed by combining optic flow information about object motion (Lee 1980) with the internal estimate of the acceleration of gravity (Lacquaniti et al. 1993). Amplitude of muscle activation is scaled with the expected momentum (mass times velocity) at impact (Lacquaniti and Maioli 1989). This prospective control scheme is adaptive; the response to the dynamic interaction predicted by the internal model is compared with the actual response of the limb (as monitored by kinesthetic and cutaneous signals), and the resulting error is used to calibrate the parameters of the neural controller of the mechanical plant, and to update the internal model.

Processes of trajectory formation have been also shown to undergo adaptive changes in response to novel force fields experienced at the hand; adaptation is based on the progressive changes of the internal models of limb dynamics initially defined in the intrinsic coordinates of the joints and muscles (Shadmer and Mussa-Ivaldi 1994). Proprioceptive information is essential to maintain these internal models (Gordon et al. 1995). Indeed, deafferented subjects (due to large fiber neuropathies) are unable to compensate for workspace anisotropies in limb inertia and produce pointing errors that are direction-dependent.

3.1.3.2 Coordinate transformations in the control of contact forces at the hand. A perturbation, such as the object impacting on the hand in interceptive tasks, is resisted by the restoring forces of the limb. In multi-jointed movements, the relation between applied displacement and resulting force is a vector field, with the direction of force generally not coincident with the direction of displacement (Hogan 1985). This vector field depends not only on the pattern of muscle activity, but also on the geometrical configuration of the limb. In other words, a given pattern of muscle activity may result in very different values of contact forces at the hand depending on the values of the joint angles. However, the CNS is able to output the specific patterns of muscle activity that produce the desired values of the contact forces at the hand (Johansson 1991; see also the chapter by Johansson). In catching, the direction of the reactive force vector at the hand tends to rotate to become aligned with the direction of the expected perturbation (Lacquaniti et al. 1993). This has been taken to suggest that the intended restoring forces at the hand are initially defined in an earth-fixed frame of reference, and are subsequently translated in the values of muscle forces and joint torques to be output in intrinsic arm coordinates. This process of coordinate transformation is contingent on the same internal models of the mechanical properties of the arm that have been considered in the previous section.

3.2 CODING MECHANISMS OF ARM MOVEMENST IN CEREBRAL CORTEX

Although many details still need to be elucidated, a broad picture of cortical representations of arm movements has emerged from the studies of the last decade (see Georgopoulos 1994; Jeannerod et al. 1995; Kalaska and Crammond 1992). This picture encompasses a number of potential correlates at the neural level of aspects of the global motor behaviour described in section 3.1. In what follows, I focus on the neuronal properties of a fronto-parietal cortical network for the control of arm reaching under visual guidance. In particular, 2 different but related problems will be considered: the nature of the spatial parameters of limb movement putatively encoded in neural activity, and the reference frames used for their encoding.

3.2.1 VECTOR CODING OF DIRECTION

3.2.1.1 Kinematic parameters. Recordings of neural activity during reaching movements in the monkey have revealed a vector code of movement direction in primary motor cortex (M1, Caminiti et al. 1990; Georgopoulos et al. 1982; Schwartz et al. 1988), dorsal premotor cortex

(PMd, Caminiti et al. 1991; Fu et al. 1993), primary somatosensory cortex (area 2, Cohen et al. 1994), and superior parietal lobule (area 5, Kalaska et al. 1983). Many neurons at all these sites are broadly tuned to the direction of the upcoming hand movement. Activity is maximal along a best direction (called preferred direction) and decreases proportionally to the cosine of the angle between the preferred direction and the direction of actual movement. Preferred directions of the ensemble of neurons tend to be distributed uniformly throughout space. The broad directional tuning indicates that each neuron participates in movements in several directions and that, by converse, a movement in a given direction engages a whole population of cells (Georgopoulos et al. 1988). A population code of movement direction has been formulated in vector terms: each neuron contributes a vector in its preferred direction with an amplitude proportional to its level of activity (Gergopoulos et al. 1983; 1988).

The time-varying changes in length and direction of the population vectors in M1 parallel the corresponding changes in the vector of tangential velocity in reaching (Georgopoulos et al. 1988) and drawing (Schwartz 1994). In drawing, the virtual changes in movement velocity predicted by the population vectors are related to path curvature by the same power law that applies to actual movement velocity (Schwartz 1994).

3.2.1.2 Dynamic parameters. The relation between M1 activity and limb force has long been known (Cheney and Fetz 1980; Dugas and Smith 1992; Evarts 1968). It has recently been shown that the same type of vector code that applies to movement direction, applies also to the coding of force direction in M1 (Georgopoulos et al. 1992). During step-tracking of 2D force vectors, the neuronal population vector points in the direction of the rate of change of the force (Georgopoulos et al. 1992).

From the studies reviewed in section 3.2.1.1, it appears that vector code of movement direction holds identically to several cortical areas. An important functional difference between frontal and parietal cortices emerges, however, when static loads pulling the arm in different directions are applied to the monkey's arm during planar arm movements (Crutcher and Alexander 1990; Kalaska et al. 1989; 1990). Load sensitivity forms a relative continuum in M1: some neurons are very sensitive to the applied load and appear to encode parameters related to movement kinetics (forces and torques), whereas other neurons are relatively insensitive to loads and appear to encode movement kinematics, and still other neurons fall in between, exhibiting both directional and load tuning (Crutcher and Alexander 1990; Kalaska et al. 1990). By contrast, the vast majority of neurons in area 5 are essentially insensitive to loads; they appear to encode only movement kinematics, not dynamics (Kalaska et al. 1990).

3.2.2 CODING OF MOTOR ERROR

The problem of whether directional activity carries a signal of motor error has been addressed by Fu et al. (1993, 1995). Because motor error is the difference vector between target and hand location, the amplitude of the hand movement needs to be specified, in addition to direction (Gordon et al. 1994).

3.2.2.1 Movement amplitude. Recent studies have shown that several neurons in M1 and PMd exhibit a significant correlation with both the direction and amplitude of the movement (Fu et al. 1993; 1995). For a given direction, neural discharge changes monotonically with movement amplitude. This modulation by amplitude, however, is significant for only 1 or 2 different directions (out of the 8 different directions investigated) in most neurons. Amplitude modulation is not associated preferentially with movements close to the preferred direction in a given neuron.

3.2.2.2 Neural activity is correlated with multiple movement parameters. Multivariate regression on cell activity demonstrates that target location, in addition to movement direction and amplitude, must be included to account for most experimental variance (Fu et al. 1993). Best correlations with these different parameters are not exactly synchronous, but tend to be sequenced in time (Fu et al. 1995): direction-related discharge tends to occur first (about 100 ms prior to movement onset), followed by target location (50 ms after movement onset), and movement amplitude (250 ms after movement onset). Ashe and Georgopoulos (1994) have examined the time-varying correlation of neural discharge in M1 and area 5 with the direction of movement, and with the instantaneous position, velocity, acceleration of the hand. They found significant correlations with all these parameters, most prominent being the correlation with direction, and least prominent that with acceleration. On average, M1 activity led movement by 90 ms, whereas area 5 activity lagged behind movement by 30 ms.

3.2.3 CODING IN BODY-CENTRED COORDINATES

3.2.3.1 Directional tuning is not invariant with arm posture. A vector code of either movement direction or motor error implies that neural activity should be the same for the same movement performed along parallel directions but starting from different initial positions. Directional tuning for arm reaching in different parts of the workspace has been studied by Caminiti and colleagues in M1 (Caminiti et al. 1990), PMd (Caminiti et al. 1991), and dorsal area 5 (Lacquaniti et al. 1995). Specifically, movements of constant amplitude were required from 3 different initial positions of the hand in 8 possible directions. They found that the mean activity of most neurons is significantly different for movements performed in parallel directions from different initial positions. In fact, the vector of preferred direction, computed for any given cell in each part of the workspace separately, rotates by an amount and direction that are highly variable among different cells. However, the population vector computed by summation of all cell contributions remains invariant for parallel movements performed in different regions of the workspace (Caminiti et al. 1990; Kettner et al. 1988). On the other hand, when pointing in a given direction is performed with the abducted arm, both the preferred directions of single neurons and the population vector summed over all recorded neurons in M1 have orientations different from those observed when pointing in the same direction is performed with the arm adducted (Scott and Kalaska 1995). In other words, similar hand trajectories associated with different arm postures are associated with different patterns of neural activity.

3.2.3.2 Evidence for positional coding in body-centred coordinates. To account for the inconsistencies described in the previous section, the concept of directional coding has recently been revisited from a different perspective (Lacquaniti et al. 1995). In particular, the hypothesis has been put forth that reaching movements might be encoded neurally in a body-centred reference frame, rather than in a hand-centred frame. Indeed, it has been shown that the activity of most neurons in area 5 (Lacquaniti et al. 1995), M1 and PMd (Guigon et al. 1996; Lacquaniti et al. 1996) is quasi-linearly related to the azimuth, elevation and distance of either the target or the hand with respect to the body. In these experiments, the origin of the frame could not be determined precisely, but head-centred coordinates yielded a slightly better fit than shoulder-centred coordinates, on average. A body-centred representation has recently been found also in area F4 of ventral premotor cortex (PMv, see the chapter by Fogassi et al.).

3.2.3.3 Classes of neurons with distinct functional properties. On the basis of the relative weight of different spatial terms, 4 main classes of body-centred neurons have been distinguished in M1, PMd, and area 5: 1) "initial position" neurons, 2) "via-point" neurons, 3) "final position" neurons, and 4) "motor error" neurons (Lacquaniti et al. 1995). Their activity is related most prominently to the starting point of the movement, to positions located along the direction of movement, to the position of the peripheral target, and to the difference vector between final and initial position, respectively. According to the hypothesis of a body-centred code, "motor error" neurons correspond to a small subset only of all directional and motor error units identified in previous studies (Fu et al. 1993; Kettner et al. 1988), namely those whose preferred direction should not change in different parts of workspace.

The activity of several neurons may reflect the current position of the limb, as signalled by peripheral feedbacks or efferent copy of motor commands. However, the activity of many other neurons is related to the final point before it is actually reached, that is during reaction-time or movement-time; this activity reflects, therefore, a preplanned desired position (Hocherman and Wise 1991) or command function (Mountcastle et al. 1975) rather than current kinematics.

On the whole, these data are compatible with the idea that central commands may specify changes in the equilibrium point of the limb (Bizzi et al. 1992; Feldman and Levin 1995). In particular, according to the model proposed by Bullock and Grossberg (1988), 3 types of signals would be computed in the brain: a target-position command (TPC) which is switched on once and for all at movement inception, a present-position command (PPC) which is continuously updated until it matches TPC, and a difference vector (DV) between TPC and PPC. "Final position" neurons, "via-point" neurons and "motor error" neurons could reflect TPC, PPC and DV, respectively.

3.2.3.4 Regional differences. None of the cortical areas investigated (M1, PMd, area 5) is functionally homogeneous, inasmuch as each contains a mixture of all 4 classes of neurons. However, the relative prevalence of these classes differs among the areas, indicating the existence of some degree of regional specialization (Guigon et al. 1996; Lacquaniti et al. 1996). Thus, during reaction time, most area 5 neurons are related to starting position, whereas most M1 neurons are related to the motor error, and most PMd neurons are related to target position. During movement time, regional differences become less pronounced, most neurons in all 3 areas being related to via-points or a combination of final position and motor error. At target-holding, most area 5 and M1 neurons, and a smaller proportion of PMd neurons, are related to final hand position.

3.2.4 DISTRIBUTION OF SPATIAL INFORMATION IN NEURONAL ENSEMBLES

A key issue in understanding distributed representations of movement in the brain is given by the statistical properties of the population ensemble in any given cortical area (cf. Georgopoulos et al. 1988). Irrespective of the specific parameter encoded (final position, via-point, motor error, movement direction etc.), one may ask whether the optimal directions of activations of the individual neurons of a given ensemble are distributed uniformly throughout space, or whether they cluster around preferred coordinate axes. The concept of population code of movement direction by coarsely tuned neurons implies that the preferred directions of single cells should be distributed uniformly (Burnod et al. 1992; Georgopoulos et al. 1988). By contrast, the positional code in body-centred coordinates described in section 3.2.3 does not appear to be uniform in three-dimensional space, at least in area 5 (Lacquaniti et al. 1995).

Tuning functions of most neurons (about 70%) cluster around the cardinal spatial axes (azimuth, elevation and distance). Thus each spatial coordinate tends to be encoded in a different subpopulation of neurons. Segregation of different spatial dimensions at given sites of the CNS are well known (Simpson 1984; Soechting and Flanders 1992), and examples will be provided in subsequent sections of this chapter.

What is the functional significance of segregating the information about distance, elevation and azimuth of hand movements in distinct subpopulations of neurons ? Neural parcellation could be a correlate of the psychophysical observation (reported in section 3.1.1.2) that these spatial parameters are processed in parallel and largely independent of each other. Projecting the spatial information onto separate axes of neural coding could be important for sensorimotor coordination, because information related to motor commands could be matched more easily with incoming sensory feedbacks. Spatial axes which result from a hybrid combination of sensory and motor axes would be best suited to accomplish this matching process (Carrozzo and Lacquaniti 1994). Although spatial information is parcellated at the level of single neurons, the complete information about limb position can be reconstructed by simple summation of the individual contributions over a population of neurons, because positive and negative spatial coefficients tend to be evenly distributed (Lacquaniti et al. 1995). This can be demonstrated by using a neural network model. A layered neural network whose intermediate layer contains positionally tuned neurons can be trained to generate the position of the hand in body-centered coordinates. It is not known whether global reconstruction of limb position takes place at a cortical level or at a subcortical level (Mountcastle 1995).

3.2.5 A CORTICAL NETWORK FOR COORDINATE TRANSFORMATIONS IN REACHING FOR A VISUAL TARGET

It has recently been shown in the monkey that the convergence of different anatomical pathways onto motor areas results in a combinaton of visually derived signals about target position with kinesthetic signals related to arm position (see the chapter by Johnson et al.). On the basis of the data reviewed in section 3.2.3.2, it has been proposed that this visuo-kinestethic integration might occur in a common body-centred frame, realizing a network for the coordinate transformations involved in the control of reaching under visual guidance (Lacquaniti et al. 1995).

3.2.5.1 Visual inputs. Objects in the visual field are processed within a network of 30 or more cortical visual areas (Felleman and VanEssen 1991). These areas are organized within two major cortico-cortical pathways or streams, each of which begins with the primary visual cortex (V1). The ventral stream is directed to the inferior temporal cortex and is important for object recognition, whereas the dorsal stream is directed into the posterior parietal cortex and is important for spatial perception and visuomotor performance (Ungerleider and Mishkin 1982; Jeannerod 1995). Here we are concerned with this dorsal stream.

Spatial visual information flows from V1, V2 to middle temporal area (MT) and middle superior temporal area (MST). Cells in these temporal areas are brodly tuned to the direction and pattern of visual motion (Snowden et al. 1992; Sakata et al. 1985). Visual-tracking neurons in MST receive convergent signals indicating current retinal image motion, eye rotation and head rotation (Thier and Erickson 1992). This convergence might result in an output encoding the direction of target motion in absolute space. Many neurons in MT and MST are selective to different combinations of the optic flow components (translation, dilation, rotation and deformation, Duffy and Wurtz 1991). Parieto-occipital area (PO), which receives direct projections from V1, V2, MT and MST, has a quasi-uniform representation of the retina without foveal magnification, and eye-position signals may contribute to define a head-centred representation of visual inputs (Galletti et al. 1993). As one moves from V1, V2 to MT, MST and PO, the receptive field size of individual neurons increases. Large receptive fields may contribute to localizing objects over retinal translation.

Visual information is transmitted from PO to the medial intraparietal area (MIP) and area 7m. MIP and 7m are connected with the dorsal premotor cortex (area PMd) and supplementary motor cortex (SMA), both of which project to primary motor cortex (M1). This then represents the candidate pathway for the transmission of visual information to motor cortex (Johnson et al. 1993).

PMd activity is also modulated by eye position signals (Boussaoud 1995). This observation, coupled with that of signal-related activity in this area (di Pellegrino and Wise 1993; Johnson et al. 1993), may explain the prominence of target-related activity in PMd during reaction-time.

3.2.5.2 Somatosensory inputs. The dorsal part of area 5 (superior parietal lobule) is a somatosensory and somatomotor area receiving its main cortical input from primary somatosensory cortex (mostly area 2), and projecting extensively to M1. Somesthetic receptive fields are generally large, often multimodal and multisegmental (Burbaud et al. 1991; Mountcastle et al. 1975; Sakata et al. 1973). Hand azimuth could be estimated by combining proprioceptive information about the horizontal rotation (yaw) of both shoulder and elbow joints, and hand elevation and distance would be similarly derived from combinations of horizontal and vertical rotations of these 2 joints. As for the source of the vertical reference that is necessary for computing arm coordinates in geotropic space, area 5 could receive this information from area 2, which in turn is recipient of vestibular inputs (in addition to parietoinsular region, see section 6.2.1). Alternatively, the reference could correspond to the trunk midsagittal axis. In this context, one may note that several neurons in area 5 have receptive fields encompassing both chest and shoulder (Burbaud et al. 1991).

Our notions on the counterpart of this network in man are still fragmentary. They are mostly based on imaging the brain (PET or MRI) during the execution of visuomotor tasks. With these techniques, functional activations of areas belonging to the dorsal cortical stream involved in the processing of spatial information have been demonstrated (Grafton et al. 1992; Kawashima et al. 1994; Decety et al. 1994). The main point of discussion concerns the nature of the possible equivalent in man of area 5 (superior parietal lobule) in the monkey.

4. Eye movements

The eye has 3 degrees of rotational freedom (horizontal, vertical and torsional), and all rotation axes pass roughly through one point (Henn 1993). Each eye is moved by six extraocular muscles, but pairs of muscles are organized antagonistically and their pulling directions define three non-orthogonal axes of rotation in the orbit (Ezure and Graf 1984a; Robinson 1982). Horizontal recti rotate the eyes about a vertical axis, whereas vertical recti rotate the eyes about an axis with a substantial torsional component, and the oblique muscles rotate the eyes about an axis which is approximately orthogonal to the vertical recti axis. Although there exist many different types of eye movements, the full range of behaviour can be accounted for by 2 main subsystems: a slow smooth system and a fast saccadic system.

4.1 GAZE STABILIZATION

It is comprised of a slow phase and a fast phase of nystagmus. During head movements in the light, the line of sight is stabilized using vestibular and visual information. This requires that the visual signals are transformed to match the vestibular ones in terms of spatial direction and physical dimension (position and velocity, Henn 1993). The main site for visual-vestibular interaction is given by the vestibular nuclei, where visual signals match vestibular signals with the common dimension of velocity, and with directions roughly aligned with those of the semicircular canals. Visual signals undergo a similar transformation also at the level of the accessory optic system.

4.1.1 VESTIBULAR ASPECTS

Vestibular organs contribute important sensory information for the maintenance of spatial orientation. Utricular and saccular otolith organs detect linear accelerations of the head, providing information about translatory head movements as well as about head orientation relative to gravity. Semicircular canals, on the other hand, are activated by head angular acceleration, and provide information about head angular velocity. Vestibular information is generally supplemented by visual and somatosensory inputs related to self-orientation and self-motion in space.

4.1.1.1 Coordinate axes at the input and output. Both types of vestibular sensors are endowed with specific coordinate systems dictated by anatomy. Thus, in each ear saccular and utricular maculae lie in roughly orthogonal planes, as do the three semicircular canals (Ezure and Graf 1984a). For each canal an optimal direction can be defined. Angular acceleration in the optimal plane leads to maximal activation, whereas acceleration in any other plane attenuates the response according to a cosine function. The null direction is orthogonal to the optimal one. The vestibulo-ocular reflex (VOR) produces compensatory eye movements in the direction opposite to head rotation, thus minimizing the retinal slip of the visual world. Compensatory eye movements are contributed by the six extraocular muscles that are organized in pairs of antagonists. Pulling directions of these muscle pairs define three axes of eye rotation in the orbit that are neither orthogonal to each other, nor aligned with the axes of the semicircular canals (Ezure and Graf 1984a; Robinson 1982).

4.1.1.2 Coordinate transformations in vestibular nuclei. The VOR then provides a clear example of coordinate transformations involved in sensorimotor coordination. Inputs (head velocity) are encoded in the canals coordinates, and outputs (eye velocity) in the muscles coordinates. Behaviorally, the coordinate transformation works almost perfectly: eye velocity vector tends to be equal and opposite to head velocity vector (Robinson 1982), although with 3D anisotropies (Straumann et al. 1993). What about the neural processing involved in the coordinate transformation ? Recordings of electrical activity from the second-order vestibular neurons that project to ocular motoneurons show that their sensitivity axes (i.e., best directions of head rotation) cluster around the axes of the canals and eye muscles, but with substantial scatter (Fukushima et al 1990). This scatter is presumably due to the extensive convergence of vestibular afferents from different canals at the level of the vestibular nuclei. Further convergence of axons of vestibular neurons occurs in the oculomotor nuclei (Ezure and Graf 1984b; Peterson and Baker 1991).

4.1.1.3 A neural network model of the VOR. Interestingly, the same type of scatter in the sensitivity axes of the interneurons is found in a three-layered neural network modelling the VOR (Anastasio and Robinson 1990). Each interneuron in the hidden layer, just as the real interneurons in the vestibular nuclei, exhibits its own idiosyncratic tuning axis. This observation has led Robinson (1992) to suggest that neural processing does not occur in any specific reference frame in the VOR. The overall coordinate transformation would then be a simple mathematical description of a system trained to minimize an error signal (the retinal slip) over a distributed network.

4.1.2 VISUAL ASPECTS

4.1.2.1 Sensor fusion. Compensatory eye movements are generated on the basis of head velocity information, that is provided by the vestibular system at higher frequencies and by the visual system at lower frequencies. How is sensor fusion accomplished and where ? As noted above, vestibular information is initially encoded along the axes of the semicircular canals. Wide-field visual motion is retinotopically encoded at the level of the retinal ganglion cells. Self-motion information is presumably derived from an analysis of local optic flow patterns. A high degree of visual-vestibular interaction occurs at the level of most second-order vestibular neurons (Henn et al. 1980), but also at thalamic and cortical levels (Buttner and Henn 1976). Neurons in the vestibular nuclei respond to whole-field visual motion along best directions that are closely aligned with those of the corresponding semicircular canal (Graf 1988).

4.1.2.2 Transformations of visual information in the accessory optic system. Transformation of visual information in canals and eye muscles coordinates occurs also at the level of the accessory optic system (AOS). In the retina of the rabbit, an afoveate lateral-eyed animal, there is a class of ON-direction-selective W-ganglion cells that are selectively tuned along one of three possible directions of motion: anterior, up with a posterior component, and down with a posterior component (Oyster et al 1972). These retinal ganglion cells project to the terminal nuclei of the AOS in the midbrain. Simpson and coworkers suggested that a simple summation of inputs from different types of directional cells to the terminal nuclei could be responsible for the generation of bipartite receptive fields organized around the best axes of the three semicircular canals (Simpson and Graf 1985). Neural encoding of different spatial directions is partially segregated: cells in the dorsal terminal nucleus respond to visual motion in the plane of the horizontal canal, whereas cells in the lateral and medial terminal nuclei respond to visual motion in the planes of the vertical canals (Sodak and Simpson 1988). Alignment in register of visual and vestibular information would allow sensor fusion when comparable optic flow patterns are created by self-motion in corresponding directions.

The issue of whether or not these findings can be generalized to other species remains controversial, however (Mustari and Fuchs 1989). Recently, Maioli and Ohgaki (1993) found that neurons in the lateral terminal nucleus of the alert cat have best directions not far from those of the vertical canals, but more closely aligned with the pulling directions of the eye muscles (vertical recti and obliques). Irrespective of the exact frame of alignment, however, this directional tuning cannot arise from simple summation of directional retinal cells in frontal-eyed animals, such as the cat or the primates, in contrast to what happens in lateral-eyed animals. In the former species, the smaller receptive fields of retinal cells must perform a local analysis of optic flow and a complex integration gives rise to the larger receptive fields found in terminal nuclei.

Visual coding in vestibular (or eye muscles) coordinates is maintained at subsequent stages of the AOS, i.e. the dorsal cap of inferior olive and the flocculo-nodular lobe of the cerebellum (Graf et al. 1988). This pathway is putatively involved in adaptive modification of the VOR (Ito 1982). Another subsystem of the AOS is comprised of the pretectal nucleus of the optic tract (NOT). This nucleus is connected to the pontine nuclei involved in eye movement generation and could subserve the slow phase of the optokinetic nystagmus as well as smooth pursuit (Hoffmann and Distler 1989). It could also contribute to visual-vestibular interactions at the level of the vestibular nuclei via indirect links (Magnin et al. 1983; Precht and Strata 1980).

4.2 ORIENTING GAZE

4.2.1 BEHAVIOUR

4.2.1.1 Listing's law. Eye rotations in the orbit have 3 degrees of freedom, corresponding to horizontal, vertical and torsional rotation, but the torsional component is constrained by the CNS during saccades. Listing's law states that, regardless of the specific trajectory that brought the eye to a given position, its orientation in 3D is uniquely determined by gaze direction. This 3D orientation can be described by the rotation vector that brought the eye from its primary direction (straight-ahead) to its present position. The rotation vectors of all eye saccades lie in a plane (Listing's plane) perpendicular to the primary eye direction, and eye torsion about this direction is about zero (Hepp 1994). Interestingly, torsion is null in a head-fixed frame, but is no longer equal to zero in coordinates systems that rotate with the eye (as in a gimbal). Thus, in the head-fixed frame, eye positions are uniquely defined by the coordinates of horizontal and vertical rotation angles. Listing's law applies to eye positions during fixations, saccades and smooth pursuit with the head erect and stationary. The law is not obeyed by eye movements during sleep, and by both slow and fast phases of eye movements evoked by vestibular stimulation.

4.2.1.2 Control of saccades: retinotopic or spatiotopic ? A long-standing problem in physiology has been whether we track targets using only the retinal error (difference vector between target image on the peripheral retina and central fovea), or whether we compute the position of the target in absolute space (with respect to the head, or the body or the external world), and then use this computed position to output motor commands (for a discussion of different views on this point, see Colby et al. 1992; Robinson 1992; Stein 1992). From a behavioral standpoint, the latter hypothesis seems more plausible. Consider the case of a visual target jumping from a starting point A to an intermediate point B, to a final point C, before the eyes can move from A. It is known that we can make saccades in darkness from A to B to C, even though the retinal error of the target at C seen from the eye at B was never available (Hallet and Lightstone 1976). As another example, consider the case of looking at a target, closing the eyes, and turning away from the target; we are still able to reach for the target with sufficient accuracy (Bloomberg et al. 1991). One can conclude that orienting movements are not driven by retinal error alone (Robinson 1992).

4.2.1.3 Local feedback loop. A well known model of saccade generation reconstructs the location of a target in space by means of a local (internal) feedback loop (Robinson 1975). According to this model, the CNS adds a copy of eye position to the retinal error, and reconstructs an internal estimate of target position with respect to the head, if the head is stationary, or with respect to external space if the head is moving (Robinson 1975). To make a saccade, the CNS sets the desired eye position equal to the estimated target position. The difference between desired eye position and current eye position then represents the motor error in retinotopic coordinates.

4.2.1.4 Multisensory integration and coordinate transformations. Gaze orientation is usually accomplished by a combination of eye and head rotations. As in the case of gaze stabilization, also orienting movements imply a coordinate transformation. At the input stage, the target may be encoded in a somatotopic map for tactile stimuli, in a retinotopic map for visual stimuli, and in a head-centered map for auditory stimuli. At the output stage, eye and head movements are specified by the coordinate axes specified by the pulling directions of the respective muscles. Neural processing at central stages intervenes in the sensorimotor transformations underlying the generation of appropriate command signals. Multimodal sensory information must be combined with signals on the current position of the eyes in the orbits and of the head in space to direct the gaze to the absolute location of the target in external space (Sparks 1988).

4.2.2 SUBCORTICAL MECHANISMS

4.2.2.1 Multimodal maps in the colliculus. The superior colliculus (SC) is a key structure for orienting movements of the eyes and head (Sparks 1988; Guitton 1992; see the chapters by Roucoux et al. and by Werner and Hoffmann). Superficial layers of this structure receive almost exclusively visual input, directly from the retina and also from the striate and peristriate visual cortex. Each SC contains a retinotopic map of the contralateral visual field, retinal inputs and cortical inputs being in register. Deeper layers, instead, receive convergent inputs from visual, auditory and somatosensory modalities, and the receptive fields for each sensory modality are approximately in register among each other (Knudsen and Brainard 1995). The visual map is retinotopically organized and represents the reference system to which the other sensory maps are normally aligned. The somatosensory map is topographically organized, and is in register with the visual map when eyes, head and body are oriented in alignment.

4.2.2.2 Auditory cues are segregated in the brainstem and recombined in SC. The auditory map in SC contains a computational map of space that is created by merging information from lower brainstem pathways (Knudsen et al. 1987). The cues for sound localization include binaural differences in the timing and intensity of spectral components of sound. These cues are frequency-specific: a given cue (such as a value of interaural intensity difference) corresponds to one location in space at one frequency, but to a different location at a different frequency. Therefore, localization cues must be evaluated in a frequency-specific manner (Knudsen et al. 1987).

Information about the timing and intensity of each frequency component is conveyed to the cochlear nuclei by frequency-tuned fibers in the auditory nerve. Neurons from cochlear nuclei project along parallel pathways to pontine nuclei specialized for comparing signals from the two ears. Interaural time delay of frequency-specific signals from each ear is processed in the medial superior olive, whereas interaural intensity difference is processed in the lateral superior olive. Thus, because the azimuth of an auditory target is mainly specified by interaural time difference, whereas target elevation is specified by interaural intensity difference, these two spatial parameters are processed in separate brainstem nuclei. Azimuth and elevation are combined together in a topographic representation of auditory space in SC by convergence of the parallel brainstem pathways.

4.2.2.3 Dynamic remapping of auditory fields. In order for the nervous system to benefit from the convergence of visual and auditory spatial information in SC, the representations of space derived from these modalities must remain aligned (Knudsen and Brainard 1995). The problem is that visual and auditory receptive fields in SC are based, respectively, on retinotopic and ear-dependent localization cues, and therefore movements of the eyes or ears relative to the head tend to disrupt the alignment of these receptive fields. However, the auditory map is kept aligned with the visual map by means of dynamic remapping of auditory receptive fields; these may shift almost in parallel with eye position (Jay and Sparks 1987). However, the influence of eye position is not constant in all units, but ranges from no effect at all in many units, to shifts that approximate closely the change in eye position in other units. Most units exhibit intermediate degrees of eye-position influence.

4.2.2.4 Collicular control of saccades. Deeper layers of SC contain also neurons that are critical for initiating and controlling eye saccades (cf. Sparks and Mays 1990). These neurons are topographically organized according to movement fields defined by the direction and amplitude of eye saccades. The static motor error (difference between initial and final position of the eye in the orbit) is computed by means of a population vector code (Lee et al. 1988). In this motor map, each neuron is broadly tuned to a best direction and amplitude of the relative change in eye position. Actual eye saccades are predicted by the vectorial average of the activity of contiguous populations of collicular neurons. This motor map is aligned with the multi-sensory (visual, auditory and somesthesic) map existing in the same structure. Motor and sensory maps are roughly retinocentric.

The colliculus is also involved in the control of orienting movements of the head. Spatial information is segregated in the optic tectum (the analogue of SC in lower vertebrates) of the owl: horizontal rotations of the head are controlled by neural circuits that are distinct from those controlling vertical rotations (Masino and Knudsen 1990).

4.2.2.5 Different cell types. It has been proposed that, instead of static motor error, dynamic motor error (i.e., instantaneous difference between desired eye position and current eye position) would be computed. Different cell types have recently been found in the SC that may support the derivation of dynamic motor error (Guitton 1992; Wurtz and Optican 1995). Fixation cells receive a constant command from cortical areas and prevent saccades by inhibiting the other types of cells. Saccades are initiated by inhibiting fixation cells and exciting burst and build up cells. Burst cells encode the desired eye displacement, and the build up cells integrate spatially the velocity feedback to determine how far the eye has already moved. The dynamic motor error would be computed downstream the SC, in the brainstem, as the difference between desired and actual eye displacement (Wurtz and Optican 1995).

4.2.2.6 Signal transformations in the brainstem. Multiple parallel pathways connect the SC with the oculomotor nuclei. For horizontal eye movements, the main path is represented by the projections of tecto-reticulo-spinal neurons to long-lead bursters (LLBs) and medium-lead bursters (MLBs) in the paramedian pontine reticular formation (Grantyn et al 1993). MLBs project to the motoneurones of the extra-ocular muscles either directly or indirectly (via the velocity-to-position integrator located largely in the nucleus prepositus hypoglossi). The pathways for the control of vertical and torsional eye movements involve the interstitial nucleus of Cajal in the mesecenphalon.

Vector coding in the SC defines relative changes in eye position and does not fully compensate for the differences in innervation required to produce invariant movement vectors from different initial positions of the eye in the orbit. Collicular movement signals must be transformed in the brainstem so as to code the position of the eye in the specific coordinate system of the motoneurons innervating eye muscles, which is organized along the horizontal and vertical meridians of the eye. This organization reflects neural constraints rather than anatomical constraints, in as much as the pulling directions of the extra-ocular muscles do not exactly act around horizontal and vertical meridians (see section 4., Hepp and Henn 1985).

Listing's law is implemented downstream of the SC, as documented by microstimulation of this structure in the monkey (van Opstal et al. 1991). The SC encodes 2D eye displacements in retinotopic coordinates, whereas the 3D orbital coordinates of the rotation vectors necessary to keep the eye in Listing's plane are derived at the level of the brainstem.

4.2.3 CORTICAL MECHANISMS

Stimulation of the SC, the cerebellum, and the frontal eye fields (FEF), yields retinotopic saccades that tend to displace the eyes in a fixed direction by a fixed amount, independently of the initial position in the orbit. However, behavioral evidence (see section 4.2.1.2) indicates that an estimate of target position in absolute space is available in the CNS. In fact, stimulation of the supplementary eye fields (SEF) drives the eyes to a fixed orbital position from any initial position, thus generating a spatially directed saccade, instead of a retinotopic saccade (Schlag and Schlag-Rey 1987). How a spatiotopic representation of target position is coded remains unknown. Two main mechanisms have been suggested: dynamic remapping of retinotopic receptive fields (Colby et al. 1992; Duhamel et al. 1992), and combination of eye-position signals with retinotopic receptive fields (Andersen et al. 1985).

4.2.3.1 Remapping of visual receptive fields. It has been shown that the retinotopic location of the receptive field of posterior parietal neurons (area LIP) shifts transiently before the onset of a saccade so as to encompass a new visual feature (Duhamel et al. 1992). The direction and amount of this shift can be explained by assuming that the origin of the visual reference frame is shifted to the location in space where the centre of gaze is expected to be after the saccade (Duhamel et al. 1992). After the saccade, the centre of gaze and the origin of the visual frame are realigned again, but the visual feature remains in the same receptive field in spite of the saccade. Moreover, these parietal neurons are activated when a saccade moves their receptive field to a spatial location where a stimulus had been presented before the saccade. Thus, these neurons can anticipate the retinal consequences of eye movements, and can also update the retinal coordinates of remembered stimuli.

4.2.3.2 Gaze-position signals. It has been suggested that neural encoding of a visual target in space results by combining retinotopic receptive fields with signals about eye position in the orbit, and of the head position relative to the external world (Andersen et al. 1993). Target location is encoded topographically on the retina and in cortical visual areas, in a reference frame centered at the eyes. Eye position signals (as derived by proprioception or by efference copy) are combined with retinocentric receptive fields in a number of cortical areas, including V1 (Trotter et al. 1992), V3a (Galletti and Battaglini 1989), PO (Galletti et al. 1991), 7a (Andersen et al. 1985), and LIP (Andersen et al. 1990). As in other systems, different spatial dimensions may be segregated across distinct populations of neurons: a proportion of eye-position neurons in area 7a are monotonically tuned along either the horizontal, the vertical or the depth axis of gaze (Andersen et al. 1990; Sakata et al. 1980).

In general, head-centered coordinates do not seem to be encoded explicitly at the level of single neurons, because visual receptive fields remain retinocentric, and only the amplitude of the visual response is modulated by eye position (Andersen et al. 1993). Instead, head-centred representations might be distributed over a population of neurons. Neural network modelling has shown that incorporating the interactions between retinocentric fields and eye-position signals results in head-centered coding at the level of output units (Zipser and Andersen 1988).

Because the head can move on the body, information about head position might be integrated centrally to modulate the visual receptive fields, much in the same way as eye position signals are used. Indeed, there is recent evidence for distributed neural representations in body-centered coordinates in areas 7a and LIP (Brotchie et al. 1995). These body-centered representations seem to coexist with the head-centered representations demonstrated at the same sites (Andersen et al. 1993).

4.2.3.3 Voluntary saccades. Eye saccades to visual or auditory or somatosensory targets are generally automatic, stereotyped responses. However, saccades can be also initiated voluntarily in the absence of an overt target or they can be oriented toward a previously presented, then memorized target. Although the central structures implicated in these intentional movements may encompass other areas (FEF, SEF, dorso-lateral prefrontal cortex, basal ganglia, see Funahashi et al. 1989; Goldberg and Bruce 1990; Hikosaka and Wurtz 1983) in addition to those listed above, the general organizational principles we have outlined remain similar, namely all such areas encode stimulus-response attributes in a variety of different modes in different reference frames.

5. Stabilization of body posture

5.1 BEHAVIORAL STUDIES

Body posture depends on a highly integrated, multisensory system of control (Massion 1992). Responses in limb and body muscles evoked by postural perturbations depend on several feedback loops that use visual, vestibular, and somatosensory inputs. Each of these loops has its own set of input and output variables to monitor and control. However the coordinated action of the whole postural system is not effected muscle by muscle, but is aimed at the control of more global goals (Massion 1992; Nashner and McCollum 1985).

5.1.1 Control of limb geometry. Stabilization of the body againts gravity requires that the vertical projection of the centre of mass of the body falls within the support area. Recent studies have shown that this goal is achieved by means of a specific strategy of control of the postural geometry (Lacquaniti 1992). In quadrupeds, the length and the angle of orientation relative to the vertical of each limb axis are maintained roughly constant when the support platform is tilted by variable amounts, and when external loads tending to shift the body centre of mass are applied (Lacquaniti et al. 1990). In humans similar kinematic strategies of stabilization apply to both the postural responses evoked by external perturbations, and to the anticipatory responses associated with the voluntary movements of the trunk involved in axial bending (Massion 1991). Moreover, these kinematic strategies remain unchanged under microgravity, that is in the absence of equilibrium constraints (Massion et al. 1993).

5.1.2 Body scheme for posture. The reference values for the geometrical variables are presumably defined on the basis of the postural body scheme. The problem of the maintenance of balance can then be solved by complying with this body scheme that predicts limited changes of the projected centre of mass under normal conditions. Even when this prediction fails under altered conditions, the animal may still attempt to adhere to the body scheme, as indicated by the absence of adaptation of postural geometry over several hours of continuous application of an external load (Lacquaniti et al. 1990). The CNS controls postural geometry directly rather than balance presumably because it has learnt that the preferred posture is stable under normal operating conditions.

5.1.3 A geotropic frame for postural control. Posture has largely evolved to oppose gravity for the maintenance of balance, and it is accordingly organized in a reference frame that is anchored to the direction of gravity. Indeed, head, trunk and limb orientation are all controlled relative to the vertical. As noted in section 3.1.1.1.3, the vertical is normally estimated using a combination of labyrinthine gravitoinertial, visual, and somatosensory cues. The head is normally stabilized in space during posture and locomotion, and provides an inertial platform for monitoring both gravity direction and visual reference lines (Pozzo et al. 1990).

5.1.4 Role of whole-field visual stimuli. Dynamic pitch or linear vection of the visual surround around a stationary animal induce an illusory change of the subjective vertical and a transient perturbation of the equilibrium (Berthoz et al. 1979; Maioli and Poppele 1991). Muscle responses are evoked at a long latency (about 100 ms), and limbs rotate in the same direction as the visual stimulus. Limb orientation returns to prestimulus values more than a second later (Maioli and Poppele 1991).

5.1.5 Role of somesthetic stimuli. They can affect the perception of body orientation in space (Young 1984). Thus aymmetric tactile stimuli spread over large contact areas of the body determine errors in psychophysical matching of the perceived vertical in humans and can evoke righting reflexes in labyrinthectomized animals. It is known that skin stimuli applied to the neck can induce changes in postural tone, with tonic flexion of the neck and forelimbs in cats. Abnormal somesthetic stimuli applied to the trunk of cats determine a profound disturbance of body posture (Lacquaniti et al. 1994a). In fact, when cats wear a vest fit to the shaved trunk, they tend to keep the limbs extended at a roughly constant orientation relative to a platform tilted in the sagittal plane, rather than relative to the vertical. This behaviour results in a progressive shift of the projected centre of mass up to the loss of balance at pronounced platform tilts. Somesthetic cues from the feet also contribute to the perception of body orientation in space, as estimated from the perceived orientation of the ground reaction forces sensed by skin receptors and proprioceptors (Dietz et al. 1992).

5.1.6 Modular organization. In contrast to limb orientation, limb length is not significantly affected by tilt of the visual surround or by the application of abnormal somesthetic stimuli. These observations and that of a differential dynamic behaviour of the changes in limb length and orientation in response to dynamic pitch suggest that these 2 geometrical variables might be controlled independently of each other.

5.1.7 Coordinate transformations for the control of posture. Length and orientation specify the position of the limb endpoint (the foot) relative to the proximal joint (the hip) in a global manner, leaving the detailed geometric configuration undetermined. There is strong evidence in favour of an intermediate processing stage that transforms endpoint coordinates (limb length and orientation) into the angular coordinates of the joints (Lacquaniti et al. 1994a; 1994b). The static geometrical configurations of the limbs during posture on a tilted table may involve variable combinations of the elevation angles of hip, knee and ankle at the hindlimbs. However, one finds that these combinations are not random, but the values of each angle covary with those of the other angles according to a planar law (Lacquaniti et al. 1994a). Also, the dynamic postural responses evoked by unpredictable table rotations describe paths that may diverge in several possible directions, but all these directions lie close to the plane of angular covariations that applies to static conditions.

A similar law of planar covariation applies also to the changes in the elevation angles of lower limb segments during locomotion, although the anatomical angles of flexion-extension do not exhibit this invariant behaviour (Borghese et al). The fact that similar laws of intersegmental coordination apply to the control of posture and locomotion is functionally significant, in as much as locomotion must assure a forward progression compatible with dynamic equilibrium, adapting to potentially destabilizing factors (e.g. uneven terrain, obstacles etc.) in an anticipatory fashion by means of co-ordinated synergies of the whole body (Hirschfeld and Forssberg 1991).

5.2 NEURAL MECHANISMS FOR THE CONTROL OF POSTURAL GEOMETRY

5.2.1 Vestibulospinal and neck reflexes. Vestibulospinal and tonic neck reflexes tend to exert antagonistic effects on limb muscles, and thus may cancel in response to head tilts about a stationary trunk (Ezure et al. 1984; Lindsay et al. 1976; Manzoni et al. 1983). The functional significance of this reflex interaction for postural control may consist in the control of the orientation of the body in space when the whole animal is moved (Mittelstaedt 1964; Roberts 1973). Signals from neck receptors encoding the orientation of the head relative to the trunk would be subtracted from vestibular signals encoding head orientation relative to the gravity vector, thus providing an error signal proportional to the deviation of the trunk from the vertical. It must be noted, however, that the gain of vestibular and neck reflexes is direction-dependent, being much lower for pitch than for roll (Wilson 1986). Moreover, the reflex gain may be substantially lower in intact, behaving animals than in decerebrate cats.

5.2.2 Convergence of vestibular and neck inputs in vestibular nuclei. The behaviour of the described reflexes is strictly correlated with that of vestibular neurons. A high degree of convergence between vestibular (otolith and canal) and neck inputs has been found in the lateral and inferior vestibular nuclei (Boyle and Pompeiano 1981; Kasper et al. 1988). Best directions of activation of neurons in these nuclei tend to cluster around roll (± 45°), the direction for vestibular stimuli being roughly opposite to that for neck stimuli. Gain and phase for vestibular and neck inputs are coherent for neurons with predominant otolith inputs, and for those with otolith plus canal inputs. Responses are more divergent in the case of neurons with canal inputs only (Kasper et al. 1988). Also the responses of pontomedullary reticulospinal neurons to vestibular stimuli are oriented around roll (Bolton et al. 1992), and the gain of vestibulospinal reflexes in roll can be modulated by the locus coeruleus (Andre et al. 1992).

5.2.3 Mechanisms for head stabilization. The vestibulocollic and cervicocollic reflexes contribute to head stabilization in all directions, including pitch (Baker et al. 1985). The type of reflex interaction depends on the experimental conditions. These reflexes add and minimize changes in head position when the head is moved on a fixed body. By contrast, the reflexes oppose each other when the whole animal is moved: the vestibulocollic reflex tends to keep the head fixed in space, but the resulting head movement evokes a cervicollic reflex that tends to keep the head fixed to the body. The vectors of maximal activation of neck muscles in response to multidirectional rotations can be predicted by assuming that the coordinate transformation from semicircular canals to neck muscle coordinates minimizes the amount of muscle coactivation (Pellionisz and Peterson 1988)

5.2.4 Role of spinocerebellar afferents. In posture and locomotion afferent information from the moving legs and efferent copy from spinal networks are sent to cortical and subcortical centres via the ventral and dorsal spinocerebellar tracts (DSCT, Arshavsky et al. 1972). DSCT neurons are broadly tuned to the length and vertical elevation of the hindlimbs (Bosco and Poppele 1993), that is the same coordinates described in behavioral studies on posture (see section 5.1.1).

5.2.5 Stretch reflexes. Although stretch reflexes from the lower limbs contribute to stabilizing body posture, their role is highly context-dependent (Nashner 1976). When calf muscles are stretched as a result of body sway determined by horizontal translation of the support platform, they are reflexly activated. When the same muscles are stretched as a result of a platform rotation under the feet, they exhibit small or no response. In fact, reflex activation compensates for body sway in the former condition, but would destabilize the body in the latter condition (Nashner 1976). Thus reflex gain is adapted according to the task.

Recent evidence indicates that changes of reflex gain can be based on a prospective control of the mechanical properties of the limbs (Lacquaniti 1992), as in the case of the reflex modulation in leg muscles during hopping and jumping (Dyhre-Poulsen et al. 1991). Limb stiffness, in particular, is a mechanical parameter that is actively controlled under many postural conditions (Lacquaniti 1996). The overall stiffness of a limb depends on both the pattern of muscle activity and the postural geometry of the limb (Hogan 1985). Therefore, muscle activation and gating of stretch reflexes must be based on a coordinate transformation of the intended stiffness defined in global limb coordinates into the actual stiffness controlled in intrinsic muscle coordinates (see section 3.1.3.2).

6. Navigation

Whole-body navigation involves the ability to move from a home location to other locations in the external world, and then return home after travelling for longer or shorter distances. This ability depends on an accurate estimate of heading direction, and on path integration (summation of all partial heading vectors). It is believed that mental maps for navigation on land are constructed in allocentric coordinates, that is in a world-based frame (McNaughton et al. 1991). Visual, vestibular, olfactory cues, can contribute to build allocentric maps in the brain, if properly calibrated and interpreted in the context of cognitive cues (Wiener and Berthoz 1993). These latter cues correspond to environmental landmarks that are recognized explicitly to belong to specific spatial locations, and that can then be used to implicitly denote these locations.

6.1 BEHAVIORAL ASPECTS

6.1.1 Vestibular contributions to navigation. Humans and animals are able to return blindfolded to a known location from which they have been passively transported away (Beritoff 1965). Vestibular-lesioned subjects, however, are often impaired in this performance, impairment that can persist even in the presence of visual cues (Miller et al. 1983). Thus, vestibular cues are important for path integration. The ability of humans to determine head displacement from vestibular information only, and to use this information to orient in space has been demonstrated for both angular motion (Bloomberg et al. 1991) and linear motion (Israel and Berthoz 1989).

As noted in section 4.1.1, linear accelerations of the head are detected by the otolith organs, whereas angular accelerations are detected by the semicircular canals. Double integration of these signals within the CNS yields positional information about distance and direction of translational displacements and about amplitude of rotations. However, this information does not indicate the absolute position in the environment, but only the position relative to the most recent point where integration of the vestibular signals was reset and started again (Wiener and Berthoz 1993). In other words, the vestibular system can only inform about relative motion between one's previous and present positions. In order to compute allocentric coordinates of movement, vestibular information must be merged with other sensory information (e.g., vision, olfaction) and must be calibrated by means of the cognitive cues mentioned above. Stabilization of head orientation (see section 5.1.3) helps to maintain the geotropic reference during navigation; in this respect, the head can be equated to an inertial platform for monitoring both gravity direction and visual reference lines, like the gyroscopes used in nautical guidance systems (Pozzo et al. 1990).

6.1.2 Visual optic flow. Recent psychophysical data on navigation using optic flow have shown that extra-retinal eye and head movement signals play an important role in extracting heading direction (Royden et al. 1992). Eye and head motion signals are probably combined in brain areas to compute eye velocity in space. For instance, it has been found that cells in MST (an area involved in optic flow analysis, see section 3.2.5.1) receive eye and head velocity signals that are usually tuned to the same direction of movement (Thier and Erickson 1992).

6.2 NEURAL SUBSTRATES

Intuitively, the complexity of sensorimotor and cognitive processing required by navigation potentially involves a vast expanse of the CNS. Nevertheless, two regions are worth of a specific mention in discussing spatial codes: the parieto-insular vestibular cortex (PIVC, Grusser et al, 1990) and the hippocampus.

6.2.1 PIVC. This region, located in the depth of the lateral sulcus near the auditory areas, receives vestibular information from the vestibular nuclei via a thalamic relay (VPLo and VPI nuclei), visual information from the accessory optic system (see section 4.1.2.2) and the pulvinar, and somatosensory information from areas 2v and 3a in primary somatosensory cortex (S1). PIVC is reciprocally connected with posterior parietal cortex (area 7), and with cingulate cortex (both anterior and posterior portions). Many neurons in PIVC respond to angular accelerations of the head, visual motion and somatosensory stimuli from neck and trunk (Grusser et al. 1990). Multisensory fusion may then result in a spatiocentric code of movement in PIVC. (For identification of a central vestibular region in man based on PET-studies, see Bottini et al. 1994).

6.2.2 Hippocampus. It receives vestibular and visual information from PIVC via the subiculum and entorhinal cortex. Visual information may also reach the hippocampus from the inferior temporal cortex via the entorhinal cortex. Pyramidal cells in the hippocampus (CA1) are active when the animal occupies specific places in the environment (*place coding*), or when his body has specific orientations in space (O'Keefe 1979; McNaughton et al. 1991). The strongest evidence for a vestibular contribution to these effects is that the directional selectivity is maintained when the animal is rotated in darkness, although the selectivity can be lost after several minutes of stimulation (Taube et al. 1990). There is a sample of cells that respond to specific directions of whole-body motion (linear vs angular motion, forward vs backward translation, clockwise vs counterclockwise rotation) independently of the specific position of the animal in space or of the specific visual view (O'Mara et al. 1994). Visual stimuli appear also to be encoded in allocentric coordinates in several cells (Feigenbaum and Rolls 1991). The responses of these units do not change when the stimuli remain in the same laboratory position but the animal is rotated or displaced to a different position. A proportion of hippocampal units respond according to the direction of the stimulus, independently of the stimulus type (visual, acoustic etc., Tamura et al. 1992). On the whole, neural encoding of an integrated vestibular signal, in association with the use of visual landmarks, results in an allocentric representation of space in the hippocampus (McNaughton et al. 1991). It has also been suggested that this kind of representation could be used in episodic memory of space and movement (Rolls 1989).

7. General conclusions

7.1 Multiple representations in the brain. The gist of this review is that sensorimotor coordination involves a multiplicity of different frames of reference within the CNS. Multimodal integration is necessary to take into account the position of a target relative to the eyes, head, body, and arm. Thus, eye-centred, head-centred, shoulder-centred, and hand-centred representations are formed at different stages of sensorimotor transformations. These different representations may be localized in different brain sites: for instance, the map of eye movements is eye-centred in the frontal eye fields, but is spatiotopic (or head-centred) in the supplementary eye fields. In other regions of the brain, instead, different types of representations may co-exist: thus, both head-centred and body-centred representations have been found in posterior parietal cortex (areas 7a and LIP, Brotchie et al. 1995). Moreover, global representations in specific reference frames may even co-exist with local, reference-free representations. In primary somatosensory cortex (SI), the topographic map of the body is replicated several times in contiguous cortical strips, and each map corresponds to a local somatotopic representation associated with specific afferent informations (skin receptors, joint and muscle receptors). In addition, however, SI is endowed with a global representation of the posture of the hand in body-centred coordinates (Helms Tillery 1994).

7.2 Explicit versus implicit spatial codes. Single neurons may encode movement explicitly in a given reference frame; thus, neurons of the superior parietal lobule are tuned to the azimuth, elevation or distance of the target and hand position (Lacquaniti et al. 1995). By contrast, head-centred and body-centred coordinates are not encoded explicitly at the level of single neurons in the inferior parietal lobule; instead, these representations are distributed over a population of neurons (Zipser and Andersen 1988).

Some researchers have suggested that the presence of scatter in the tuning functions of single neurons does not merely indicate biological noise in the encoding of a given frame, but reflects the way of operating of the learning algorithm which determined the tuning, an algorithm inherently free of geometrical coordinates, just as that implemented in artificial neural networks (Robinson 1992).

Irrespective of the specific mechanism implemented at the level of single neurons, spatial representations in reference frames are an emergent property at the systems level. Examples of idiosyncrasies in the overt behaviour, that are revealing of specific frames of reference, have been provided for all types of sensorimotor coordination considered in this review. Thus, Listing's law for eye saccades predicts that torsion is null in a head-fixed frame, but not-null in an eye-centred frame. In pointing, the spatial distributions of hand positions may cluster around coordinate axes centred at the head in a given task performed under given conditions, but may cluster around hand-centred or shoulder-centred axes in a different task or under different conditions. The angles of elevation of lower limb segments covary according to a planar law in posture and locomotion. By contrast, the anatomical angles of flexion-extension do not exhibit this invariant behaviour in locomotion (Borghese et al.). Finally, allocentric representations have been documented in the case of whole-body navigation in the environment.

7.3 Spatial representations in the brain are flexible. Spatial frames should not be viewed as abstract, rigid entities. In real life, often a target is not a single point in an otherwise homogeneous background, but it is a more or less complex object belonging to a textured environment cluttered with many other objects. This fact has several implications. First, it might be inefficient to use a single frame of reference for representing the full course of an orienting movement to the target. Instead, movement planning could be broken in segments for which different frames are used (Berthoz 1991). Whereas the initial phase could be planned in body-centred coordinates, the final phase of homing on the target could be planned more usefully in object-centred coordinates.

A second consequence of the complexity of visual arrays of targets has to do with the processes of selective attention (Posner and Petersen 1990). The capacity of the visual system to process in parallel several objects that are simultaneously present in the visual field is limited (Desimone and Duncan 1995). Only few of them can be attended, due to neural competition, and even less of them can become targets of a motor action. Therefore, local, fine-grained representations focused on the selected target must be embedded within the global, low-resolution representation of the whole visual field. Neural mechanisms for spatial selection of visual targets involve gating of the receptive field of single neurons; the receptive field can shrink around the target excluding a closely spaced distractor that would normally fall inside the field (Moran and Desimone 1985). This spatial bias is created by simple instructions administered prior to the trial, and presumably involves spatial working memory.

7.4 Reference frames linked to the sensorimotor periphery versus computational frames of reference. The frame of reference used for neural coding at certain brain sites is determined directly by the functional anatomy of the corresponding sensory epithelium or muscular apparatus at the periphery: for instance, the retinocentric coordinates in visual cortex, the coordinates of semicircular canals in the vestibular nuclei, or the coordinates associated with the pulling directions of muscles at the level of spinal motoneurones. However, many other frames are computational (Knudsen et al. 1987), resulting from complex neural processing along the central pathways. Thus, the map of auditory space in the colliculus is organized in retinocentric coordinates that are unrelated to the tonotopic organization of the cochlea, and the terminal nuclei of the accessory optic system encode optic flow along the coordinate axes of the vestibular semicircular canals, that are unrelated to the retinotopic map.

7.5 Spatial alignment of multimodal informations. A crucial problem of multisensory integration is that spatial information derived by different sensors is encoded in different frames of reference. However, multimodal information can be matched at a given brain site by means of computational representations. A particularly clear example is provided by the multimodal maps of the superior colliculus (see section 4.2.3). The SC contains convergent spatial maps from visual, auditory and somatosensory modalities. The receptive fields for each such modality are approximately in register among each other, and in register with the motor fields present in the same structure; both sensory and motor fields are organized in retinotopic coordinates. This spatial alignment is established during development and is maintained dynamically afterwards (Knudsen and Brainard 1995).

During development, the visual modality dominates in the process of adjusting and aligning the topography of the other modalities. The visual space map is based on retinotectal projections that follow genetically determined molecular markers (Sanes 1993). In dark-reared animals the visual map is normal, but the auditory and motor maps are not, the precision and topography of both these maps being severely disrupted (Knudsen and Brainard 1995). Presumably, the patterns of intermodal connectivity that are necessary to enable a neuron to respond to a particular constellation of spatial cue values are established by activity-dependent mechanisms during development: for instance, increases in synaptic efficacy resulting from synchronous input of converging afferent axons. Alignment, once established during development, must be maintained by taking into account the consequences brought about by the relative motion of the different effectors and, therefore, of the sensors attached to them. Thus, the auditory receptive fields in the SC are dynamically remapped according to eye position (Jay and Sparks 1987).

Multimodal information needs to be spatially aligned not only within a given brain region, but also across different brain regions that are concurrently active in sensorimotor coordination. Thus, we have seen that visual information is transformed in the spatial coordinates of the semicircular canals all along the brainstem pathways involved in visuo-vestibular integration: the vestibular nuclei, the terminal nuclei of the accessory optic system, the inferior olive, and the flocculo-nodular lobe of the cerebellum (see section 4.1.2).

7.6 Geotropic reference. In the cases considered in the previous section, spatial alignment is obtained by transforming the representation derived from one sensory modality into that of another sensory modality. However, this is not the only principle used for spatial alignment in the brain. There exist also neural representations in hybrid reference frames, in which spatial parameters are referred at the same time to the body and to an external reference (Carrozzo and Lacquaniti 1994). In many cases, the external reference is provided by the vertical, that is by the direction of the geotropic referential, gravity, within which all animals have been moulded (Paillard 1971). Thus, neurons in the superior parietal lobule receive convergent somatosensory information (skin receptors, muscle and joint receptors), and are active prior to and during arm movements. Many such neurons are tuned to the geotropic coordinates (azimuth, elevation and distance) of target and hand position, even though these coordinates are not directly related to those of somatosensory and motor apparati. From a behavioral standpoint, sound localization in space (Kundsen et al. 1987), proprioceptive sense of limb position (Soechting and Ross 1984), orienting movements of the head (Masino and Knudsen 1990), hand (Lacquaniti et al. 1986), and body (Masino and Grobstein 1989), postural stabilization of the body (Lacquaniti and Maioli 1994), locomotion (Borghese et al), and navigation (Wiener and Berthoz 1993) all involve a geotropic frame of reference. Because the head is normally stabilized in space, it provides an inertial platform for monitoring both gravity direction and visual reference lines (Pozzo et al. 1990). Thus, the head serves also to define an internal reference, the relative straight-ahead axis, which is used for both eye and hand orienting. The geotropic frame is also interiorized to some extent, as demonstrated by the persistence of standard postural adjustments of the body relative to the estimated vertical even in the absence of gravity (Massion et al 1993).

7.7 Internal models. The geotropic frame is used to build not only a scheme of the spatial relations of body segments, but also to build internal models of the dynamical properties of objects. Thus, in tasks that require the interception of moving objects, the timing and the amplitude of anticipatory activity in arm muscles are based on internal estimates of the dynamics of the object (path, law of motion, and momentum) that are dictated by the gravitational field (Lacquaniti 1996). Moreover, the contact forces between the hand and object are optimized in an earth-fixed frame, thanks to an internal model of the mechanical properties of the limb. Also, the processes of trajectory formation undergo adaptive changes in response to novel force fields experienced at the hand, based on the progressive changes of the internal models of limb dynamics (Shadmer and Mussa-Ivaldi 1994). Proprioceptive information has been shown to play an essential role in maintaining and re-calibrating these internal models (Gordon et al. 1995).

7.8 Cognitive representations. Although reference frames have been mainly considered in the context of sensorimotor coordination in this chapter, we know that the same frames are also used to process spatial information for mental representations that are not sensory-driven. Thus, spatial information is stored in the buffer of working memory in the same coordinates (direction and distance) that are used for visually guided reaching movements (Rossetti and Lacquaniti). Unilateral cerebral lesions may determine neglect of contralateral space for either visual stimuli or visual imagery (Bisiach and Vallar 1988; Guariglia et al. 1993). Both forms of neglect may be based on body-centred, object-centred or allocentric coordinates (Rafal 1994). Directional biases in neglect are reduced by means of experimental manipulations of gravicentric references, such as vestibular stimulation (Cappa et al. 1987) or reclination of the body in the supine posture (Pizzamiglio et al. 1995).

7.9 Modular organization. Modularity is pervasive in brain functions (see Heiligenberg 1991; Livingston and Hubel 1988). At early stages of processing within the visual system, separate subsystems are dedicated to analysing information about different stimulus attributes (brightness, colour, texture, depth, motion), although interdependencies among different modules are known to exist (colour can contribute to motion perception, for instance). At later levels of processing, a ventral cortical stream is concerned with extracting an object's form, and a dorsal cortical stream with tracking its position and motion (Ungerleider and Mishkin 1982). At a high level of processing within the ventral stream (inferotemporal cortex), analysis of object form is accomplished by neurons that are selective for different 2D-views of the object (Tanaka 1993). Contrary to the theoretical prediction by Marr (1982), recent psychophysical evidence suggests that an internal 3D model of an object for view-independent recognition is never formed (Bulthoff and Edelman 1992). Instead, multiple 2D views are stored for each object, and recognition is achieved by matching to the most similar 2D view or by interpolation between the neighboring views.

In addition to the visual system, other systems are also organized along modular lines. In the auditory system, frequency, phase, and intensity are processed in different modules (Knudsen et al. 1987; Konishi 1986). The reconstruction of spatial information of sound is also segregated anatomically, azimuth and elevation being processed in distinct nuclei of the brainstem auditory pathway (Carr and Konishi 1990). Parcellation of spatial information is also apparent in sensorimotor transformations. Thus, in the optokinetic system, different spatial directions of visual motion are processed in distinct terminal nuclei of the accessory optic system (Simpson and Graf 1985). In the saccadic system, eye or head rotations in the horizontal plane are controlled by neural circuits that are distinct from those controlling vertical rotations

(Masino and Knudsen 1990; Henn 1993). In the postural control system, limb length is controlled independently of limb orientation in response to whole-body disturbances (Lacquaniti and Maioli 1994a).

In reaching for a visual target, azimuth, elevation and distance of the hand are encoded in largely distinct subpopulations of neurons in the superior parietal lobule (Lacquaniti et al. 1995). We do not know whether global 3D-reconstruction occurs at some other cortical stage, or whether parcellation is maintained all the way down to the subcortical centers involved in movement execution. However, parallel processing of the different spatial parameters of hand reaching is evident at the behavioral level. Thus, psychophysics and chronometric studies have shown that programming the direction of a hand movement is distinct from programming its distance (Gordon et al. 1994; Soechting and Flanders 1989a). Movement direction and distance are also stored in separate channels in visuospatial working memory (Rossetti and Lacquaniti).

Modular organization and parcellation of neural processing of different attributes of sensory stimuli and motor responses probably reflect both evolutionary reasons (Heiligenberg 1991) and principles of computational parsimony to handle multidimensional information in a hugely distributed, parallel system such as the brain.

Acknowledgments
Work from the author's laboratory was partially supported by grants from CNR, IRCCS S. Lucia, MURST, and Ministero della Sanità.

References

Anastasio, T.J., Robinson, D.A. Distributed parallel processing in the vertical vestibulo-ocular reflex: learning networks compared to tensor theory. *Biol. Cybern.* 63:161-167, 1990.

Andersen, R.A., Bracewell, R.M., Barash, S., Gnadt, J.W., Fogassi, L. Eye position effects on visual, memory, and saccade-related activity in areas LIP and 7a of macaque. *J. Neurosci.* 10:1176-1196, 1990.

Andersen, R.A., Essick, G.K., Siegel, R.M. Encoding of spatial location by posterior parietal neurons. *Science* 230:456-458, 1985.

Andersen, R.A., Snyder, L.H., Li, C-S., Stricanne, B. Coordinate transformations in the representation of spatial information. *Curr. Opin. Neurobiol.* 3:171-176, 1993.

Andre, P., D'Ascanio, P., Ioffe, M., Pompeiano, O. Microinjections of vasopressin in the locus coeruleus complex affect posture and vestibulospinal reflexes in decerebrate cats. *Pflügers Arch.* 420:376-388, 1992.

Arbib, M.A. Interaction of multiple representations of space in the brain. In: *Brain and Space* (ed. J. Paillard) Oxford University Press, 1991.

Arshavsky, Yu. I., Berkinblit, M.B., Fukson, O.I., Gel'fand, I.M. Recordings of neurons of the dorsal spinocerebellar tract during evoked locomotion. *Brain Res.* 43:272-275, 1972.

Ashe, J., Georgopoulos, A.P. Movement parameters and neurla activity in motor cortex and area 5. *Cerebr. Cortex* 4:590-600, 1994.

Baker, J., Goldberg, J., Peterson, B. Spatial and temporal responses of the vestibulocollic reflex in decerebrate cats. *J. Neurophysiol.* 54:735-756, 1985.

Berkinblit, M.B., Feldman, A.G., Fukson, O.I. Adaptability of innate motor patterns and motor control mechanisms. *Behav. Brain Sci.* 9:585-638, 1986.

Beritoff, J.S. *Neural Mechanisms of Higher Vertebrate Behavior*. Brown, New York, 1965.

Berthoz, A. Reference fames for the perception and control of movement. In: *Brain and Space* (ed. J. Paillard) Oxford University Press, Oxford, UK, pp. 81-111, 1991.

Berthoz, A., Lacour, M., Soechting. J.F., Vidal, P.P. The role of vision in the control of posture during linear motion. *Prog. Brain Res.* 50:197-209, 1979.

Biguer, B., Donaldson, I.M.L., Hein, A., Jeannerod, M. Neck muscle vibration modifies the representation of visual motion and direction in man. *Brain* 111:1405-1424, 1988.

Bisiach, E., Vallar, G. Hemineglect in humans. In: *Handbook of Neuropsychology*. (Eds. F. Boller, J. Grafman) Elsevier, Amsterdam pp.195-222, 1988.

Bizzi, E., Hogan, N., Mussa-Ivaldi, F.A., Giszter, S. Does the nervous system use equilibrium-point control to guide single and multiple joint movements ? *Behav. Brain Sci.* 15:603-613, 1992.

Bizzi, E., Mussa-Ivaldi, F.A., Giszter, S. Computations underlying the execution of movement - A biological perspective. *Science* 253:287-291, 1991.

Bloomberg, J., Melvill Jones G., Segal, B. Adaptive modification of vestibularly perceived rotation. *Exp. Brain Res.* 84:47-56, 1991.

Bolton, P.S., Goto, T. Schor, R.H., Wilson, V.J., Yamagata, Y., Yates, B.J. Response of pontomedullary reticulospinal neurons to vestibular stimuli in vertical planes. Role in vertical vestibulospinal reflexes of the decerebrate cat. *J. Neurophysiol.* 67:637-647, 1992.

Borghese, N.A., Bianchi, L., Lacquaniti, F. Kinematic determinants of human locomotion. submitted

Bosco, G., Poppele, R.E. Broad directional tuning in spinal projections to the cerebellum. *J. Neurophysiol.* 70:863-866, 1993.

Bottini G, Sterzi R, Paulesu E, Vallar G, Cappa SF, Erminio F, Passingham RE, Frith CD, Frackowiak RSJ Identification of the central vestibular projections in man: a positron emission tomography activation study. *Exp. Brain Res.* 99:164-169, 1994.

Boussaoud, D. Primate premotor cortex: modulation of preparatory neuronal activity by gaze angle. *J. Neurophysiol.* 886-890, 1995.

Boyle, R., Pompeiano, O. Convergence and interaction of neck and macular vestibular inputs of vestibulospinal neurons. *J. Neurophysiol.* 45:852-868, 1981.

Bridgeman, B. Complementary cognitive and motor image processing. In: *Presbyopia Research*. (Eds. G. Obrecht, L.W. Stark) Plenum, New York, NY, pp.189-198, 1991.

Brotchie, P.R., Andersen, R.A., Snyder, L.H., Goodman, S.J. (1995) Head position signals used by parietal neurons to encode locations of visual stimuli. Nature 375:232-235.

Bullock, D., Grossberg, S. Neural dynamics of planned arm movements: emergent invariants and speed-accuracy properties during trajectory formation. *Psychol. Rev.* 95: 49-90, 1988.

Bülthoff, H.H., Edelman, S. Psychological support for a two.dimensional interpolation theory of object recognition. *Proc. Natl. Acad. Sci. USA* 89:60-64, 1992.

Burbaud, P., Doegle, C., Gross, C., Bioulac, B. A quantitative study of neuronal discharge in area 5, 2, and 4 of the monkey during fast arm movements. *J. Neurophysiol.* 66:429-443, 1991.

Burnod, Y., Grandguillaume, P., Otto, I., Ferraina, S., Johnson, P.B., Caminiti, R. Visuo-motor transformations underlying arm movements toward visual targets: a neural network model of cerebral cortical operations. *J. Neurosci.* 12:1435-1453, 1992.

Buttner, U., Henn, V. Thalamic unit activity in the alert monkey during natural vestibular stimulation. *Brain Res.* 103:127-132, 1976.

Caminiti, R., Johnson P.B., Galli, C., Ferraina, S., Burnod, Y. Making arm movements within different parts of space: the premotor and motor cortical representation of a coordinate system for reaching to visual targets. *J. Neurosci.* 11:1182-1197, 1991.

Caminiti, R., Johnson, P.B., Ferraina, S., Bianchi, L. The source of visual information to the primate frontal lobe: a novel role for the superior parietal lobule. *Cerebr. Cortex* in press

Caminiti, R., Johnson, P.B., Urbano, A. Making arm movements within different parts of space: Dynamics aspects in the primate motor cortex. *J. Neurosci.* 10:2039-2058, 1990.

Cappa, S.F., Sterzi, R., Vallar, G., Bisiach, E. Remission of hemineglect and anosognosia after vestibular stimulation. *Neuropsychologia* 25:775-782, 1987.

Carr, C.E. Konishi, M. A circuit for the detection of interaural time differences in the brain stem of the barn owl. *J. Neurosci.* 10:3227-3246, 1990.

Carrozzo, M., Lacquaniti, F. A hybrid frame of reference for visuo-manual coordination. *NeuroReport* 5:453-456, 1994.

Cheney, P.D., Fetz, E.E. Functional classes of primate corticomotoneuronal cells and their relation to active force. *J. Neurophysiol.* 44:773-791, 1980.

Cohen, D.A.D., Prud'homme, M.J.L., Kalaska, J.F. Tactile activity in primate primary somatosensory cortex during active arm movements: correlation with receptive field properties. *J. Neurophysiol.* 71:161-172, 1994.

Colby, C.L., Duhamel, J-R., Goldberg, M.E. Posterior parietal cortex and retinocentric space. *Behav. Brain Sci.* 15:727-728, 1992.

Crutcher, M.D., Alexander, G.E. Movement-related neuronal activity selectively coding either direction or muscle pattern in three motor areas of the monkey. *J. Neurophysiol.* 64:151-163, 1990.

Decety J, Perani D, Jeannerod M, Bettinardi V, Tadary B, Woods R, Mazziotta JC, Fazio F Mapping motor representations with positron emission tomography. *Nature* 371:600-602, 1994.

Desimone, R., Duncan, J. Neural mechanisms of selective visual attention. *Annu. Rev. Neurosci.* 18:193-222, 1995.

di Pellegrino, G., Wise, S.P. Visuospatial versus visuomotor activity in the premotor and prefrontal cortex of a primate. *J. Neurosci.* 13:1227-1243, 1993.

Dietz, V., Gollhofer, A., Kleiber, M., Trippel, M. Regulation of bipedal stance: dependency on "load" receptors. *Exp. Brain Res.* 89:229-231, 1992.

Duffy, C.J., Wurtz, R.H. Sensitivity of MST neurons to optic flow stimuli. I. A continuum of response selectivity to large-field stimuli. *J. Neurophysiol.* 65:1329-1345, 1991.

Dugas, C., Smith, A.M. Responses of Cerebellar Purkinje Cells to Slip of a Hand-Held Object. *J. Neurophysiol.* 67:483-495, 1992.

Dyhre-Poulsen, P., Simonsen, E.B., Voigt, M. Dynamic Control of Muscle Stiffness During Hopping and Jumping in Man. *J. Physiol. (London)* 437:237-304, 1991.

Evarts, E.V. Relation of pyramidal tract activity to force exerted during voluntary movement. *J. Neurophysiol.* 31:14-27, 1968.

Ezure, K., Graf, W. A quantitative analysis of the spatial organization of the vestibulo-ocular reflexes in lateral- and frontal eyed animals. I. Orientation of semicircular canals and extraocular muscles. *Neuroscience* 12:85-94, 1984a.

Ezure, K., Graf, W. A quantitative analysis of the spatial organization of the vestibulo-ocular reflexes in lateral- and frontal eyed animals. II. Neuronal networks underlyng vestibulo-oculomotor coordination. *Neuroscience* 12:95-110, 1984b.

Ezure, K., Wilson, V.J. Interaction of tonic neck and vestibular reflexes in the forelimb of the decerebrate cat. *Exp. Brain Res.* 54:289-292, 1984.

Feigenbaum, J.D., Rolls, E.T. Allocentric and egocentric spatial information processing in the hippocampal formation of the behaving primate. *Psychobiology* 19:21-40, 1991.

Feldman, A.G., Levin, M.F. Positional frames of reference in motor control. The origin and use. *Behav. Brain Sci.* in press

Felleman, D.J., Van Essen, D.C. Distributed hierarchical processing in the primate cerebral cortex. *Cereb. Cortex* 1:1-47, 1991.

Flanders, M., Helms Tillery, S.I., Soechting, J.F. Early Stages in a Sensorimotor Transformation. *Behav. Brain Sci.* 15:309-362, 1992.

Flash, T., Hogan, N. The coordination of arm movements: an experimentally confirmed model. *J. Neurosci.* 5:1688-1703, 1985.

Fookson, O.I., Smetanin, B., Berkinblit, M.B., Adamovich, S., Feldman, G., Poizner, H. Azimuth errors in pointing to remembered targets under extreme head rotations. *NeuroReport* 5:885-888, 1994.

Fu, Q.-G., Flament, D., Coltz, J.D., Ebner, T.J. Temporal encoding of movement kinematics in the discharge of primate primary motor and premotor neurons. *J. Neurophysiol.* 73:836-854, 1995.

Fu, Q.-G., Suarez, J.I., Ebner, T.J. Neuronal specification of direction and distance during reaching movements in the superior precentral premotor area and primary motor cortex of monkeys. *J. Neurophysiol.* 70:2097-2116, 1993.

Fukushima, K., Perlmutter, S.J., Baker, J.F., Peterson, B.W., Spatial properties of second-order vestibulo-ocular relay neurons in the alert cat. *Exp. Brain Res.* 81:462-478, 1990.

Funahashi, S., Bruce, C.J., Goldman-Rakic, P.S. Mnemonic coding of visual space in the monkey's dorsolateral prefrontal cortex. *J. Neurophysiol.* 61:331-349, 1989.

Galletti, C., Battaglini, P.P. Gaze-dependent visual neurons in area V3A of monkey prestriate cortex. *J. Neurosci.* 9:1112-1125, 1989.

Galletti, C. Battaglini, P.P., Fattori, P. Functional properties of neurons in the anterior bank of the parieto-occipital sulcus of the macaque monkey. *Eur. J. Neurosci.* 3:452-461, 1991.

Galletti, C., Battaglini, P.P., Fattori, P. Parietal neurons encoding spatial locations in craniocentric coordinates. *Exp. Brain Res.* 96:221-229, 1993.

Gandevia, S.C., Burke, D., Does the nervous system depend on kinesthetic information to control natural limb movements? *Behav. Brain Sci.* 15:614-632, 1992.

Georgopoulos, A.P. On reaching. *Annu. Rev. Neurosci.* 9:147-170, 1986.

Georgopoulos, A.P. Higher Order Motor Control. *Annu. Rev. Neurosci.* 14:361-377, 1991.

Georgopoulos, A.P. New concepts in generation of movement. Neuron 13:257-268, 1994.

Georgopoulos, A.P., Ashe, J., Smyrnis, N., Taira, M. The motor cortex and the coding of force. *Science* 256:1692-1695, 1992.

Georgopoulos, A.P., Caminiti, R., Kalaska, J.F., Massey, J.T. Spatial coding of movement: a hypothesis concerning the coding of movement direction by motor cortical populations. *Exp. Brain Res.* Suppl. 7:327-336, 1983.

Georgopoulos, A.P., Kalaska, J.F., Caminiti, R., Massey, J.T. On the relations between the direction of two-dimensional arm movements and cell discharge in primate motor cortex. *J. Neurosci.* 2:1527-1537, 1982.

Georgopoulos, A.P., Kettner, R.E., Schwartz, A.B. Primate motor cortex and free arm movements to visual targets in three-dimensional space. II. Coding of the direction of movement by a neuronal population. *J. Neurosci.* 8:2928-2937, 1988.

Goldberg, M.E., Bruce, C.J. Primate frontal eye fields: III. Maintenance of a spatially accurate saccade signal. *J. Neurophsyiol.* 64:489-508, 1990.

Gordon, J., Ghilardi, M.F., Ghez, C. Accuracy of planar reaching movements. I. Independence of direction and extent variability. *Exp. Brain Res.* 99:97-111, 1994.

Gordon, J., Ghilardi, M.F., Ghez, C. Impairments of reaching movements in patients without proprioception. I. Spatial errors. *J. Neurophysiol.* 73:347-360, 1995.

Graf, W. Motion detection in physical space and its peripheral and central representations. *Ann. NY Acad. Sci.* 545:154-169, 1988.

Graf, W., Simpson, J.I., Leonard, C.S. Spatial organization of visual messages of the rabbit's cerebellar flocculus. II. Complex and simple spike responses of Purknie cells. *J. Neurophysiol.* 60:2091-2121, 1988.

Grafton ST, Mazziotta JC, Woods RP, Phelps ME Human functional anatomy of visually guided movements. *Brain* 115:565-587, 1992.

Grantyn, A., Berthoz, A., Olivier, E., Chat, M. Control of gaze by tectal and reticular projection neurons. In: *Multisensory control of movement* (eds. A. Berthoz et al.) Oxford University Press, Oxford, pp. 185-200, 1993.

Grusser, O.J., Pause, M., Schreiter, U. Localization and responses in the parieto-insular vestibular cortex of awake monkeys (Macaca fascicularis). *J. Physiol. (London)* 430:537-557, 1990.

Guariglia, C., Padovani, A., Pantano, P., Pizzamiglio, L. Unilateral neglect restricted to visual imagery. *Nature* 364:235-237, 1993.

Guitton, D. Control of eye-head coordination during orienting gaze shift. *Trends Neurosci.* 15:174-179, 1992.

Hallet, P.E., Lightstone, A.D. Saccadic eye movements towards stimuli triggered by prior saccades. *Vision Res.* 16:99-106, 1976.

Heiligenberg, W. The neural basis of behavior: a neuroethological view. *Annu. Rev. Neurosci.* 14:247-267, 1991.

Henn, V. Neuronal control of eye movements. In: *Multisensory control of movement* (eds. A. Berthoz et al.) Oxford University Press, Oxford, pp. 7-26, 1993.

Henn, V., Cohen, B., Young, L.R. Visual-vestibular interaction in motion perception and the generation of nystagmus. *Neurosci. Res. Prog. Bull.* 18:457-651, 1980.

Hepp, K. Oculomotor control: Listing's law and all that. *Curr. Op. Neurobiol.* 4:862-868, 1994.

Hepp, K., Henn, V. Iso-frequency curves of oculomotor neurons in the rhesus monkey. *Vision Res.* 25:493-499, 1985.

Hikosaka, O., Wurtz, R.H. Visual and oculomotor functions of monkey substantia nigra pars reticulata. III. Memory-contingent visual and saccade responses. *J. Neurophysiol.* 49:1268-1284, 1983.

Hirayama, M., Kawato, M., Jordan, M.I. The cascade neural network model and a speed-accuracy trade-off of arm movement. *J. Mot. Behav.* 25:162-174, 1993.

Hirschfeld, H., Forssberg, H. Phase-dependent modulations of anticipatory postural activity during human locomotion. *J. Neurophysiol.* 66:12-19, 1991.

Hocherman, S., Wise, S.P. Effects of hand movement path on motor cortical activity in awake, behaving rhesus monkeys. *Exp. Brain Res.* 83:285-302, 1991.

Hoffmann, K.-P., Distler, C. Quantitative analysis of visual receptive fields of neurons in nucleus of the optic tract and dorsal terminal nucleus of the accessory optic tract in macaque monkey. *J. Neurophysiol.* 62:416-428, 1989.

Hogan, N. The mechanics of multi-joint posture and movement control. *Biol. Cybern.* 52:315-332, 1985.

Hogan, N. Planning and execution of multijoint movements. *Can. J. Physiol. Pharmacol.* 66:508-517, 1988.

Hollerbach, J.M., Atkeson, C.G. Deducing planning variables from experimental arm trajectories: pitfalls and possibilities. *Biol. Cybern.* 56:279-292, 1987.

Israel, I., Berthoz, A. Contributions of the otoliths to the calculation of linear displacement. *J. Neurophysiol.* 62:247-263, 1989.

Jay, M.F., Sparks, D.L. Sensorimotor integration in the primate superior colliculus. II. Coordinates of auditory signals. *J. Neurophysiol.* 57:35-55, 1987.

Jeannerod M *The Neural and Behavioural Organization of Goal-directed Movements.* Oxford: Clarendon Press, 1988.

Jeannerod, M. A neurophysiological model for the directional coding of reaching movements. In: *Brain and Space* (ed. J. Paillard) Oxford University Press, Oxford, U.K., pp. 49-69, 1991.

Jeannerod, M., Arbib, M.A., Rizzolatti, G., Sakata, H. Grasping objects: the cortical mechanisms of visuomotor transformation. *Trends Neurosci.* 18:314-320, 1995.

Jeannerod, M., Biguer, B. The directional coding of reaching movements. A visuomotor conception of spatial neglect. In: *Neurophysiological and Neuropsychological Aspects of Spatial Neglect.* (Jeannerod, M. Ed.) Elsevier, Amsterdam, pp.87-113, 1987.

Johansson, R.S. How is grasping modified by somatosensory input ? In: *Motor Control: Concepts and Issues.* (D.R. Humphrey, H.-J. Freund, Eds.) Wiley, New York, pp. 331-355, 1991.

Johnson, P.B., Ferraina, S., Caminiti, R. Cortical networks for visual reaching. *Exp. Brain Res.* 97:361-365, 1993.

Kalaska, J.F., Caminiti, R., Georgopoulos, A.P. Cortical mechanisms related to the direction of two-dimensional arm movements: Relations in parietal area 5 and comparison with motor cortex. *Exp. Brain Res.* 51:247-260, 1983.

Kalaska, J.F., Cohen, D.A.D., Hyde, M.L., Prud'homme, M. A comparison of movement direction-related versus load direction-related activity in primate motor cortex, using a two-dimensional reaching task. *J. Neurosci.* 9:2080-2102, 1989.

Kalaska, J.F., Cohen, D.A.D., Prud'homme, M., Hyde, M.L. Parietal area 5 neuronal activity encodes movement kinematics, not movement dynamics. *Exp. Brain Res.* 80:351-364, 1990.

Kalaska, J.F., Crammond, D.J. Cerebral cortical mechanisms of reaching movements. *Science* 255:1517-1523, 1992.

Karnath, H.O., Christ, K., Hartje, W. Decrease of contralateral neglect by neck muscle vibration and spatial orientation of trunk midline. *Brain* 116:383-396, 1993.

Kasper, J., Schor, R.H., Wilson, V.J. Response of vestibular neurons to head rotations in vertical planes. II. Response to neck stimulation and vestibular-neck interaction. *J. Neurophysiol.* 60:1765-1778, 1988.

Kawashima R, Roland PE, O'Sullivan BT Fields in human motor areas involved in preparation for reaching, actual reaching, and visuomotor learning: a positron emission tomography study. *J. Neurosci.* 14:3462-3474, 1994.

Knudsen, E.I., Brainard, M.S. Creating a unified representation of visual and auditory space in the brain. *Annu. Rev. Neurosci.* 18:19-43, 1995.

Knudsen, E.I., Du Lac, S., Easterly, S. Computational maps in the brain. *Annu. Rev. Neurosci.* 10:41-65, 1987.

Konishi, M. Centrally synthesized maps of sensory space. *Trends Neurosci.* 9:163-168, 1986.

Lacquaniti, F. Central representations of human limb movements as revealed by studies on drawing and handwriting. *Trends Neurosci.* 12:287-291, 1989.

Lacquaniti, F. Automatic control of limb movement and posture. *Curr. Opin. Neurobiol.* 2:807-814, 1992.

Lacquaniti, F. Neural control of hand mechanics for visuomanual coordination. In: *Hand and Brain* (A. Wing, P. Haggard, R. Flanagan eds.) Academic Press, 1996

Lacquaniti, F., Borghese, N.A., Carrozzo, M. Transient reversal of the stretch reflex in human armmuscles. *J. Neurophysiol.* 66:939-954, 1991.

Lacquaniti, F., Borghese, N.A., Carrozzo, M. Internal models of limb geometry in the control of hand compliance. *J. Neurosci.* 12:1750-1762, 1992.

Lacquaniti, F., Carrozzo, M., Borghese, N.A. Time-varying mechanical behavior of multi-jointed arm in man. *J. Neurophysiol.* 69:1443-1464, 1993.

Lacquaniti, F., Guigon, E., Bianchi, L., Ferraina, S., Caminiti, R. Representing spatial information for limb movement: role of area 5 in the monkey. *Cereb. Cortex* 5:391-409, 1995.

Lacquaniti, F., LeTaillanter, M., Lopiano, L., Maioli, C. The control of limb geometry in cat posture. *J. Physiol. (London)* 426:177-192, 1990.

Lacquaniti, F., Maioli, C. The role of preparation in tuning anticipatory and reflex responses during catching. *J. Neurosci.* 9:134-148, 1989.

Lacquaniti, F., Maioli, C. Independent control of limb position and contact forces in cat posture. Journal of Neurophysiology 72:1476-1495, 1994a.

Lacquaniti, F., Maioli, C. Coordinate transformations in the control of cat posture. *J. Neurophysiol.* 72:1496-1515, 1994b.

Lacquaniti, F., Soechting, J.F. Coordination of arm and wrist motion during a reaching task. *J. Neurosci.* 2:399-408, 1982.

Lacquaniti, F., Soechting, J.F., Terzuolo, C. Some factors pertinent to the organization and control of arm movements. *Brain Res.* 252:394-397, 1982.

Lacquaniti, F., Soechting, J.F., Terzuolo, C. Path constraints on point-to-point arm movements in three-dimensional space. *Neuroscience* 17:313-324, 1986.

Lacquaniti, F., Terzuolo, C., Viviani, P. The law relating the kinematic and figural aspects of drawing movements. *Acta Psychol.* 54:115-130, 1983.

Lacquaniti, F., Terzuolo, C., Viviani, P. Global metric properties and preparatory processes in drawing movements. In: *Preparatory States and Processes* (eds. S. Kornblum and J. Requin) Erlbaum, Hillsdale, NJ, pp.357-370, 1984.

Lee, C., Rohrer, W.H., Sparks, D.L. Population coding of saccadic eye movements by neurons in the superior colliculus. *Nature* 332:357-360, 1988.

Lee, D.N. Visuo-motor coordination in space-time. In: *Tutorials in Motor Behavior* (Eds. G.E. Stelmach and J. Requin) North-Holland Publishing, p. 281-295, 1980.

Lindsay, K.W., Roberts, T.D.M., Rosenberg, J.R. Asymmetric tonic labyrinthine reflexes and their interaction with neck reflexes in the decerebrate cat. *J. Physiol. (London)* 261:583-601, 1976.

Livingston, M., Hubel, D. Segregation of form, color, movement, and depth: anatomy, physiology and perception. *Science* 240:740-749, 1988.

Magnin, M., Courjon, J.H., Flandrin, J.M. Possible visual pathways to the cat vestibular nuclei involving the nucleus prepositus hypoglossi. *Exp. Brain Res.* 51:298-303, 1983.

Maioli, C., Ohgaki, T. Role of the accessory optic system for sensorimotor integration. The problem of reference frames. In: *Multisensory control of movement* (eds. A. Berthoz et al.) Oxford University Press, Oxford, pp. 73-92, 1993.

Maioli, C., Poppele, R.E. Parallel processing of multisensory information concerning self-motion. *Exp. Brain Res.* 87:119-125, 1991.

Manzoni, D., Pompeiano, O., Srivastava, U.C., Stampacchia, G. Responses of forelimb extensors to sinusoidal stimulation of macular labyrinth and neck receptors. *Arch. Ital. Biol.* 121:205-214, 1983.

Marr, D. *Vision.* Freeman, San Francisco, 1982.

Masino, T., Grobstein, P. The organization of descending tectofugal pathways underlying orienting in the frog, Rana pipiens. I. Lateralization, parcellation and an intermediate spatial representation. *Exp. Brain Res.* 75:227-244, 1989.

Masino, T., Knudsen, E.I. Horizontal and vertical components of head movement are controlled by distinct neural circuits in the barn owl. *Nature* 345:434-437, 1990.

Massion, J. Movement, posture and equilibrium: interaction and coordination. *Prog. Neurobiol.* 38:35-56, 1992.

Massion, J., Gurfinkel, V., Lipshits, M., Obadia, A., Popov, K. Axial synergies under microgravity conditions. *J. Vestib. Res.* 3:275-287, 1993.

Matelli, M., Camarda, R., Glickstein, M., Rizzolatti, G. Afferent and efferent projections of the inferior area 6 in the macaque monkey. *J. Comp. Neurol.* 251:281-298, 1986.

McNaughton, B.L., Chen, L., Markus, E.J. 'Dead reckoning', landmark learning, and the sense of direction: a neurophysiological and computational hypothesis. *J. Cogn. Neurosci.* 3:190-202, 1991.

Miller, S., Potegal, M., Abraham, L. Vestibular involvement in a passive transport and return task. *Physiol. Psychol.* 11:1-10, 1983.

Mittelstaedt, H. Basic control patterns of orientational homeostasis. *Symp. Soc. Exp. Biol.* 18:365-385, 1964.

Mittelstaedt, H. A new solution to the problem of the subjective vertical. *Naturwissenschaften* 70:272-281, 1983.

Moran, J., Desimone, R. Selective attention gates visual processing in the extrastriate cortex. *Science* 229:782-784, 1985.

Mountcastle, V.B., Lynch, J.C., Georgopoulos, A., Sakata, H., Acuna C Posterior parietal association cortex of the monkey: command functions for operations within extrapersonal space. *J. Neurophysiol.* 38:871-908, 1975.

Mustari, M.J., Fuchs, A.F. Response properties of single units in the lateral terminal nucleus of the accessory optic system in the behaving primate. *J. Neurophysiol.* 61:1207-1220, 1989.

Nashner, L.M. Adapting reflexes controlling the human posture. *Exp. Brain Res.* 26:59-72, 1976.

Nashner, L.M., McCollum, G. The organization of human postural movements: a formal basis and experimental synthesis. *Behav. Brain Sci.* 8:135-172, 1985.

O'Keefe, J. A review of the hippocampal place cells. *Prog. Neurobiol.* 13:419-439, 1979.

O'Mara, S.M., Rolls, E.T., Berthoz, A., Kesner, R.P. Neurons responding to whole-body motion in the primate hippocampus. *J. Neurosci.* 14:6511-6523, 1994.

Oyster, C.W., Takahashi, E., Collewijn, H. Direction selective retinal ganglion cells and control of optokinetic nystagmus in the rabbit. *Vision Res.* 12:183-193, 1972.

Paillard, J. Les déterminants moteurs de l'organisation spatiale. *Cahiers Psychol.* 14:261-316, 1971.

Paillard, J. The contribution of peripheral and central vision to visually guided reaching. In: *Analysis of Visual Behavior* (eds. D.J. Ingle, M.A. Goodale and R.J. Mansfield) MIT Press, Cambridge, MA, p. 367-385, 1982.

Paillard, J. Motor and representational framing of space. In: *Brain and Space* (Paillard J, ed), pp 163-182. New York: Oxford University Press, 1991.

Pellionisz, A., Llinas, R. Tensorial approach to the geometry of brain function: cerebellar coordination via a metric tensor. *Neuroscience* 5:1125-1136, 1980.

Pellionisz, A., Peterson, B.W. A tensorial model of neck motor activation. In: *Control of Head Movement.* (Eds. B.W. Peterson, F. Richmond), Oxford Univ. Press, Oxford, pp. 178-186, 1988.

Peterson, B.W., Baker, J.F. Spatial transformations in vestibular reflex systems. In *Motor Control: Concepts and Issues.* Edited by Humphrey DR, Freund H-J. Chichester: Wiley 1991 pp. 121-135, 1991.

Pizzamiglio, L., Vallar, G., Doricchi, F. Gravity and hemineglect. *NeuroReport* in press

Posner, M.I., Petersen, S.E. The attention system of the human brain. *Annu. Rev. Neurosci.* 13:25-42, 1990.

Pozzo, T., Berthoz, A., Lefort, L. Head stabilization during various locomotor tasks in humans. *Exp. Brain Res.* 82:97-106, 1990.

Precht, W., Strata, P. On the pathway mediating optokinetic responses in vestibular nuclear neurons. *Neuroscience* 5:777-787, 1980

Rafal, R.D. Neglect. *Curr. Opin. Neurobiol.* 4:231-236, 1994.

Roberts, T.D.M. Reflex balance. *Nature* 244:156-158, 1973.

Robinson, D.A. Oculomotor control signals. In: *Basic mechanisms of ocular motility and their clinical implications* (Eds. G. Lennerstrand, P. Bach-y-Rita), Pergamon, Oxford, pp. 337-374, 1975.

Robinson, D.A. The use of matrices in analyzing the three dimensional behavior of the vestibulo-ocular reflex. *Biol. Cybern.* 46:53-66, 1982.

Robinson, D.A. Implications of neural networks for how we think about brain function. *Behav. Brain Sci.* 15:644-655, 1992.

Roll, R., Velay, J.L., Roll, J.P. Eye and neck proprioceptive messages contribute to the spatial coding of retinal input in visually oriented activities. *Exp. Brain Res.* 85:423-431, 1991.

Rolls, E.T. Functions of neuronal networks in the hippocampus and neocortex in memory. In: *Neural Models of Plasticity: Experimental and Theoretical Approaches* (Eds. J.H. Byrne, W.O. Berry) Academic Press, San Diego, pp. 240-265, 1989.

Rosenbaum, D.A. Human movement initiation: specification of arm, direction, and extent. *J. Exp. Psychol.: General.* 109:444-474, 1980.

Roucoux, A., Crommelinck, M. Orienting gaze: a brief survey. In: *Multisensory control of movement* (eds. A. Berthoz et al.) Oxford University Press, Oxford, pp. 130-149, 1993.

Royden, C.S., Banks, M.S., Crowell, J.A. The perception of heading during eye movements. *Nature* 360:583-585, 1992.

Sakata, H., Shibutani, H., Kawano, K. Spatial properties of visual fixation neurons in posterior parietal association cortex of the monkey. *J. Neurophysiol.* 43:1654-1672, 1980.

Sakata, H., Shibutani, H., Kawano, K., Harrington, T.L. Neural mechanisms of space vision in the parietal association cortex of the monkey. *Vision Res.* 25:453-464, 1985.

Sakata, H., Takaoka, Y., Kawarasaki, A., Shibutani, H. Somatosensory properties of neurons in the superior parietal cortex (area 5) of the rhesus monkey. *Brain Res.* 64:85-102, 1973.

Sanes, J.R. Topographic maps and molecular gradients. *Curr. Opin. Neurobiol.* 3:67-74, 1993.

Schlag, J., Schlag-Rey, M. Evidence for a supplementary eye field. *J. Neurophysiol.* 57:179-200, 1987.

Schwartz, A.B. Direct cortical representation of drawing. *Science* 265:540-542, 1994.

Schwartz, A.B., Kettner, R.E., Georgopoulos, A.P. Primate motor cortex and free arm movements to visual targets in three-dimensional space. I. Relations between single cell discharge and direction of movement. *J. Neurosci.* 8:2913-2927, 1988.

Scott, S.H., Kalaska, J.F. Changes in motor cortex activity during reaching movements with similar hand paths but different arm postures. *J. Neurophysiol.* 73:2563-2567, 1995.

Shadmehr, R., Mussa-Ivaldi, F.A. Adaptive representation of dynamics during learning of a motor task. *J. Neurosci.* 14:3208-3224, 1994.

Simpson, J.I. The accessory optic system. *Annu. Rev. Neurosci.* 7:13-41, 1984.

Simpson, J.I., Graf, W. The selection of reference frames by nature and its investigation. *Rev. Oculomot. Res.* 1:3-20, 1985.

Snowden, R.J., Treue, S., Andersen, R.A. The response of neurons in areas V1 and MT of the alert rhesus monkey to moving random dot patterns. *Exp. Brain Res.* 88:389-400, 1992.

Sodak, R.E., Simpson, J.I. The accessory optic system of rabbit. I. Basic visual response properties. *J. Neurophysiol.* 60:2037-2054, 1988.

Soechting, J.F., Flanders, M. Sensorimotor representations for pointing to targets in three-dimensional space. *J. Neurophysiol.* 62:582-594, 1989a.

Soechting, J.F., Flanders, M. Errors in pointing are due to approximations in sensorimotor transformations. *J. Neurophysiol.* 62:595-608, 1989b.

Soechting, J.F., Flanders, M. Arm movements in three-dimensional space: computation, theory, and observation. *Exerc. Sports Sci. Rev.* 19:389-418, 1991.

Soechting, J.F., Flanders, M. Moving in three-dimensional space: frames of reference, vectors, and coordinate systems. *Annu. Rev. Neurosci.* 15:167-191, 1992.

Soechting, J.F., Lacquaniti, F. Invariant characteristics of a pointing movement in man. *J. Neurosci.* 1:710-720, 1981.

Soechting, J.F., Lacquaniti, F., Terzuolo, C. Coordination of arm movements in three-dimensional space. Sensorimotor mapping during drawing movement. *Neuroscience* 17:295-311, 1986.

Soechting, J.F., Ross, B. Psychophysical determination of coordinate representation of human arm orientation. *Neuroscience* 13:595-604, 1984.

Sparks, D.L. Neural cartography: sensory and motor maps in the superior colliculus. *Brain Behav. Evol.* 31:49-56, 1988.

Sparks, D.L., Mays, L.E. Signal transformations required for the generation of saccadic eye movements. *Annu. Rev. Neurosci.* 13:309-336, 1990.

Stein, J.F. The representation of egocentric space in the posterior parietal cortex. *Behav. Brain Sci.* 15:691-700, 1992.

Straumann, D., Yué, Q., Kawachi, N., Henn, V. Three-dimensional anisotropy of vestibulo-ocular reflex in rhesus monkey. In: *Multisensory control of movement* (eds. A. Berthoz et al.) Oxford University Press, Oxford, pp 42-50, 1993.

Tamura, R., Ono, T., Fukuda, M., Nakamura, K. Spatial responsiveness of monkey hippocampal neurons to various visual and auditory stimuli. *Hippocampus* 2:307-322, 1992.

Tanaka, K. Neuronal mechanisms of object recognition. *Science* 262:685-688, 1993.

Taube, J.S., Muller, R.U., Ranck, J.B. Head-direction cells recorded from the postsubiculum in freely moving rats. I. Description and quantitative analysis. *J. Neurosci.* 10:420-435, 1990.

Thier, P., Erickson, R.G. Responses of visual-tracking neurons from cortical area MST-1 to visual, eye and head motion. *Eur. J. Neurosci.* 4:539-553, 1992.

Trotter, Y., Celebrini, S., Stricanne, B., Thorpe, S., Imbert, M. Modulation of neural stereoscopic processing in primate area V1 by the viewing distance. *Science* 257:1279-1281, 1992.

Ungerleider, L.G., Mishkin, M. Two cortical visual systems. In: *Analysis of Visual Behavior* (Eds. Ingle, D.J., Goodale, M.A., Mansfield, R.J.W.) MIT press, Cambridge, pp.549-586, 1982.

Uno, Y., Kawato, M., Suzuki, R. Formation and control of optimal trajectory in human multijoint arm movement. Minimum torque-change model. *Biol. Cybern.* 61:89-101, 1989.

Van Opstal, A.J., Hepp, K., Hess, B.J.M., Straumann, D., Henn, V. Two- rather than three-dimensional representation of saccades in monkey superior colliculus. *Science* 252:1313-1315, 1991.

Vanden Abeele, S., Crommelinck, M., Roucoux, A. Frames of reference used in goal-directed arm movement. In: *Multisensory control of movement* (eds. A. Berthoz et al.) Oxford University Press, Oxford, pp. 363-378, 1993.

Viviani, P., Flash, T. Minimum-jerk, two-thirds power law and isochrony: converging approaches to movement planning. *J. Exp. Psychol: Hum. Percept. Perf.* 21:32-53, 1995.

Viviani, P., Terzuolo, C. Space-time invariance in learned motor skills. In: *Tutorials in Motor Behavior* (eds. G.E. Stelmach & J. Requin) North-Holland, Amsterdam, pp.525-533, 1980.

Viviani, P., Terzuolo, C. Trajectory determines movement dynamics. *Neuroscience* 7:431-437, 1982.

Wiener, S., Berthoz, A. Forebrain structures mediating the vestibular contribution during navigation. In: *Multisensory control of movement* (eds. A. Berthoz et al.) Oxford University Press, Oxford, pp. 427-456, 1993.

Wilson, V.J. Vestibulospinal and neck reflexes: interaction in the vestibular nuclei. *Arch. Ital. Biol.* 129:43-52, 1991.

Wilson, V.J., Schor, R.H., Suzuki, I., Park, B.R. Spatial organization of neck and vestibular reflexes acting on the forelimbs of the decerebrate cat. *J. Neurophysiol.* 55:514-526, 1986.

Worringham, C.J., Stelmach, G.E., Martin, Z.E. Limb segment inclination sense in proprioception. *Exp. Brain Res.* 66:653-658, 1987.

Wurtz, R.H., Optican, L.M. Superior colliculus cell types and models of saccade generation. *Curr. Opin. Neurobiol* 4:857-861, 1994.

Young, L.R. Perception of the body in space: mechanisms. In: *Handbook of Physiology. The Nervous System. Motor Control.* Section 1, Vol. 3, Part 2 (Brookhart, J.M., Mountcastle, V.B. Eds.) Am. Physiol. Soc. Bethesda, MD, pp. 1023-1066, 1984.

Zipser, D., Andersen, R.A. A back-propagation programmed network that simulates response properties of a subset of posterior parietal neurons. *Nature* 331:679-684, 1988.

THE NEURAL CONTROL OF OBJECT-ORIENTED ACTIONS*

M. JEANNEROD
Vision et Motricité, INSERM U94
16, avenue du Doyen Lépine
F-69500 Bron, FRANCE

A large part of human actions are directed toward objects. Fundamental aspects of our behavior, like the ability to use tools, for example, originate from neural specialization for perceiving, grasping, recognizing and categorizing objects. These operations correspond to adaptive acquisitions in primates and some are unique to man. In this chapter, an ensemble of mechanisms for behaving with objects will be described, with emphasis on the anatomical and physiological arguments that allow to delineate a specific neural system devoted to object-oriented movements.

1. Visuomotor coordination as a dissociable visual function

Visuomotor coordination requires specific processing of visual information for transfering it into a motor code. In this section, the neural systems which account for

* Note: This Chapter is an abbriged version of a Chapter to appear in: M. Jeannerod, "Representations for action. Neural coding and cognitive structure". Oxford, Blackwell (In press).

F. Lacquaniti and P. Viviani (eds.), Neural Bases of Motor Behaviour, 41–76.

visuomotor transformation will be described, starting with an analysis of the relevant aspects of the anatomy of the visual system (see Jeannerod and Rossetti, 1993).

1.1. THE TWO-VISUAL-SYSTEMS HYPOTHESIS

The notion that retinofugal fibers are distributed along several central visual pathways has a long history in neuroscience. The visuomotor function of the subcortical pathway relaying in the superior colliculus was particularly advocated by Cajal (1909). He described a "descending" or "motor" pathway arising from the 4th layer of the superior colliculus and terminating in the ocular motor nuclei and in the adjacent reticular formation. Later on, Hess et al (1946) based on stimulation experiments in cats, concluded that the visual grasp reflexes were organized in the superior colliculi. Accordingly, extensive lesions of this structure (Sprague and Meikle, 1965) were shown to produce severe impairment in eye movements and visuomotor behaviour.

The modern version of this hypothesis re-emerged in the late 1960s. The distribution of retinofugal fibers between retinogeniculate and retinotectal pathways was thus interpreted within the framework of a dichotomy between two visual systems endowed with complementary functions. Schneider (1969) proposed that the geniculostriate pathway was a system essential in the learning of pattern discrimination, and the retinotectal pathway, a system for mediating spatial orientation. Using hamsters (a species with a large and easily accessible tectum) to demonstrate his point, Schneider showed that, following ablation of visual areas 17 and 18 animals became unable to learn simple pattern discriminations (e.g., vertical vs horizontal stripes), although they remained able to orient toward stimuli presented across their visual field. By contrast, following large undercuttings through the midbrain tectum, the spatial orientation ability was lost, whereas pattern discrimination was still possible. This anatomical and functional duality became known as the now classical opposition between a system specialized for answering the question "What is it?", and another one specialized for answering the question "Where is it?" (Schneider, 1969).

A similar version of the parallel visual functions model was used for explaining the mechanisms of visuomotor coordination. Hein and Held (1967) showed that kittens reared without sight of their forelimbs presented a defective placing reaction when approached from the edge of a table. Whereas they were still able to extend their forelimbs prior to contact with the table, their response was inaccurate: if the edge was made discontinuous to render the task more difficult, they were unable to guide their forelimbs at the appropriate location, and contacted the table at random. Hein and Held thus concluded that the placing reaction was in fact dissociable into components, one of which, elicited extension, developed without sight of the forelimbs, and the other, guided placing, required a prolonged viewing of the forelimbs to be present. This dissociation of visuomotor behaviour into components implicitly suggested separate substrates for the visual control of each of them. Indeed, a similar dissociation was observed in kittens reared in complete darkness (a condition known to severely affect the functioning of visual cortex) (Vital-Durand et al, 1974). Accordingly, accurate guidance would be preferentially controlled by the visual cortex, whereas elicited extension would be a more primitive response arising from stimulation of the peripheral retina and processed outside visual cortex. Another example of such a dissociation is illustrated in Figure 1.

The introduction of primates in visuomotor research led to one of the most striking outcomes of the two-visual-system hypothesis, namely the (re)discovery of residual visual function following lesion of visual cortex. Pathological destruction of visual cortex in man was classically thought to produce total blindness, except for pupillar response to light and very crude visual perception limited to sudden changes in illumination. This conventional opinion, however, was called into question on the basis of experimental findings in monkeys. Although destriated monkeys also appeared to be profoundly impaired in their ordinary visual behavior, they were still able to avoid obstacles and to generate motor responses for reaching objects appearing in, or moving across their visual field (Humphrey and Weiskrantz, 1967) These residual visual abilities were attributed to retinal ganglion cells surviving ablation of visual cortex and projecting subcortically (see Dineen and Hendrickson, 1981). These anatomical findings

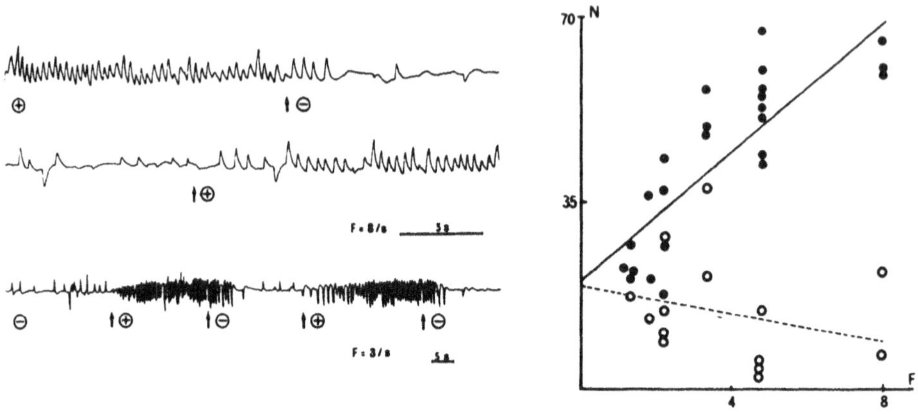

Figure 1.

An example of an experiment prompted by the "Two visual systems" hypothesis. Kittens reared in the dark from birth were exposed 1 hour per day, between the 4th and the 10th week of age, to a visual environment which consisted exclusively of vertical edges moving horizontally at a constant speed and always in the same direction. Total exposure varied between 10 and 60 hours. At the 12th week of age optokinetic nystagmus (OKN) was tested. Displacements of a striped visual pattern in the direction of exposure elicited a normal OKN with a positive correlation between beat frequency and velocity of displacement. In the direction oposite to exposure, OKN was recorded only for slow displacements and did not correlate to displacement velocity. The left part of the figure shows OKN in response to a pattern moving at the frequency of 8 stripes per second, in the direction of exposure (+ signs), or in the opposite direction (- signs) in one kitten exposed to unidirectional motion for 25 hours. The right part shows a plot of OKN frequency (in beats per second) as a function of stimulus velocity (in stripes per second) in one animal: dark symbols and solid bar show responses in the direction of exposure; open symbols and dashed bar, responses in the opposite direction. This result suggests the existence of two components in OKN: a reponse to slow motion pre-existing visual experience, and a response built up by visual experience allowing adaptation to increasing velocities (From Vital-Durand and Jeannerod, 1974).

represented a strong argument for the role of subcortical structures in mediating residual visual function in destriated monkeys. Mohler and Wurtz (1977) showed that partially destriated monkeys, which were able to orient visually toward stimuli presented within their scotoma, lost this ability after subsequent destruction of the retinotopically corresponding zones of the superior colliculi. Thus in monkey, superior colliculi, and possibly other subcortical areas receiving input from the retina, might play a critical role either in mediating pure "subcortical vision", or in relaying visual input to other structures on which they project, including extra-striate cortex. The role of extrastriate visual cortex in residual vision is also demonstrated by experiments with transient inactivation of striate cortex. In this condition, normal visual responses can still be recorded from neurons in visual areas (Girard et al, 1991). In man, clinical obervations suggestive of incomplete or "relative" blindness within scotomata of cortical origin had been mentioned previously by several authors (the now classical "blindsight" phenomenon, see Weiskrantz, 1986 for review).

1.2. TWO CORTICAL VISUAL SYSTEMS

The two-visual-systems model as it was heralded by Schneider (1969) was rapidly considered as unsatisfactory. Lesions limited to the superficial layers of the superior colliculus in the tree shrew, sparing the deep layers and the adjacent midbrain, were shown to produce little or no visuomotor deficit (Casagrande et al., 1972). In rodents, lesions of striate cortex appeared to affect orientation toward targets located within the rostral visual field, whereas this ability was spared after collicular lesion. Superior colliculus, by constrast, was necessary for orienting toward targets placed in the far peripheral visual field (see Goodale, 1983). Thus, orientation seems a more complex function than suggested by the Schneider's results, and cannot be completely dissociated from pattern discrimination, specially for what concerns the most central parts of the visual field.

Another school of thought, more influenced by human neuropsychology than by phylogenetic considerations, and using monkeys (rather than rodents) as subjects, came to the conclusion that the subcortical route was of little importance in spatial vision. Instead, it was postulated that both modes of vision were mediated by two diverging corticocortical pathways (Ungerleider and Mishkin, 1982, Mishkin et al, 1983; for a recent version of these pathways, see Boussaoud et al, 1990). One pathway was the ventral occipitotemporal route linking striate cortex to prestriate areas and from there, reaching inferotemporal cortex on both sides via callosal connections. Interruption of this pathway abolished object discrimination without affecting perception of spatial relations between objects. The other, dorsal, pathway diverged from the previous one by linking the prestriate areas to the posterior part of the parietal lobe. Interruption of this pathway produced visual spatial disorientation characterized, not only by misperception of the relative positions of spatial landmarks, but also by localization deficits during object-oriented action (for review, see Ungerleider and Mishkin, 1982).

These "cortical" systems in fact expand to subcortical structures: Baleydier and Morel (1992) showed that groups of neurons in the posterior pulvinar complex projecting to cortical areas belonging to the ventral and the dorsal routes, respectively, are entirely segregated. In spite of both receiving separate subcortical inputs, however, the two systems seem to have different degrees of dependence with respect to these inputs. This was demonstrated in experiments with functional exclusion (by cooling) or lesion of area V1, at the origin of the two cortico-cortical pathways. One should expect that this exclusion will abolish, or greatly diminish, visual responsiveness in areas downstream to V1 in both systems. In fact, according to Girard et al (1991, 1992) visual responses are completely abolished only in the occipito-inferotemporal pathway, whereas altered responses can still be recorded in the occipitoparietal areas. Bullier et al's (1994) conclusion is that subcortical pathways responsible for visual responses in parietal cortex in the absence of V1 are part of a primitive organization of the visual system, common to primate and non-primate species. These pathways would be responsible for fast and unconscious processing of visual signals. By contrast, the more recently acquired ventral system would be exclusively dependent on V1 input (Figure 2).

In order to account for these different functional properties for processing visual input, it has been suggested that the two cortico-cortical systems relay different types of visual input. The ventral system would relay input from the P retino-geniculate pathway (which originates from the P retinal ganglion cells and projects on the parvocellular layers of the lateral geniculate), a pathway specialized for processing color and form; the dorsal system would relay input from the M pathway (from M ganglion cells through magnocellular geniculate layers), specialized for processing motion (see Merigan and Maunsell, 1993 for review). This dichotomy, in spite of its obvious heuristic value, now seems far too simplistic. First, many of the cortical areas pertaining to the ventral or to the dorsal system appear to be strongly interconnected, as it indeed appears on Figure 1 (see also Morel and Bullier, 1990). Second, there are probably more than two pathways originating from V1. Finally, some areas, like V3 receive inputs from both P and M pathways. This same area V3, where most cells are orientation selective and which should be classified as a "form" area, in fact projects mostly to parietal cortex (for a review of these points, see Zeki, 1993). Based on these arguments, it would seem more appropriate to draw a functional separation, not between a "form" (or "object") system (the ventral system) and a "space" system (the dorsal system), but rather between two "form" systems, the ventral one more concerned with form in association to color and the dorsal one with "dynamic" form in relation to motion (Zeki, 1993). These points will become highly relevant in the discussion about form processing in relation to object-oriented action.

The dorsal system thus appears as the place where to look for the neural correlates of visuomotor transformations, for at least two reasons. First, the aspects of visual information processed in this system, like motion, are implicitly more relevant to motor control than static form or color; second, as will be emphasized below, parietal lesions, not inferotemporal lesions, produce visuomotor impairments. Accordingly, the dorsal system will be considered here as a visuomotor system specialized for dealing

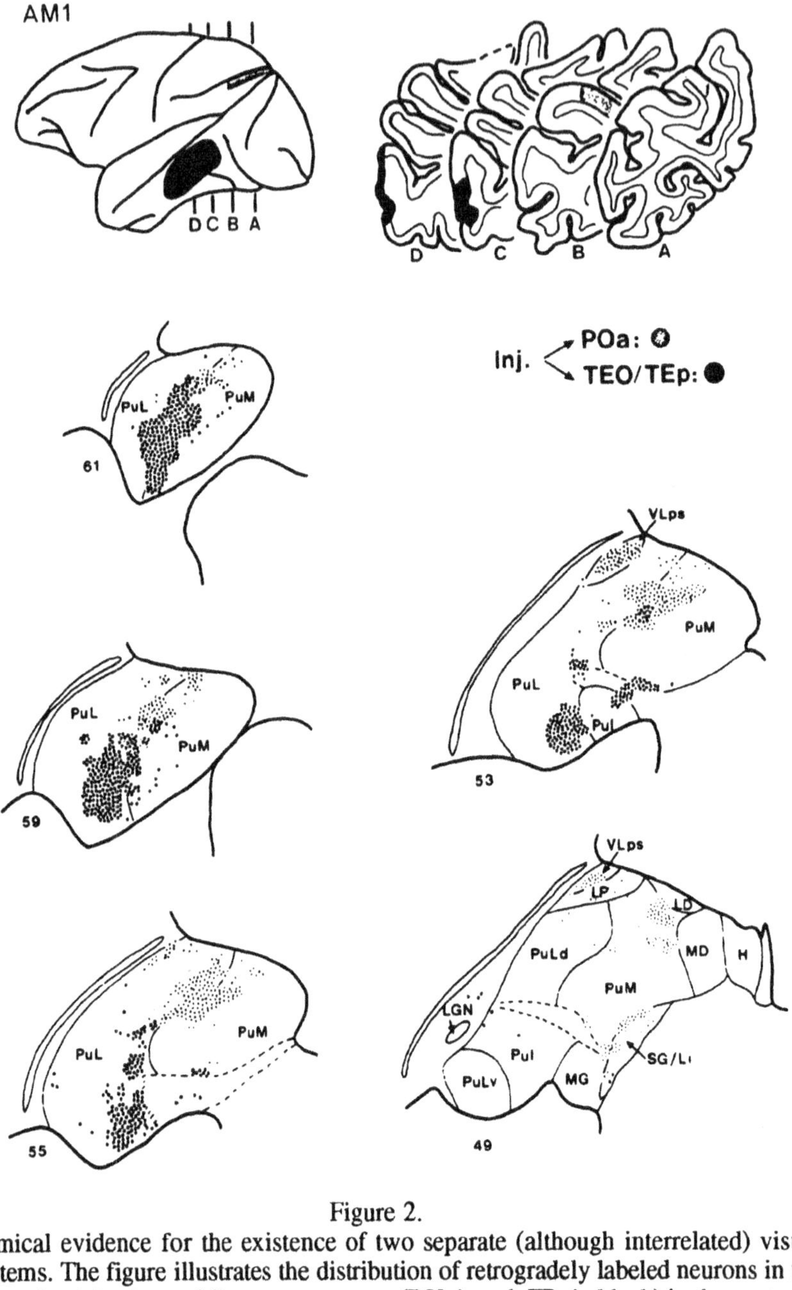

Figure 2.
Anatomical evidence for the existence of two separate (although interrelated) visual subsystems. The figure illustrates the distribution of retrogradely labeled neurons in the pulvinar after injections of fluorescent tracers (DY, in red; FB, in black) in the posterior parietal and the inferotemporal cortices. The two populations of neurons do not overlap with each other (from Baleydier and Morel, 1992; with permission).

with all aspects of object-oriented actions, which includes, not only directing action in the proper direction, but also interacting locally with objects.

1.3. VISUOMOTOR CHANNELS

This renewed conception of parallel visual systems coincided with the formulation of another visuomotor model, no longer based on the modalities of visual coding of the movement, but rather on the modes of representation of the goal of the movement. Objects in space afford two main types of interactions, which correspond to two different visuomotor channels (Jeannerod, 1981). One channel deals with extrinsic properties of these objects (their location in space with respect to the body, their velocity of motion, etc). Its function is to carry the hand at a desired location within extrapersonal action space. It must include mechanisms for computing distance and direction of the point in space where to move. The other channel deals with intrinsic object properties (like shape or size). Its function is to acquire objects with the purpose of manipulating, identifying, or transforming them. The difference in biological significance between the two behaviors should reflect at the level of their neural substrates and modes of control. Reaching is mostly effected by the proximal joints of the arm, it operates within a body-centered system of coordinates. By contrast, grasping relates to the specialization of the hand in primates and man (see below). The location in space where the interaction between the hand and the object takes place is therefore immaterial, at least within the limits imposed by biomechanical constraints.

2. Neural coding in the visuomotor (dorsal) pathway

Reaching movements

The action of reaching toward a location in extrapersonal space (be it pointing to a spot of light or transporting the hand for grasping an object) has been extensively studied, both in animals and man. The point here is not to review the kinematic aspects of reaching, nor its modalities of control by visual or proprioceptive feedbacks: these aspects have been extensively treated in several reviews (e.g., Jeannerod, 1988). The present description will concentrate on the cortical mechanisms for coding the direction of reaching movements.

2.1. REACHING NEURONS IN THE PARIETAL CORTEX

The first arguments as to a role of parietal cortex in coding the direction of reaching were drawn from lesion experiments. Studies focusing on the effects of inferior parietal lobule lesions on visuomotor behavior in monkey have revealed highly specific impairments, including a reaching deficit, first mentionned by Peele (1944). The reaching deficit following this lesion in monkey is characterized by the fact that the animals misreach with their arm contralateral to the lesion in either part of the visual field (the so-called "hand effect") (Hartje and Ettlinger, 1973, Faugier-Grimaud et al, 1978, 1985, Lamotte and Acuna, 1978). Their ipsilesional arm is usually not affected. Misreaching following parietal lesion consistently involves a systematic bias of reaches toward the side of the lesion, a fact that was reported by all the above mentioned authors. Finally, the reaching deficit is more severe in the absence of visual feedback from the limb (e.g., in darkness) than under visual guidance.

Neurons related to reaching were first described by Hyvarinen and Poranen (1974) and by Mountcastle et al (1975) within areas 5 and 7. These neurons belong to a broader category of neurons selectively activated during various aspects of visuomotor behavior, also including manipulation, visual fixation or eye movement. The cells described by Mountcastle et al in both areas have no receptive field or other sensory properties in the visual or somatosensory modalities: instead, they discharge during active reach of objects of motivational interest (a piece of food for example), or during active manipulation of these objects. Hence the terminology of "projection neurons" and of "command functions" used by these authors to designate an unspecified reaching system. By contrast, Hyvarinen and Poranen (1974) considered that the cells responding to arm movements in area 7 had directional properties, in specifying reaching in a given direction of extrapersonal space. In their experiments, activation of such neurons required association of a visual stimulus and of a movement toward it. Neither presentation of the visual stimulus, nor execution of the movement alone were found sufficient conditions for firing these neurons. A population of neurons coding for movement direction was also found in parietal area 5 (Kalaska et al, 1983).

Hyvarinen and Poranen (1974) and Mountcastle et al (1975) both tended to assign the parietal cortex a role in generating neural commands for movements directed at extrapersonal space. Other authors (e.g., Robinson et al,1978), however, considered that properties of the visuomotor neurons (specially a category of eye movement-related cells, the visual-fixation cells) could well be explained by "passive" responses to visual stimuli. Accordingly, they proposed that area 7 is specialized in high-order sensory (visual) processing and has a predominant role in visual and visuospatial functions. In order to account for the fact that discharges of neurons in area 7 do increase when movements are actively produced by the animal during presentation of the stimuli, they proposed the intervention of an additional "attentional" mechanism. The cells fulfilling these criteria are concentrated in subarea 7a. It remains, as Lynch (1980) argued, that the fact that a given cell receives a sensory input does not preclude its participation to sensorimotor or even motor processes: quite the reverse, it is indeed critical that these parietal "reaching neurons" receive the relevant inputs (visual, tactile, kinesthetic) for elaborating the representation of the movement and controlling its execution. More recently, MacKay (1992) showed that "reach" neurons in area 7a were activated during arm movements towards targets located in specific parts of the workspace, including when the movements were performed in the dark. In some cases, this directional preference held for movements performed with either arm. Reach neurons were only related to the execution, not to the preparation of a movement in the preferred direction. This is a surprising finding, because directional anticipatory activities were commonly observed in area 5 (Crammond and Kalaska, 1989), which is considered somewhat "closer" from execution than area 7a. Finally, although reach related activity was clearly influenced by visual cues, no detectable receptive fields were recorded. Moreover, MacKay's reach neurons were not influenced by eye movements and position, which suggests that they did not receive any eye position signal, and could not encode the position of visual stimuli in space relative to the body. This result contrasts with those obtained in other posterior parietal areas area. In area LIP (an area buried into the intraparietal sulcus at the rostral edge of area 7a), Andersen et al (1985) found a population of cells with distinct receptive fields, the response of which was influenced by the position of the eyes in the orbits. In area PO in the superior parietal lobule also, Galetti et al (1993) found neurons responding to visual stimuli placed at a given spatial location, regardless of eye position. These neurons would thus be suited for detecting the position of visual targets in head-centered coordinates.

The precise role of reach neurons in posterior parietal cortex still remains uncertain. MacKay (1992) considers as doubtful that reach neurons in area 7a directly participate in motor output: the reason is that the recorded part of area 7a projects to prefrontal areas, not to premotor cortex. By contrast, these neurons would have a prominent role in monitoring ongoing movements, and in providing signals for the generation of corrections, a suggestion which would account for the lesion effects. In parietal lesioned monkeys spatial errors at the end of reaching movements tend to remain uncorrected. Alternatively, as also proposed by MacKay (1992), the parietal output to frontal cortex could exert a role in coordinating reach and grasp. This explanation would be consistent with the fact that area 7a neurons receive not only visual but also somatosensory inputs. Disruption of somatosensory input at lower levels of the nervous system (e.g., brainstem or spinal cord) is known to affect the coordination of reach and grasp movements (Glendinning et al, 1992; Jeannerod, 1986). There are areas in the inferior parietal lobule, outside area 7a, however, that project to premotor cortex. Matelli et al (1986) found that the rostralmost part of area 7a, in the intraparietal sulcus, does project to area 6. Johnson et al (1993) also found direct connections between area MIP and the dorsal premotor area, where they recorded reaching neurons. These connections could provide a link between area PO where the location of visual targets is coded in head-centred coordinates, and the directional output generated by the motor system (Figure 3).

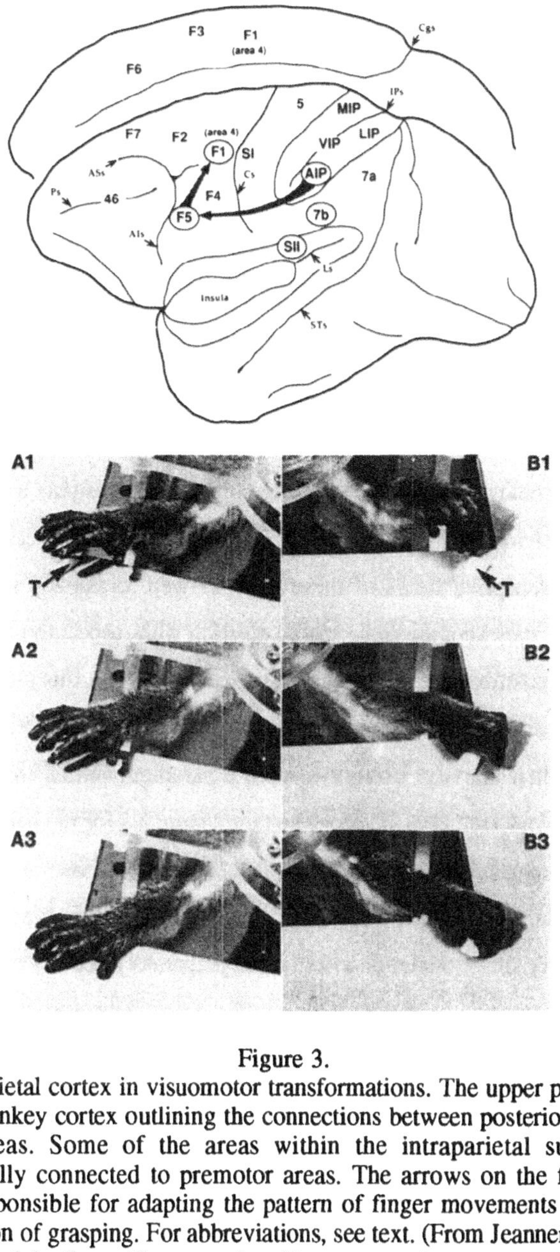

Figure 3.

The role of parietal cortex in visuomotor transformations. The upper part of the figure is a map of monkey cortex outlining the connections between posterior parietal cortex and motor areas. Some of the areas within the intraparietal sulcus (IPs) are monosynaptically connected to premotor areas. The arrows on the figure show the connection responsible for adapting the pattern of finger movements to object shape during the action of grasping. For abbreviations, see text. (From Jeannerod et al, 1995). The lower part of the figure illustrates the effect, on grasping movements, of ablation of the inferior parietal lobule in one monkey. On the left (A1-A3): successive frames drawn from a film of the contralesional hand during a reach-to-grasp movement directed at food target T. Note stretched fingers, as compared to the preshaped posture of the non-affected hand (B1-B3, right). (From Faugier-Grimaud et al, 1978, with permission).

1991). The role of SMA is to be understood within the framework of a parallelism of pathways controlling goal-directed actions.

Parietal, premotor and motor cortical mechanisms combine for selecting the proper movement direction as a function of the visual goal. The timing of movement-related neuronal discharges in these cortical areas suggests that motor and premotor cortex areas (including SMA) relate to initiation of goal-directed movements, whereas parietal neurons relate to their monitoring during execution.

3. Neural coding in the visuomotor (dorsal) pathway

Grasping movements

Hand movements have been the focus of an increasing number of studies in the last fifteen years. Their description, however, is mostly available in man. This section will first provide a review of the motor patterns of grasping in humans and will include the monkey data whenever possible.

The hand is one of the highest achievements of motor function in primates and man. By using the criterion of stability of the grasp as a prerequisite for handling objects, Napier (1956) considered that human prehensile movements can be described along only two main motor patterns. "If, Napier stated, prehensile activities are to be regarded as the application of a system of forces in a given direction then the nature of prehensile activity can be resolved into two concepts -that of precision and that of power" (p 906). The precision and the power grip patterns can be used alternatively or in combination for almost every object.

Precision grip with true opposition of the pulpar surfaces of the thumb and the index finger is considered as the top attribute of dextrous hands. The problem of whether this attribute is specific to the human hand or not, is a matter of discussion. The Heffner and Masterton scale for ranking digital dexterity (Heffner and Masterton, 1975), based on anatomy of the hand, includes only man in the topmost category, that with opposable thumb and precision grip (See also Napier, 1961).

3.1. THE PATTERN OF GRIP FORMATION

The type of grip that is formed by the hand in contact with the object represents the end result of a motor sequence which starts well ahead of the action of grasping itself. The fingers begin to shape during transportation of the hand at the object location. This process of grip formation is therefore important to consider, because it shows dynamically how the static posture of the hand is finally achieved. No systematic investigation of this aspect of grasping (preshaping) seems to have been made until the film study by Jeannerod (1981).

Preshaping first involves a progressive opening of the grip with straightening of the fingers, followed by a closure of the grip until its matches object size. The point in time where grip size is the largest (maximum grip size) is a clearly identifiable landmark which occurs within about 60% to 70% of the duration of the reach, that is, well before the fingers come in contact with the object (Jeannerod, 1981, 1984; Wallace and Weeks, 1988; Wing et al, 1986). This biphasic opening-closure motor pattern is not unique to man: obervations based on films during prehension in rhesus monkeys have revealed a closely similar opening of the grip followed by closure before contact with the object.

The amplitude of grip aperture during grip formation covaries with object size (Jeannerod, 1984; Wallace and Weeks, 1988; Wing et al, 1986). Marteniuk et al (1990) found that for an increase of 1 cm in object size, the maximum grip size increases by 0.77 cm. One possible explanation for the biphasic opening-closure pattern of grip formation relates to the thumb-index finger geometry. Because the index finger is longer than the thumb, the finger grip has to open wider than required by object size, in order for the index finger to turn around the object and to achieve the proper orientation of the grip. Indeed, the movement of the index contributes the most to grip formation, whereas the position of the thumb with respect to the wrist tends to remain invariant (Wing and Fraser, 1983). The extra-opening of the grip during preshaping might also represent a safety margin for compensating the effects of the variability of the reach. Indeed, maximum grip size tends to become larger than required by object size in a number of conditions where the variability of the reach is likely to be increased (e.g., lack of visual control, movements directed at targets in the peripheral visual field, etc).

As already stated, the pattern of finger movements that arises prior to, and during grasping reflects the activity of higher order visuomotor mechanisms for detecting the shape of the object and generating appropriate motor commands. The problem is for the motor system of the hand to build an "opposition space" which would take into account both the shape of the object and the biomechanics of the hand (Arbib, 1985; Iberall and Arbib, 1990; Iberall et al, 1986). Observations like those of Stelmach et al (1994), for example, show that different positions of the same object in the workspace may yield different types of grips or different position of the fingers on the object. This strongly suggests the existence of a higher order coordination mechanism which couples the different components of prehension.

These findings demonstrate that hand movements and postures during object-oriented behaviour are largely determined at the representational level. Indeed, visual feedback signals seem of very little importance during the movement itself, as both the pattern of grip formation and the coordination of the reaching and grasping components are correctly achieved in situations where the hand remains invisible to the subject. Similarly, the size of the maximum grip aperture correlates with the size of the object in the absence of visual feedback from the hand (Jeannerod, 1984). This is not to say that the correct representation of object properties must not be reactivated by visual input arising from the central retina. Prehension movements directed at objects presented within the peripheral visual field are not only slower and less accurate; the grip formation is incomplete, the fingers do not shape properly (Sivak and MacKenzie, 1992). Similarly, grasping directed at memorized objects involve larger grip apertures than grasping directed at visible objects (e.g., Wing et al, 1986).

Grasping cannot be reduced to its visuomotor aspects, it is the motor counterpart of a broader function. During handling and manipulation, for which grasping is a pre-condition, signals for object identification arising from sight and touch are co-processed. The fingerpads have been considered by some authors (e.g.,

Sherrington) as the somatosensory "macula". Thus, the hand brings objects to be manipulated within the central field of vision, so that "the finest movements of the fingers must be under simultaneous control from the very centers of the visual and tactual maculae" (see Phillips, 1985). Touch, and particularly active touch, is not subject to these constraints, as it involves direct assessment of size and volume. In addition, touch is critical for perceiving object properties like hardness, compliance, texture, temperature, weight, etc, which can hardly be accessed by sight alone. This is one of the reasons why reaching and grasping must be functionally interrelated. Reaching is a precondition for grasping, its finality is to bring the specialized area of the hand at contact with objects, for acquiring and using object-related information. Reaching is therefore a mere transition between visual processing prior to contact and haptic processing during manipulation.

Finally, another aspect of grip accuracy is specification of grip force. This parameter also has to be, at least partly, specified in advance, during the preshaping phase, in order for the adequate force to be applied on the object at the onset of the grasp. Lifting an object implies a sequence of coordinated events where the grip force (to grasp the object) and the load force (to lift the object) vary in parallel. The grip force/load force ratio must exceed the slip ratio, itself determined by the coefficient of friction between the skin and the object surface. Changing the coefficient of friction (by using an object coated with sandpaper, suede or silk, for instance) changes the grip force, so that the load force remains invariant and the grip force/load force ratio increases when frictional forces decrease. By contrast, increasing the weight of the object results in an increase of both grip force and load force and an invariant grip force/load force ratio (Johansson and Westling, 1987).

The respective contributions of anticipatory mechanisms and of reflex adjustments to the accuracy of grip force have been extensively studied. It appears that the adaptive changes in grip force are strongly dependent on tactile afferent signals. A demonstration of this point is that adaptation of the grip force to friction disappears if the fingertips are anaesthetized. The duration of the initial, isometric, phase of lifting movements (the preload phase, ca 100 ms) is sufficient for tactile afferents from the fast adapting receptors to come into play, the latency between the onset of the slip and the change in the force ratio being in the range of 75 ms. In addition, these fast adapting receptors are very sensitive to slip signals (Westling and Johansson, 1984, Johansson and Westling, 1987). These signals may be used for updating the coding of initial forces based on internal representation of object properties, and for sensorimotor learning. It is likely that visual cues related to object size will also be used for building this representation (see Johansson and Westling, 1988). Several experiments have shown that available information about object weight, compliance, texture (e.g., based on visual size cues) can accurately determine grip and load forces in advance with respect to the grasp itself (Gordon et al, 1991).

These data are essential for accessing the content of the representation that the subject has formed about the object up representations of objects. The initial stage of the grasp (the preshape that occurs during the reach) expresses the visuomotor "decisions" that have been made for the motor apparatus to comply with object properties. Fingers shape in anticipation to object size and shape, the wrist rotates in anticipation to object orientation to give the optimal stability to the grasp. During this stage, also forces are generated which will be applied immediately at the time of contact, in order to grasp and lift the object.

3.2. NEURAL MECHANISMS INVOLVED IN THE CONTROL OF VISUALLY GUIDED GRASPING

The mechanism of how finger movements are coordinated for producing an anticipatory hand configuration corresponding the shape of an object is therefore a fascinating one, as it directly impinges upon the mechanisms of neural representation of object-oriented action. For a recent review, see Jeannerod et al (1995).

3.2.1. *Motor cortex*

The role of motor cortex in controlling finger movements was extensively studied by classical authors, using cortical stimulation. Stimulation experiments, however, may give a false idea of cortico-motoneuronal connectivity. Instead, recent results suggest a large degree of flexibility of the relations between motor cortex and distal muscles. It can be shown, for example, that a monkey corticospinal neuron which fires during a movement of a given type (e.g., a precision grip) will not fire during a movement of another type (e.g., a force grip), in spite of the fact that the same muscles are implied in both cases (Muir and Lemon, 1983; Lemon et al, 1986). This result indicates that a given motoneuron can be connected to several sets of cortical cells, and that each set becomes activated in relation to a certain type of movement, not to the contraction of a certain muscle. The logical consequence of these findings is that there are several cortical representations for the same muscle, each representation coming into play as a function of the type of movement to be performed. Microstimulation experiments in motor cortex have in fact shown large overlaps between cortical zones which control individual motoneurons (see Wiesendanger, 1986). The picture of cortical representation of finger movements gets even more complicated if one considers the fact that cortical pyramidal cells are usually connected to more than one motoneuron (up to two or three, Buys et al, 1986). This divergence was directly demonstrated by intracellular labelling of corticospinal axons (Shinoda et al, 1981).

Experimental data in man also confirm a "task-dependent" mode of organization of cortical commands, rather than a somatotopical organization in the usual sense. Experiments using transcranial magnetic stimulation illustrate this point. The response of a given hand muscle (the first dorsal interosseus, for example) to a stimulus applied at the same cortical locus was tested during performance of different types of movements. The response was found to be different according to whether the muscle was contracted in isolation or in association with other muscles (during the performance of a precision grip, for example). Accordingly, each task would be characterized by a given pattern of activity of cortical cells, with the possibility of selecting populations limited to a small number of neurons. A normal human subject can, in certain conditions, activate selectively one motoneuron: this striking effect can be obtained in showing the subject the EMG recording of his own finger muscle, and instructing him to focuse his attention on one of the motor units appearing on the record (Kato and Tanji, 1972).

These experiments make it difficult to conceive a simple model for the production of independent finger movements. It could be that the movements of each finger are controlled by a specific population of neurons. To obtain combined movements of several fingers, the corresponding populations would be activated. Alternatively, the synergy between fingers could be achieved by another type of neuronal populations, controlling the muscles of several fingers. According to Scheiber (1990), however, these explanations are probably incorrect, for the reason that finger muscles are in fact not independent from each other (they are mechanically coupled at the level of their tendons). The generation of independent movements of one finger would thus rely on inhibitory mechanisms preventing activation of muscles of other fingers, so that isolated finger movements would be "extracted" from coarser synergies (like open or close the hand, for example). If this model revealed true, it would require more neurons to produce the movement of a single finger than those of several fingers (Scheiber, 1990).

3.2.2. *Parietal cortical areas*

Parietal cortex is known to be concerned with the visual control of hand movement from the effects of posterior parietal lesions in animals. Monkeys with lesions in the inferior parietal lobule, in addition to the typical misreaching with the contralesional arm already described in the previous section, fail to shape their contralesional hand and make awkward grasps (Faugier-Grimaud et al, 1978) (Figure 3).

Neurons related to active arm movements were first recorded in the inferior parietal lobe by Mountcastle and his colleagues who, besides the "arm projection" neurons related to reaching, also identified a group of "hand manipulation" neurons (Mountcastle et al, 1975). More recently, H. Sakata and his group found hand movement-related neurons to be concentrated in a small zone (which they called area AIP) within the rostral part of the posterior bank of intraparietal sulcus. Neurons from this area were recorded in monkeys trained to manipulate various types of switches, which elicited from the animal different motor configurations of the hand. Most of them were selectively activated during grasping one or two of these objects among the four routinely used ones (Taira et al, 1990). Neuron activity was not influenced by changing the position of the object in space, which shows that they were related to distal hand and finger movements rather than to proximal movements of the arm.

Many of these visually responsive neurons were also activated by the sight of objects during fixation without grasping. Using a broader variety of graspable objects, including primitive shapes such as spheres, cubes, cones, cylinders, rings and plates of different sizes, Sakata et al (1992) found that AIP neurons which were activated during both grasping and fixation were selective for the same object whether the monkey looked at it or grasped it. Some of them were also sensitive to the size or the orientation of the objects. Finally other neurons were not activated during the fixation of objects ("non-object" type) but seemed to require other visual stimuli, such as the view of the moving hand to be activated. "Non-object" type responses were usually elicited after the initiation of the hand movement, and were likely to be concerned with the interaction of the hand with the object.

The Sakata hand neurons cannot be dissociated from other groups of neurons, identified in close-by parietal areas, and which are related to processing complex visual stimuli. This is the case of neurons in the posterior part of area 7, specifically sensitive to rotation of a visual stimulus, regardless of its shape (Sakata et al, 1986). Another group of neurons sensitive to the 3-D orientation of the longitudinal axis of visual stimuli were recently found in the caudal part of the IPS posterior bank. It is therefore likely that the 3-D characteristics of the object are processed in parietal areas different from AIP and the output of such processing is then sent to AIP.

An interesting confirmation of the role of the hand neurons in representing grasping movements was provided by an experiment of Gallese et al (1994). Transient inactivation of AIP, by injecting a GABA agonist (muscimol) in the rostral IPS posterior bank under electrophysiological control, produced a subtle change in the performance of visually guided movements during grasping tasks. Grasping errors were observed in tasks requiring a precision grip or during sticking out the index finger to insert it in a groove. These errors were due to a lack of preshaping of the hand during the approach phase of the movement. In addition, there was a clearcut dissociation of the muscimol effects on grasping and reaching. Whereas the alteration of preshaping was consistently obtained after injection in the rostral part of the posterior bank of the sulcus, misreaching occurred after injection within its more caudal part. This result provides the unambiguous demonstration that the hand-related parietal neurons play a specific role in the visuomotor transformation used for grasping objects. Indeed, the deficit is a visuomotor one, not a motor one, as it can be corrected under tactile guidance: as soon as the monkey awkwardly touches the target-object, the hand shapes in accordance with object shape. Area AIP inactivated by Gallese et al is only a couple of synapses ahead of executive areas in frontal cortex. This result obviously confirms and expands those obtained by Faugier-Grimaud et al (1978). Interestingly, the latter authors had observed, immediately following ablation of area 7, a transient "paralysis" of the contralateral hand, which appeared to remain unused, unless the monkey was forced to do so.

3.2.3. *Premotor cortex neurons*

Area AIP is directly connected with a limited zone of premotor cortex. This zone corresponds to one of the cytochrome-oxidase subfields (field F5) of area 6, identified by Matelli et al (1985). In addition, F5, which forms the rostral part of inferior area 6, is itself directly connected with that part of area 4 (field F1) which corresponds to the hand primary motor field. Intracortical microstimulation and single neuron studies showed that F5 is specifically related to distal movements (Rizzolatti et al, 1988). It is thus particularly interesting to examine the properties of neurons in this cortical zone in relation to visually-guided grasping. Because F1 neurons have only very limited access to visual information (visually responsive neurons in this area are rare and have visual properties -brisk, transient responses to abrupt stimulus presentation (Wannier et al, 1989)- that hardly fit those one would expect for grip formation) the visuomotor transformations required for grasping movements have to occur upstream in motor control, in areas more closely connected to the visual system. The connections linking parietal areas like AIP to the "distal" area of premotor cortex (F5) and then to F1 could thus represent another, parallel specialized visuomotor system for encoding object primitives and generating the corresponding hand configurations (Figure 3). It would be interesting to repeat the experiment of transient inactivation by injecting muscimol in the premotor neurons where AIP directly projects. The pattern of deficit observed during object-oriented behavior should closely resemble that observed during AIP inactivation.

Rizzolatti and his colleagues recorded single neurons from F5 in behaving monkeys tested during object oriented motor actions (Rizzolatti et al, 1988). These experiments showed that most neurons located in the upper part of F5 are related to grasping and other object-related motor actions (holding, tearing, manipulating). "Grasping" neurons discharge in relation with finger and hand movements during the action of grasping an object. The temporal relation of this discharge with grip movements changes from neuron to neuron. Some fire during the last part of grasping, that is, during finger flexion. Others start firing with finger extension and continue during finger flexion. Others are activated in advance of finger movements and often cease discharging only when the object is grasped. An important property of most F5 neurons is their selectivity for different types of hand prehension. Eighty five per cent of grasping neurons show selectivity for one of three basic types of grip: precision grip (the most represented type), finger prehension and whole hand prehension. There is specificity for different finger configurations, even within the same grip type. Thus, the prehension of a sphere, which requires the opposition of all fingers, is encoded by different neurons than the prehension of a cylinder, for which a palm opposition grip is used.

Visual responses were observed in about 20-30% of F5 neurons. Two types of responses can be distinguished. Neurons of the first class respond to presentation of graspable objects. Often, there is a relation between the type of prehension coded by the cell and the size of the stimulus effective in triggering the neurons. This is particularly clear for the precision grip neurons which are activated only by small visual objects. Neurons of the second class are of a particular relevance for the study of motor representations. These neurons (termed "mirror neurons" by di Pellegrino et al, 1992) respond when the monkey sees hand movements executed by the experimenter or another monkey. They also respond when the recorded monkey performs hand movements of a particular type. The important point is that, in order to fire a mirror neuron, the observed hand movements have to be the same as those which would activate that neuron if the monkey performed them. For example, many mirror neurons fire when the monkey grasps a piece of food and also when the experimenter or the other monkey does so. They do not fire, however, when the experimenter makes a grasping movement without food, or when the food is grasped with a tool, etc. In other words, mirror neurons encode object-oriented actions, whether they are performed or they are observed by the recorded monkey, which suggests that they might be better defined as "representation" neurons. At any rate, these F5 neurons are clearly distinct from those recorded in the superior temporal sulcus by Perrett et al (1986), which are sensitive to observed hand and body movements, but have no motor counterpart. Thus, the F5 neurons can be activated visually in two ways: by objects and by events. In both cases the eliciting stimuli address specifically the F5 neurons coding the grip congruent with them.

If one compares the properties of parietal neurons with those of F5, striking similarities, but also important differences emerge. Visual responses to 3-D objects are found more frequently in parietal cortex than in F5. By contrast, "mirror neurons" responding to the view of hand action of other individuals were not found in AIP. As for the motor properties, parietal "motor dominant" neurons also code elementary motor acts, such as precision grip, whole hand prehension, wrist rotation, etc. However, most of the parietal neurons appear to represent the entire action, since they start to discharge with the hand shaping and continue to fire while the monkey is holding the object. This property contrasts with those of F5 neurons, which were commonly related to a particular segment of the action. Indeed, in primary motor cortex, on which F5 heavily projects, neurons code even more fragmental movements.

Conclusion

The above behavioral and neurophysiological results suggest that actions are driven by implicit "knowledge" of object attributes, such that stable and decomposable motor patterns arise to cope with external reality (Figure 4). In addition, the elementary units that compose these motor patterns appear to match identifiable neuronal subpopulations. It is therefore tempting to speculate that motor elements that are neurally precoded, can be assembled to form the motor representation responsible for a given action. A model for this simplified form of motor representation was proposed by M. Arbib (Arbib, 1981, 1985; Iberall et al, 1986; Arbib and Hesse, 1986; Iberall and Arbib,1990). Arbib's view is that motor representations (which he calls "coordinated control programs") are composed of elementary units (the "motor schemas") which interact for controlling motor output. A motor schema is a predetermined set of commands (a micro-action), not an independent or an individual command: in that sense it already represents a certain degree of coordination of motor output for achieving a limited goal. When assembled with other schemas, a coordinated control program (in fact a higher order schema) arises, which in turn can be assembled with others, and so on. This recursive property of schemas is a convenient one for modeling their neural implementation at different levels of analysis (e.g., neuronal populations, single neurons, etc.) of the nervous system.

Motor schemas are assumed by Arbib to be activated by perceptual schemas which serve encoding of visual input (for review, see Jeannerod et al, 1995). Again, perceptual schemas do not deal with perceptual primitives, they encodes configurations of the external environment which, in the visual domain, may correspond to one of the attributes of an object. During visually-guided prehension, perceptual schemas activate the corresponding motor schemas for the subactions of "reach", "preshape", "enclose", "rotate forearm", etc. They are the basis for "visuomotor decisions" which are implicitly taken during actions towards visual objets, and which account for movement smoothness and accuracy.

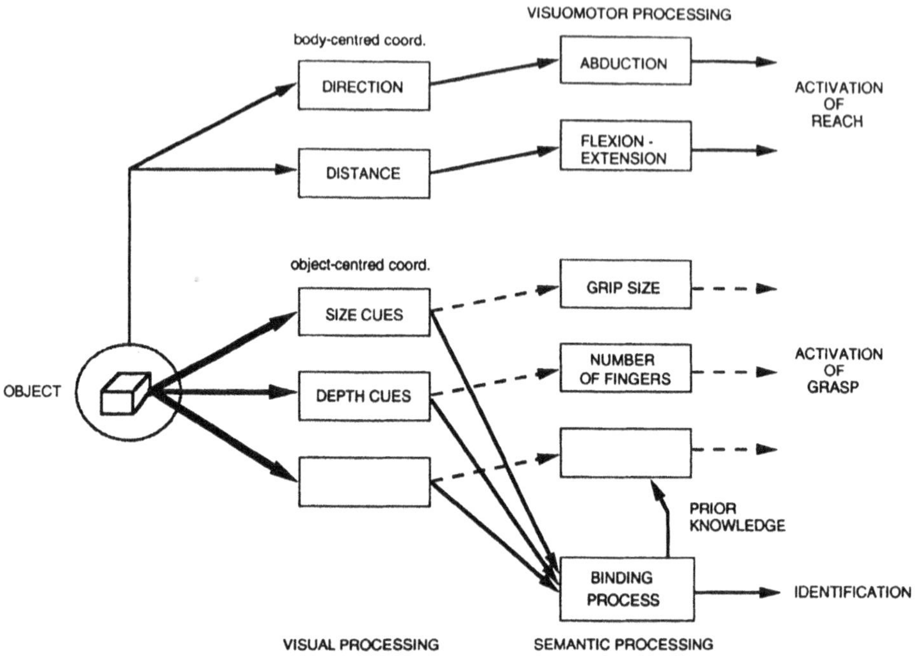

Figure 4.
Diagrammatic representation of central processes involved in object-oriented behavior. Extrinsic object properties (related to its spatial position) are processed in body-centred coordinates, intrinsic object properties in object-centred coordinates. Activation of reach and activation of grasp are thought to rely on separate, parallel processes, both pertaining to the dorsal visual pathway and to the posterior parietal areas. Semantic processing, in spite of using the same object primitives as visuomotor processing, is thought to occur in the ventral visual pathway. Semantic knowledge stored in the ventral pathway could improve visuomotor performance using connections between the two pathways (from Jeannerod, 1994).

References

Andersen, R.A., Essik, G.K., and Siegel, R.M. (1985) Encoding of spatial location by posterior parietal neurons. *Science* **230**, 456-458.

Arbib, M.A. (1985) Schemas for the temporal organization of behavior. *Human Neurobiology* **4**, 63-72.

Arbib, M.A. (1981) Perceptual structures and distributed motor control, in V.B. Brooks (ed.), *Handbook of Physiology, Section I: The nervous system, Vol.2: Motor control,* Williams et Wilkins, Baltimore, pp. 1449-1480

Arbib, M.A and Hesse, M.B. *The construction of reality.* Cambridge University press, Cambridge.

Baleydier, C. and Morel, A. (1992) Segregated thalamo-cortical pathways to infeior parietal and inferotemporal cortex in macaque monkey. *Visual Neuroscience* **8**, 391-405.

Boussaoud, D. (1995) Primate premotor cortex. Modulation of preparatory neural activity by gaze angle. *Journal of Neurophysiology* **73**, 886-890.

Bullier, J., Girard, P., and Salin, P.A. (1994) The role of area 17 in the transfer of information to extrastriate visual cortex. *Cerebral cortex* **10**, 301-330.

Buys, E.J., Lemon, R.N., Mantel, G.W.H., and Muir, R.B. (1986) Selective facilitation of different hand muscles by single corticospinal neurons in the conscious monkey. *Journal of Physiology* **381**, 529-549.

Cajal, S.R. (1909) *Histologie du système nerveux de l'homme et des vertébrés,* Maloine, Paris.

Caminiti, R., Johnson, P.B., and Urbano, A. (1990) Making arm movements within different parts of space: dynamic aspects in the primate motor cortex. *The Journal of Neuroscience* **10**, 2039-2058.

Caminiti, R., Johnson, P.B., Burnod, Y., Galli, C., and Ferraina, S. (1990) Shift of preferred directions of premotor cortical cells with arm movements performed across the workspace. *Experimental Brain Research* **83**, 228-232.

Casagrande, V.A., Harting, J.K., Hall, W.C., Diamond, I.T., and Martin, G.F. (1972) Superior colliculus of the Tree Shrew: a structural and functional subdivision into superficial and deep layers. *Science* **177**, 444-447.

Chen, D.F., Hyland, B., Maier, V, Palmeri, A., and Wiesendanger, M. (1991) Comparison of neural activity in the supplementary motoor area and in the primary motor cortex in the monkey. *Somatosensory and Motor Research* **8**, 27-44.

Crammond, D.J. and Kalaska, J.F. (1990) Cortical neuronal activity recorded in a delay task that dissociates location of cue stimulus and movement end-point. *Society for Neuroscience Abstracts* **16**, 423.

Dineen, J.J. and Hendrickson, A.E. (1981) Age-correlated differences in the amount of retinal degeneration after striate cortex lesions in monkeys. *Investigative Ophthalmology & Visual Science* **21**, 749-752.

Di Pellegrino, G., Fadiga, L., Fogassi, L., Gallese, V., and Rizzolatti, G. (1992) Understanding motor events: A neurophysiological study. *Experimental Brain Research* **91**, 176-180.

Faugier-Grimaud, S., Frenois, C., and Stein, D.G. (1978) Effects of posterior parietal lesions on visually guided behavior in monkeys. *Neuropsychologia* **16**, 151-168.

Faugier-Grimaud, S., Frenois, C., and Peronnet, F. (1985) Effects of posterior parietal lesions on visually guided movements in monkeys. *Experimental Brain Research* **59**, 125-138.

Fogassi, L., Gallese, V., di Pellegrino, G., Fadiga, L., Gentilucci, M., Luppino, G., Matelli, M., Pedotti, A., and Rizzolatti, G. (1992) Space coding by premotor cortex. *Experimental Brain Research* **89**, 686-690.

Gallese, V., Murata, A., Kaseda, M., Niki, N., and Sakata, H. (1994) Deficit of hand preshaping after muscimol injection in monkey parietal cortex. *Neuroreport* **5**, 1525-1529.

Galletti, C., Battaglini, P.P., and Fattori, P. (1993) Parietal neurons encoding spatial locations in craniotopic coordinates. *Experimental Brain Research* **96**, 221-229.

Gentilucci, M., Fogassi, L., Luppino, G., Matelli, M., Camarda, R., and Rizzolatti, G. (1988) Functional organization of inferior area 6 in the macaque monkey. 1. Somatotopy and the control of proximal movements. *Experimental Brain Research* **71**, 475-490.

Georgopoulos, A.P., Kalaska, J.F., Caminiti, R, and Massey, J.T. (1982) On the relations between the direction of two-dimensional arm movements and cell discharge in primate motor cortex. *Journal of Neuroscience* **2**, 1527-1537.

Georgopoulos, A.P., Schwartz, A.B., and Kettner, R.E. (1986) Neuronal population coding of movement direction. *Science* **233**, 1416-1419.

Girard, P., Salin, P., and Bullier, J. (1991) Visual activity in areas V3A and V3 during reversible inactivation of area V1 in the macaque monkey. *Journal of Neurophysiology* **66**, 1493-1503.

Girard, P., Salin, P., and Bullier, J. (1992) Response selectivity in neurons in area MT of the macaque monkey during reversible inactivation of area V1. *Journal of Neurophysiology* **67**, 1-10.

Glendinning, D.S., Cooper, B.Y., Vierck, C.J., and Leonard CM (1992) Altered precision grasping in stumptail macaques after fasciculus cuneatus lesions. *Somatosensory and Motor Research* **9**, 61-73.

Goodale MA (1983) Neural mechanisms of visual orientation in rodents: targets versus places, in A. Hein and M. Jeannerod (eds)*Spatially oriented behavior*, Springer-Verlag, New-York, pp 35-62.

Gordon, A.M., Forssberg, H., Johansson, R.S., and Westling, G. (1991) Visual size cues in the programming of manipulative forces during precision grip. *Experimental Brain Research* **83**, 477-482.

Graziano, M.S.A., Yap, G.S., and Gross, C.G. (1994) Coding of visual space by premotor neurons. *Science* **266**, 1054-1057.

Hartje, W. and Ettlinger, G. (1973) Reaching in light and dark after unilateral posterior parietal ablations in the monkey. *Cortex* **9**, 346-354.

Heffner, R. and Masterton, B. (1975) Variation in form of the pyramidal tract and its relationship to digital dexterity. *Brain, Behavior and Evolution*v **12**, 161-200.

Hein, A. and Held, R. (1967) Dissociation of the visual placing response into elicited and guided components. *Science* **158**, 190-192.

Hess, W.R., Bürgi, S., and Bucher, V. (1946) Motorische function des tectal und tegmentalgebietes. *Monatsschrift für psychiatriche Neurologie* **112**, 1-52.

Humphrey, N.K. and Weiskrantz, L. (1967) Vision in monkeys after removal of the striate cortex. *Nature* **215**, 595-597.

Hyvarinen, J. and Poranen, A. (1974) Function of the parietal associative area 7 as revealed from cellular discharges in alert monkeys. *Brain* **97**, 673-692.

Iberall, T. and Arbib, M.A. (1990) Schemas for the control of hand movements: an assay on cortical localization, In: M.A. Goodale (ed),*Vision and action. The control of grasping.*, Norwood, Ablex, pp. 204-242.

Iberall, T., Bingham, G. , and Arbib, M.A. (1986) Opposition space as a structuring concept for the analysis of skilled hand movements, in H. Heuer and C. Fromm (eds), *Generation and modulation of action pattern, Experimental Brain Research Series* **15**: 158-173.

Jeannerod, M. (1981) Intersegmental coordination during reaching at natural visual objects, in J. Long and A. Baddeey (eds), *Attention and Performance IX*, Erlbaum, Hillsdale, pp. 153-168.

Jeannerod, M. (1984) The timing of natural prehension movements. *Journal of Motor Behaviour* **16**, 235-254.

Jeannerod, M. (1986) The formation of finger grip during prehension. a cortically mediated visuomotor pattern. *Behavioural Brain Research* **19**, 99-116.

Jeannerod, M. (1988) *The neural and behavioural organization of goal-directed movements* Oxford University Press, Oxford.

Jeannerod M. (1994) The hand and the object. The role of posterior parietal cortex in forming motor representations. *Canadian Journal of Physiology and Pharmacology* **72**, 525-534.

Jeannerod, M. and Biguer, B. (1982). Visuomotor mechanisms in reaching within extrapersonal space, in M.A. Goodale & R. Mansfield (eds.), *Advances in the analysis of visual behaviorr,* MIT Press, D. Ingle, Boston, pp. 387-409

Jeannerod, M. and Prablanc, C. (1983) The visual control of reaching movements, in: J. Desmedt (ed), *Motor control mechanisms in man,* Raven, New-York, pp 13-29.

Jeannerod, M. and Rossetti, Y. (1993). Visuomotor coordination as a dissociable visual function: experimental and clinical evidences, in C. Kennard (ed.), *Visual Perceptual Defects. Baillière's Clinical Neurology* , Vol.2 No.2, Baillière Tindall, pp. 439-460.

Jeannerod, M., Arbib, M.A., Rizzolatti, G., and Sakata, H. (1995) Grasping objects. The cortical mechanisms of visuomotor transformation. *Trends in Neuroscience* **18**, 314-320.

Johansson, R.S. and Westling, G. (1988) Coordinated isometric muscle commands adequately and erroneously programmed for the weight during lifting task with precision grip. *Experimental Brain Research* **71**, 59-71.

Johansson, R.S. and Westling, G. (1987) Signals in tactile afferents from the fingers eliciting adaptive motor responses during precision grip. *Experimental Brain Research* **66**, 141-154.

Johnson, P.B., Ferraina, S., and Caminiti, R. (1993) Cortical networks for visual reaching. *Experimental Brain Research* **97**, 361-365.

Kalaska, J.F., Caminiti, R., and Georgopoulos, A.P. (1983) Cortical mechanisms related to the direction of two dimensional arm movements. Relations in parietal area 5 and comparisonwithmotor cortex. *Experimental Brain Research* **51**, 247-260.

Kato, M. and Tanji, J. (1972) Conscious control of motor units of human finger muscles, in: G.G. Samjen (ed), *Neurophysiology studied in man,* Excerpta Medica, Amsterdam.

Lamotte, R.H. and Acuna, C. (1978) Defects in accuracy of reaching after removal of posterior parietal cortex in monkeys. *Brain Research* **139**, 309-326.

Lemon, R.N., Mantel, G.W.H., and Muir, R.B. (1986) Corticospinal facilitation of hand muscles during voluntary movements in the conscious monkey. *Journal of Physiology* **381**, 497-527.

Lynch, J.C. (1980) The functional organization of posterior parietal association cortex. *Behavioral Brain Science* **3**, 485-498.

MacKay, W.A. (1992). Properties of reach related neuronal activity in cortical area 7a. *Journal of Neurophysiology* **67**, 1335-1345.

Marteniuk, R.G., Leavitt, J.L., MacKenzie, C.L., and Athenes, S. (1990). Functional relationships between grasp and transport components in a prehension task. *Human Movement Science* **9**, 149-176.

Matelli, M., Camarda, R., Glickstein, M., and Rizzolatti, R. (1986) Afferent and efferent projections of the inferior area 6 in the Macaque Monkey. *The Journal of Comparative Neurology* **251**, 281-298.

Merigan, W.H. and Maunsell, S.H.R. (1993) How parallel are the primate visual pathways? *Annual Reviews of Neuroscience* **16**, 369-402.

Mishkin, M. and Ungerleider, L.G. (1982) Contribution of striate inputs to the visuospatial functions of parieto-preoccipital cortex in monkeys. *Behavioural Brain Research* **6**, 57-77.

Mohler, C.W. and Wurtz, R.H. (1977) Role of striate cortex and superior colliculus in visual guidance of saccadic eye movements in monkeys. *Journal of Neurophysiology* **4O**, 74-94.

Morel, A. and Bullier, J. (1990) Anatomical segregation of two cortical visual pathways in the macaque monkey. *Visual Neuroscience* **4**, 555-578.

Mountcastle, V.B., Lynch, J.C., Georgopoulos, A. Sakata, H., and Acuna, C. (1975) Posterior parietal association cortex of the monkey: command functions for operations within extra-personal space. *Journal of Neurophysiology* **38**, 871-908.

Muir, R.B. and Lemon, R.N. (1983) Corticospinal neurons witha special role in precision grip. *Brain Research* **261**, 312-316.

Napier, J.R. (1956). The prehensile movements of the human hand. *Journal of Bone and Joint Surgery* **38B**, 902-913.

Napier, J.R. (1961) Prehensility and opposability in the hands of primates. *Symp. Zool. Soc. London* **5**, 115-132.

Peele, T.L. (1944) Acute and chronic parietal lobe ablations in monkeys. *Journal of Neurophysiology* **7**, 269-286.

Perrett, D.I., Harris, M.H., Bevan, R., Thomas, S., Benson, P.J., Mistlin, A.J., Citty, A.J., Hietanen, J.K., and Ortega, J.E. (1989) Framework of analysis for the neural representation of animate objects and actions. *Journal of Experimental Biology* **146**, 87-113.

Phillips, C.G. (1985) *Movements of the hand.* Liverpool University Press, Liverpool.

Rizzolatti, G., Camarda, R., Fogassi, L., Gentilucci, M., Luppino, G. and Matelli, M. (1988) Functional organization of area 6 in the macaque monkey. II. Area F5 and the control of distal movements. *Experimental Brain Research* **71**, 491-507.

Robinson, D.L., Goldberg, M.E., and Stanton, G.B. (1978) Parietal association cortex in the primate. Sensory mechanisms and behavioural modulation. *Journal of Neurophysiology* **41**, 910-932.

Sakata, H., Shibutani, H., Ito, Y., and Tsurugai, K. (1986) Parietal cortical neurons responding to rotary movement of visual stimulus in space. *Experimental Brain Research* **61**, 658-663.

Sakata, H., Taira, M., Mine, S., and Murata, A. (1992) Hand-movement-related neurons of the posterior parietal cortex of the monkey: their role in the visual guidance of hand movements, in R. Caminiti, P. B. Johnson and Y. Burnod, (Eds), *Control of arm movement in space: neurophysiological and computational approaches,* Heidelberg: Springer, Berlin, pp. 185-198.

Schieber, M.H. (1990) How might the motor cortex individuate movements, *TINS* **13**, 440-445.

Schneider GE (1969) Two visual systems. *Science* **163**, 895-9O2.

Shinoda, Y., Yokota, J.I., and Futami, T. (1981) Divergent projections of individual corticospinal axons to motoneurons of multiple muscles inn the monkey. *Neuroscience Letters* **23**, 7-12.

Sivak, B. and MacKenzie, C.L. (1992) The contribution of peripheral vision and central vision to prehension, In L. Proteau and D. Elliott (Eds)*Vision and motor control,* Elsevier, Amsterdam.

Sprague, J.M. and Meikle, T.H. (1965) The role of the superior colliculus in visually guided behavior. *Experimental Neurology* **11**, 115-146.

Stelmach, G.E., Castiello, U., and Jeannerod, M. (1994) Orienting the finger opposition space during prehension movements. *Journal of Motor Behavior* **26**, 178-186.

Taira, M., Mine, S., Georgopoulos, A.P., Murata, A., and Sakata, H. (1990) Parietal cortex neurons of the monkey related to the visual guidance of hand movements. *Experimental Brain Research* **83**, 29-36.

Ungerleider, L. and Mishkin, M. (1982) Two cortical visual systems, in: D.J. Ingle, M.A. Goodale and R.J.W. Mansfield (Eds),*Analysis of visual behavior*, MIT Press, Cambridge, pp. 549-586.

Vital-Durand, F. and Jeannerod, M. (1974) Role of visual experience in the development of optokinetic responses in kittens. *Experimental Brain Research* **20**, 297-302.

Vital-Durand, F., Putkonen, P.T.S., and Jeannerod M. (1974) Motion detection and optokinetic responses in dark reared kittens.*Vision Research* **14**, 141-142.

Wallace, S.A. and Weeks, D.L. (1988). Temporal constraints in the control of prehensive movements. *Journal of Motor Behavior* **20**, 81-105.

Wannier, T.M.J., Maier, M.A., and Hepp-Reymond, M.C. (1991) Contrasting properties of monkey somatosensory and motor cortex neurons activated during the control of force in precision grip. *Journal of Neurophysiology* **65**, 572-589.

Weiskrantz, L. (1986) *Blindsight. A case study and implications,* Oxford University Press, Oxford.

Westling, G. and Johansson, R.S. (1984) Factors influencing the force control during precisiongrip. *Experimental Brain Research* **53**, 277-284.

Wing, A.M. and Fraser, C. (1983). The contribution of the thumb to reaching movements. *Quaterly Journal of Experimental Psychology* **35A**, 297-309.

Wing, A.M., Turton, A., and Fraser, C. (1986). Grasp size and accuracy of approach in reaching. *Journal of Motor Behavior* **18**, 245-260.

Zeki, S. (1993) *A vision of the brain*. Blakwell, Oxford.

FUNCTIONAL ARCHITECTURE OF CORTICAL NETWORKS UNDERLYING VISUAL REACHING

P. B. JOHNSON[1], S. FERRAINA[2], R. CAMINITI[2]
[1]*Division of Neurosurgery, Box 3807, Duke University Medical Center, Durham, NC 27710, USA, and* [2]*Istituto di Fisiologia umana, Università degli Studi di Roma "La Sapienza", Piazzale Aldo Moro 5, 00185 Rome, Italy*

We exist within a intricate three-dimensional world where most of our movements are made with respect to targets and references rooted to objects which are external to our own bodies. The complexities of successfully navigating one's limbs through this extrapersonal space have only recently become fully appreciated. Planning and execution of movements in extrapersonal space require the combination of information regarding the locations of targets of interest and information concerning the geometry of our own body with respect to those targets. The act of reaching to visual targets provides a good model for the study of this general problem encountered in motor control.

The primary computation in the control of reaching is the transformation of information concerning target location and initial arm configuration into a motor command appropriate for carrying the hand to the target. Most reaching movements are toward targets whose locations are derived from visual inputs which are originally represented in the nervous system in a retinotopic frame of reference. Proprioceptive information, on initial arm configuration, originally exists in a coordinate system based on muscle lengths and joint angles. The brain regions contributing to the control of reaching must combine, transform, and transmit these disparate input sources in a way which leads to a pattern of muscle activations. Recent approaches to the study of the control of reaching have stressed these concepts of information representation and coordinate system transformations [46, 16, 17, 35, 59, 38]. This emphasis reflects a general trend in the study of sensory-motor function [44, 23, 37, 50, 51, 56].

There is mounting evidence for the existence of interactions between visual and somatic inputs in the central nervous system. Psychophysical studies have indicated an interaction between visual and proprioceptive information for the construction of a coherent internal representation of visual space [36, 53, 60, 45]. Interactions of retinal and extraretinal signals have also been shown at the level of the single neuron. The responses of neuron in the inferior parietal lobule [6] to visual stimuli have been shown to be influenced by the angle of gaze [4, 3]. Similar dependencies on gaze angle have also been shown in the visual responses of neurons in areas V6 [27] and area V3A [27] and in both dorsal [12] and ventral [63] premotor cortex. It is not known how commonly, or how early in the visual pathway, effects of non-retinal inputs appear within the visual

F. Lacquaniti and P. Viviani (eds.), Neural Bases of Motor Behaviour, 77–97.

system. However, the existence of extraretinal signals have been observed even in the activity of neurons in V1 [61].

There is also psychophysical evidence for the interaction of visual information and kinesthetic input from the arm in the perception of arm configuration [37]. Furthermore, it is well known that deafferented patients make significantly more errors in the performance of visual reaching tasks than do normal subjects [54, 34, 11]. There have been no quantitative studies examining interactions between visual responses and arm proprioception at the single neuron level, however, some preliminary evidence of such interactions has been found. Recordings of neuronal activity in parietal area 5 indicate that within the medial wall of the intraparietal sulcus (IPS) there are neurons which can be driven by both visual and somatosensory stimuli [21, 42, 43]. During reaching tasks to visual targets, neurons in frontal lobe motor areas demonstrate a shift in their directional tuning as a result of changes in arm configuration, reflecting the combination of inputs that we know must occur [15, 16, 17, 9]. Similar results have been obtained in parietal area 5 [26], where a body-centered positional code has been successfully used to combine static postural and movement-related aspects of reaching [48].

In this paper we will provide an outline of the relationships between the representations of sensorimotor information and the anatomical substrates by which this information is combined and transformed in those cortical areas which are most directly involved in the control of movement: the primary motor cortex (MI) and dorsal premotor cortex (PMd) in the frontal lobe, dorsal area 5 and area MIP (medial intraparietal) [42, 43] in the parietal lobe. The data discussed in this paper were obtained during studies [42, 43] aiming at assessing the relationships between the physiological properties and the cortical connectivity of these regions.

1. The Reaching-related Region of the Frontal Lobe

The region of the frontal lobe rich in neurons related to proximal arm motion and therefore active during a reaching task (Fig 1A), includes, in the flat exposed part of the cortex (Fig. 1B), a zone spanning from the central sulcus (CS) to the spur and posterior part of the medial limb of the arcuate sulcus (AS). Mediolateraly, it extends approximately from the Superior Precentral Sulcus (SPcS) to the imaginary line represented by the "projection" of the arcuate spur on the central sulcus.

2. Functional Properties of Frontal Neurons

The flow of information set in motion by a visual stimulus serving as target of a reaching movement can be studied in this as in other cortical regions, through the analysis of the correlations between the dynamic properties of cortical neurons and behavior. This study can be significantly enriched by the analysis of the spatial distribution of these functional properties within the cortical tissue.

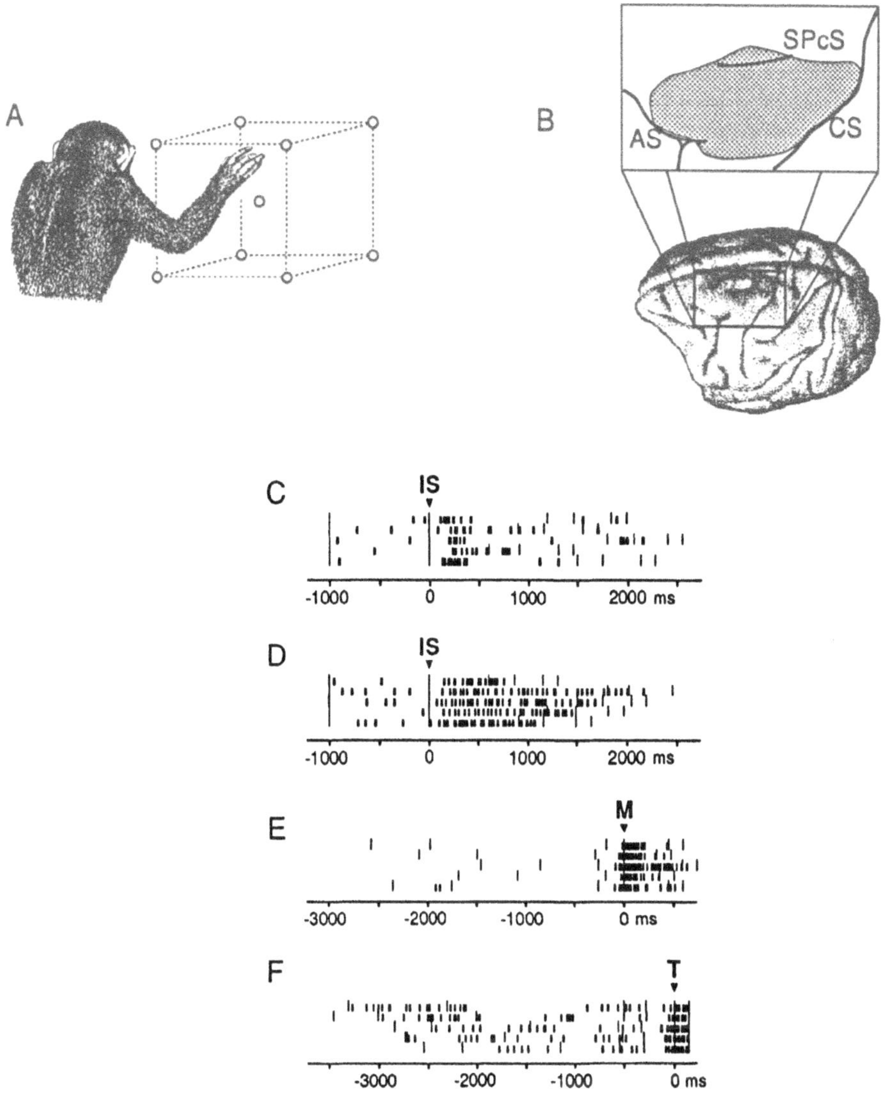

Fig 1. Behavioral apparatus and summary of types of neuronal activity studied in the frontal lobe. **A.** Circles represent the locations of LED/pushbuttons situated at the ends of metal rods protruding from a rear panel. The central LED/pushbutton was the origin of all movements. **B.** Grayed area outlines frontal lobe proximal arm representation studied by single unit recording methods. **C.** Raster showing example of signal-related activity. Rasters from 5 trials to the same target are aligned to the presentation of the IS. Thick tick marks represent single action potentials. Longer, thin tick marks correspond to behavioral events during the evolution of the trial; from left, beginning of neuronal recording, presentation of the IS, presentation of the GS, onset of movement, target acquisition, end of trial/reward. The time scale is in milliseconds. **D.** Example of cell displaying set-related activity, conventions as in C. **E.** Example of neuron with movement-related activity. Rasters are aligned to movement onset, other conventions as in C. **F.** Example of neuron with position-related activity. Rasters are aligned to the beginning of the target hold period.

If one wants to dissociate in time the early events concerning target location from the later ones related to movement generation and execution, instructed-delay directional reaching tasks can be used. In these tasks, a visual "cue" instructs the animal where the next movement has to be made after an appropriate waiting period at the end of which a "go" signal for reaching is presented. Four main neuronal activity types have been observed in the frontal lobe reaching-related region, during four temporally distinct epoches of this task: phasic signal-related activity occurring immediately after the presentation of the instruction signal (IS; Fig 1C); tonic set-related activity occurring during the interval between IS and go-signal (GS; Fig 1D); movement-related activity which usually begins before the initiation of movement (Fig 1E); and position-related activity which occurs while the monkey hold its hand at the different target positions (Fig 1 F). Defined time epoches (DT1, DT2, RTMT, THT) are used to distinguish these 4 types of activity.

The majority of neurons in the reaching-related region of the frontal lobe exhibits more than one of these activity-types. Some neurons display all 4 of the above-defined activity-types. Signal- and set-related activities are broadly tuned to the direction of the upcoming movement [42, 43] as has been previously reported for movement-related activity [57, 17].

2.1. INTRACORTICAL DISTRIBUTION OF FUNCTIONAL PROPERTIES

Our understanding of cortical representation of reaching can be significantly improved by studying the spatial organization of functional properties of neurons in both the the tangential and radial domains of the cortex.

2.1.1. *Tangential Organization of Frontal Cortex*

The distribution of the activity-types above mentioned is not uniform in the tangential domain of dorsolateral frontal cortex. This can be assessed by using different approaches. The incidence of neurons with significant directional modulation during a given epoch varies with rostrocaudal level (Fig. 2 A-D). Signal- and set modulation are common rostral to the 4/6 border (the border between primary motor cortex and dorsal premotor cortex) and become less frequent moving toward the CS. Here, in between the central sulcus (CS) and the 4/6 border, significant movement-related modulation is more common than rostral to the 4/6 border. Neurons displaying significant position-related modulation can be found throughout the rostrocaudal extent of frontal lobe in a rather uniform fashion. The 4/6 border region is very active during the task and display a relatively high incidence of all types of neuronal activities.

A different complementary way to study this aspect of the cortical organization consists in deriving a measure of the degree to which each cortical neurons displays one or more of the directional activity-types observed and to compute mean levels of this measure for each penetration [see 43 for details]. Since most neurons display various degrees of directional modulation during both IDT and movement epoches of the task, the ratios of these activity can be computed and expressed as a function of the rostrocaudal location in the cortex. Plots of these ratios (Fig. 2 E-F) show a significant trends ($p < 0.0001$, regression F test) indicating how directional movement-related activity increases relatively as one moves caudally through the cortex.

Thus, moving in the rostrocaudal tangential domain of the cortex, the functional properties of neurons progressively change in a trend-like fashion from those which are expression of the neural phenomena leading to target location to those more linked to movement generation.

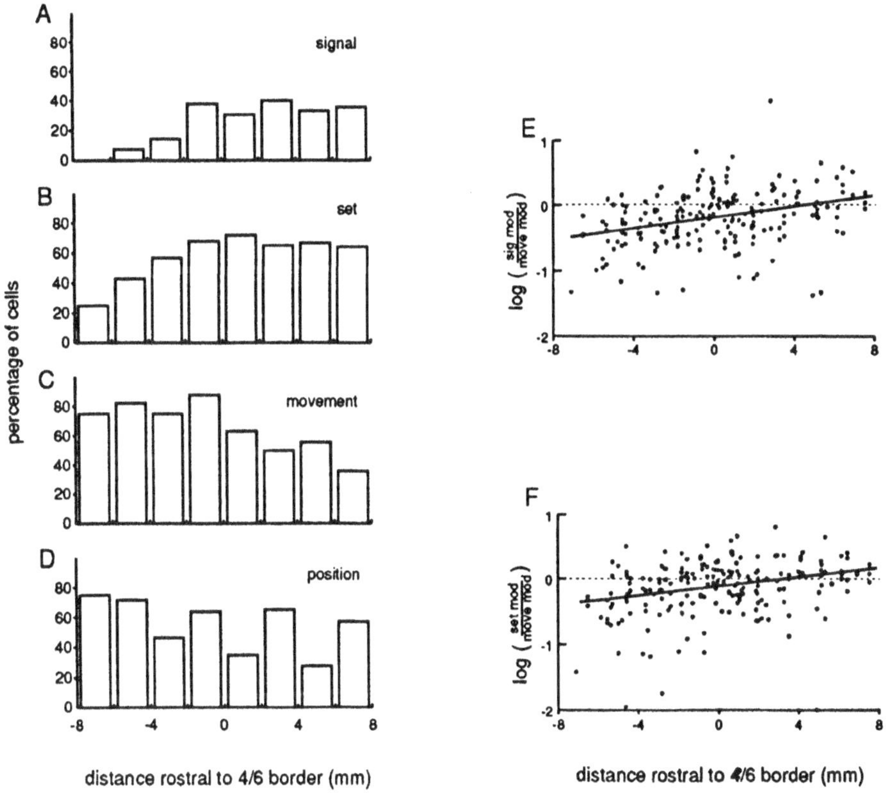

Fig 2. Left. Percentages of neurons with significant directional modulation ($p < 0.01$, ANOVA) as a function of rostrocaudal location within frontal cortex. Percentages were calculated for 2 mm wide rostrocaudal bins relative to the area 4/6 border. **A.** Signal-related directional modulation. **B.** Set-related directional modulation. **C.** Movement-related directional modulation. **D.** Position-related directional modulation. Right. Plots of directional activity ratios as functions of rostrocaudal position. **E.** Plot of the logarithm of the ratio of signal- to movement-related activity as a function of cell position relative to the area 4/6 border. Positive values indicate a high signal-related activity relative to movement-related activity. Negative values indicate relatively more movement-related activity. Solid line is a linear regression with slope significantly different from zero ($p < 0.0001$, regression F test; $r = 0.322$). **F.** Plot of the logarithm of the ratio of set- to movement-related activity. Conventions as in A. Regression slope is significantly different from zero ($p < 0.0001$, regression F test; $r = 0.295$). (From Johnson et al., 1996, reprinted with modification)

2.1.2. Encoding of movement direction across task epoches.

When the arm moves in different directions toward a visual target, cell activity in the motor [31, 57] and premotor [15, 17] cortices is highest for a particular preferred direction of movement (PD). The spatial orientation of the neuronal preferred directions (PDs) throughout the evolution of the behavioral trial is of interest since it probably relates to the nature of the intracortical information processing within the frontal lobe reaching-related region during the composition of commands for reaching to a visible target. The stability of the orientation of the PD can be measured by comparing, for each directionally tuned neuron, the angular differences between PDs calculated during different epoches (Fig 3).

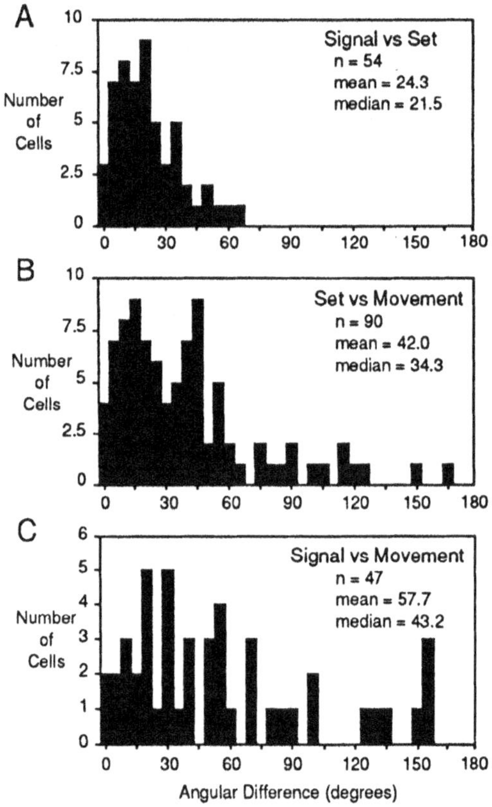

Fig 3. Histograms of angular differences between PDs of different task epoches. For each directional neuron, preferred directions were calculated for signal-, set-, and movement-related activities separately. Angular differences are expressed as the inverse cosine of the inner product between vectors. **A.** Distribution of angular differences between signal and set PDs of the same neuron. **B.** Distribution of angular differences between set and movement PDs. **C.** Distribution of angular differences between signal and movement PDs.

For many neurons, the PD remains relatively invariant across task epoches, for others it changes significantly. PDs from epoches DT1 and DT2 are the most similar (median angular difference = 21.5 degrees), those from epoches DT1 and MT are the most different (median angular difference = 43.2 degrees). In all cases the spatial orientation of PDs across task epoches displays little stability, suggesting that once the visually-derived input about target location becomes available to the frontal lobe, further computation is necessary for the final shaping of the command for reaching.

2.1.3. *Distribution in depht of preferred direction vectors of movement-related activity*

The radial distribution of the orientations of PDs within the frontal cortex is of interest since it relates to the problem of cortical coding of movement direction within vertical assemblies of cortical neurons. This distribution is represented graphically in Fig 4. No obvious organization in the distribution of these orientations across the cortex can be observed. However, it can be seen that, in some cases, neurons recorded in close proximity to one another within the same penetration have similar PDs.

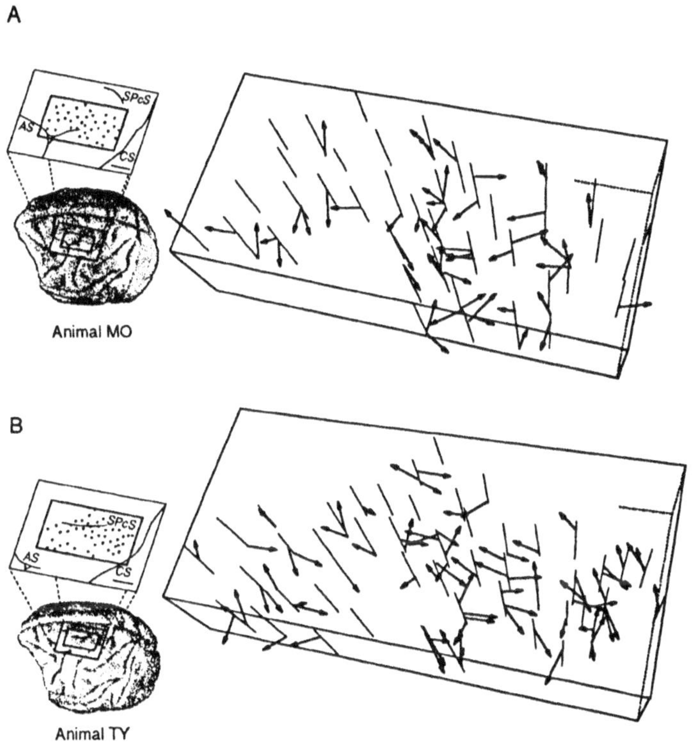

Fig 4. Representation of the distribution of neuronal PDs within the reaching-related zone of dorsolateral frontal cortex. Solid rectangle represents the slab of cortex indicated on the brain figurine. Lines represent the locations of microelectrode recording penetrations through the cortex. The PDs of neurons during the RTMT task epoch are indicated by arrows. The origin of each PD arrow is drawn at the depth at which the neuron was recorded along the penetration.

The relationship, for pairs of individual neurons, between similarity in orientation of PD and distance between neurons can be examined by at least 2 methods. First, for pairs of neurons which are recorded in different penetrations, the distance between them is approximated by the tangential cortical distance between the microelectrode entry points of the two penetrations. In this analysis, neurons recorded in the same penetration are assigned a cortical separation of zero. The angular difference is then computed between PDs of all pairs of neurons directionally tuned during the RTMT. Angular differences and cortical distances are shown for a total of 4725 pairs of directionally tuned neurons (Fig 5). Interneuron distances for neurons recorded in differing penetrations range from 0.481 to 14.263 mm. There is no relationship between angular difference of PDs and interneuron distance. As to be expected with a uniform distributions of PDs, the median angular difference between pairs of PDs is close to 90 degrees (90.1 degrees).

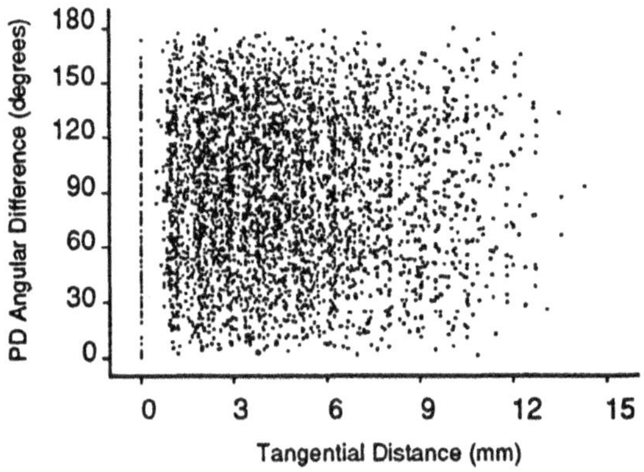

Fig 5. Plot of angular difference between PDs of pairs of neurons as a function of the tangential cortical distance between the neurons. Pairs of neurons with positive tangential distances were recorded in different microelectrode penetrations.

The relationship between angular difference and cortical distance can also be examined within individual microelectrode penetrations (Fig 6). Interneuron distances are approximated using the depth of the recording electrode as a measure. Pairs of neurons recorded without moving the recording electrode are assigned a cortical separation of zero. In the analysis of 93 pairs of directionally tuned neurons, interneuronal distances range from 0 to 3521 mm. The median angular difference between pairs of PDs within single penetrations is less than for pairs in differing penetrations (median = 76.4 degrees). The tendency for within-penetration angular differences to be less than between-penetration differences is statistically significantly ($p < 0.05$ Mann-Whitney test). Neurons very close to one another within the same penetration tend to be even more similar. Pairs of neurons recorded within 300 μm of each other in the same penetration have a median angular difference of 64.8 degrees. However, as with the between-penetration data, there is no obvious linear relationship of difference in PD to interneuron distance within single penetrations.

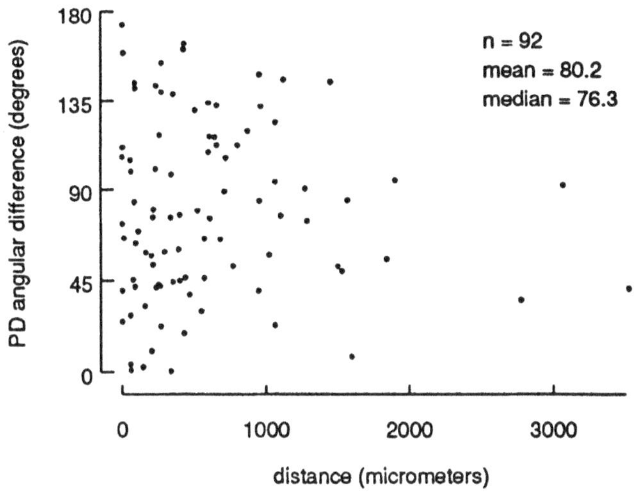

Fig 6. Plot of angular difference between PDs of pairs of neurons which were recorded in the same recording penetration. Angular differences are plotted as a function of the difference in depth along the recording penetration between the 2 neurons.

3. Anatomical Organization of the Reaching-related Zone of the Frontal Lobe. Instrinsic and Estrinsic Cortico-cortical Connectivity

The frontal lobe reaching-related zone is characterized by an uneven distribution of functional properties which are partially segregated in the tangential cortical domain. This suggests a spatial segregation of inputs to the frontal lobe and postulates the existence of significant frontal intracortical processing leading to composition of motor commands for reaching. It becomes therefore important to see whether the features of intrinsic local connectivity and the spatial organization of the cortico-cortical inputs provide coherent anatomical *substrata* to the segregation observed.

3.1. INTRINSIC ANATOMICAL ORGANIZATION OF THE FRONTAL LOBE REACHING-RELATED REGION

The intrinsic connectivity of the frontal lobe arm-related region is primarily local in nature. This pattern of connectivity will be illustrated by using two prototypical experimental cases in which different retrograde tracers were injected, as shown in Figure 7 [43].

Within the local area of the proximal arm representation, regions of near-continuous labelling of neurons are found immediately surrounding the injection sites for all types of tracers (Fig 7). These regions, in general, extend for approximately 1.5 mm from the injection site. Beyond this 1.5 mm range, the distribution of labelling begins to be more "patchy". Even excluding the region of continuous labelling within 1.5 mm of the injections, the vast majority of labelled neurons are located within 5 mm of the injection sites. Labelled neurons are found deep in the CS, SPcS, and the AS.

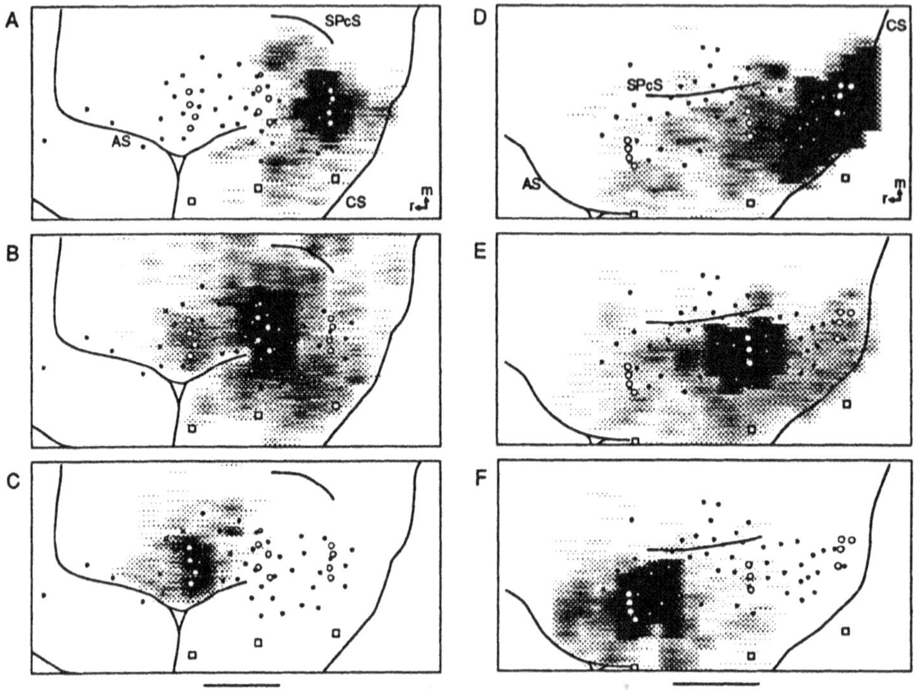

Fig 7. Maps of labelled cell density in the frontal lobe reaching-related region. Coronal sections were flattened by a computer-based procedure (see Methods) and the density of labelled neurons (neurons per 100 x 100 µm bin) is indicated on a gray scale. Darker regions contain more labelled neurons. Relative locations of recording penetrations (small open circles), tracer injections (large open circles), and marking dye injections (open squares) are also indicated. A-C. Label resulting from injections in MI (RLM), MI/PMd (WGA-apoHRP-Au), and PMd (GLM) of animal MO. D-F. Distribution of labelled neurons following injections in MI (RLM), MI/PMd (CTB-Au), and PMd (GLM) of animal TY. Scale bars are 5 mm.

3.1.1. *Neurons Projecting to the MI Reaching-related Region*

In general, neurons projecting to the caudal part of MI can be identified within the task-related area. However, they are absent in the most rostral portion of this region. Additional projections arise from more lateral regions which probably correspond to the MI distal forelimb representation and from the rostral bank of the CS. A projection originates in the fundus and medial bank of the SPcS.

When retrograde tracers injections (Fig 7 A) are in a focus of high position-related directional activity and low directional activity during the other epoches, cells labelled by this injection are restricted to the caudalmost portion of the proximal arm representation, as is the distribution of position-related activity. When the MI injection (Fig 7 D) is in a region of high position- and movement-related directional activities, the distribution of labelled cells extends further rostrally, resembling the distributions of position- and movement-related directional activities.

3.1.2. *Neurons Projecting to the MI/PMd Border Region*

As with the projections to caudal MI, those projecting to the MI/PMd border originates also within the physiologically identified task-related area. These projections however originates from regions displaced rostrally. More projecting neurons are found in PMd, fewer in caudal MI. In addition, more projecting neurons are found in the lateral regions corresponding to the MI distal forelimb representation than following the caudal injections. Patches of labelled cells are also found within and medial to the SPcS.

When the MI/PMd (Fig 7 B) injections are in a zone of high directional activity related to signal, set, and movement, the pattern of labelling shows features in common with the distributions of each of these activity types. When the MI/PMd border region injections (Fig 7 E) are in a zone of moderate to high directional activity for all epoches, the resulting pattern of labelling from these injections covers most of the cortical regions which display moderate to high directional activity in any of the four epoches.

3.1.3. *Neurons Projecting to the PMd Reaching-related Region*

The most rostral task-related area receives projections from a relatively small portion of this region, including the medial bank of the superior limb of the AS.

The rostral injections shown in Figure 7 C, are in regions of generally lower levels of both mean and directional activity in the task. In general, the overlap between labelling and patterns of neuronal activity in these cases is low and the distributions of labelled cells does not resemble the distributions of any of the defined activity-types.

3.2. EXTRINSIC PARIETAL PROJECTION TO THE FRONTAL LOBE REACHING-RELATED REGION

Parietal projections to ipsilateral frontal lobe reaching-related regions originate almost exclusively in the SPL. Here, the cells of origin of these projections are located in the exposed cortex of the lobule and extend caudoventrally into the medial wall of the intraparietal sulcus (IPS). Additional projections originate from the caudalmost tip of the SPL and within the caudal portions of the cortex located on the medial aspect of the hemisphere.

3.2.1. *Parietal projections to MI*

Parietal projections to the caudal regions of the frontal lobe reaching area, originate from cells located in the exposed cortex of the SPL and in the medial crown of the IPS. Parieto-frontal association cells can be found from the level of the postcentral dimple to the crown of the IPS and into the dorsalmost part of the medial wall of the IPS (Fig 8). It is probable that these projections originate from both areas 2 and 5. The border between these areas is in fact difficult to determine.

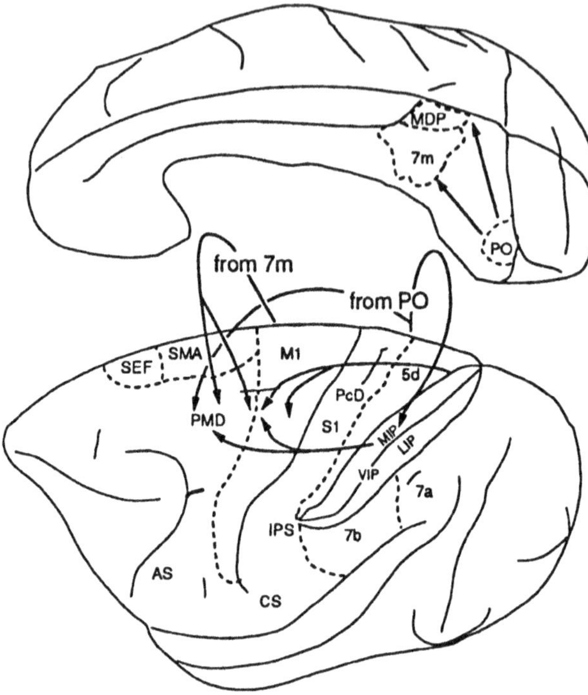

Fig 8. A summary of parieto-frontal association connectivity. Top: medial view of the cerebral hemisphere. Bottom: lateral view of the cerebral hemisphere. Abbreviations: AS, arcuate sulcus; CS, central sulcus; PcD, postcentral dimple; IPS, intraparietal sulcus; MIP, LIP, and VIP, respectively the medial, lateral and ventral intraparietal areas; MDP, medial dorsal parietal area; VIP, ventral intraparietal area; PO, parieto-occipital area; MDP, medial dorsal parietal area

3.2.2. *Parietal projections to the MI/PMd border region*

Parietal projections to the MI/PMd border originate from neurons located within the SPL at intermediate locations between the cells projecting to MI and those projecting to PMd. When the pattern of injections is more caudal in M1 the cells of origin of parietal projections are located on the exposed cortical surface and extend into the dorsal part of the medial wall of the IPS . When injections are placed more rostrally in M1 (Fig 8), projecting cells can be found over much of the medial wall of the IPS (area MIP) as well as over a small region of the ventral wall of the posterior cingulate sulcus. Some projection cells are also located onto the exposed surface of the SPL.

3.2.3. *Parietal projections to PMd.*

Parietal projections to the PMd proximal arm area originate from separate regions of the SPL. These are primarily area MIP (Fig 8) in the medial wall of the IPS, with some additional projections coming from the most caudal and dorsal extent of the SPL, from area MDP (Fig. 8). The projections coming from the medial SPL generally originate from a region extending from the ventral wall of the most caudal portion of the CiS to the mesial surface of the hemisphere, in area 7m [18] (Fig 8).

Interestingly, projections to a more rostral region of the frontal lobe lying immediately anterior to the proximal arm representation of PMd originate from the most posterior locations within the SPL, including visual area area PO.

4. Physiological Trends in the Parietal Lobe

The spatial correspondence between tangential distribution of activity types in the frontal lobe and pattern of association parietal projections suggests that these functional trends may be, at least in part, imposed be the set of parietal association projections. These in turn, must reflect, the general functional organization of the cortex of origin.

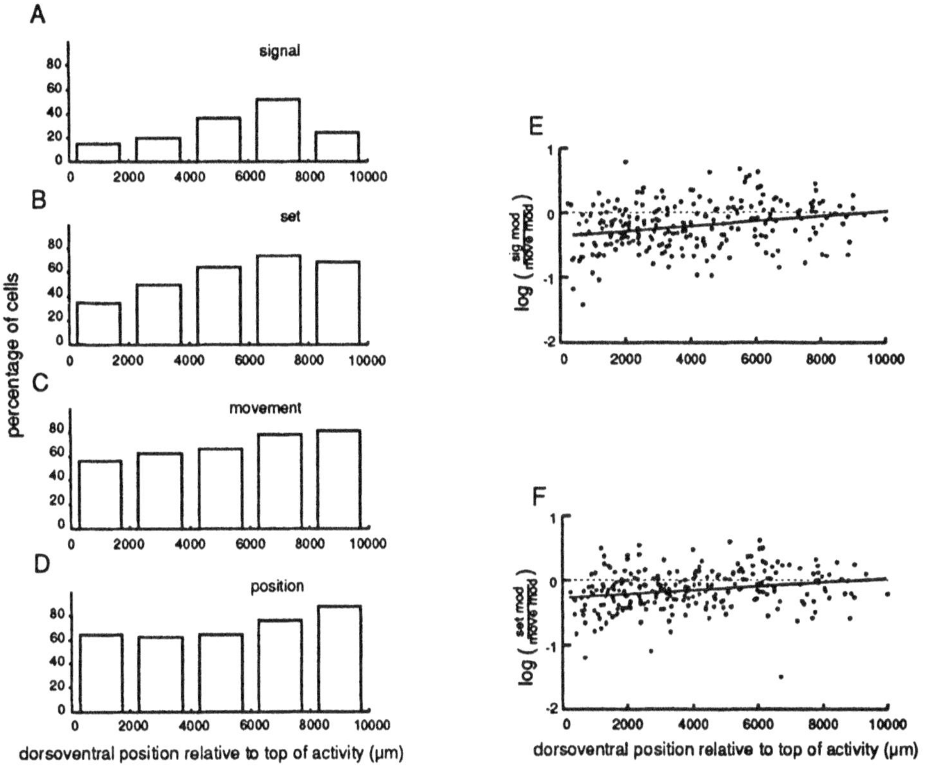

Fig 9. Left. Percentages of parietal neurons with significant directional modulation ($p < 0.01$, ANOVA) as a function of dorsoventral location. Percentages were calculated for 2 mm wide dorsoventral bins along each penetration relative to the position of the first neuronal activity encountered. **A.** Signal-related directional modulation. **B.** Set-related directional modulation. **C.** Movement-related directional modulation. **D.** Position-related directional modulation. Right. Plots of directional activity ratios as functions of dorsoventral position. **E.** Plot of the logarithm of the ratio of signal- to movement-related activity as a function of cell position relative to the top of neuronal activity. Positive values indicate a high signal-related activity relative to movement-related activity. Negative values indicate relatively more movement-related activity. Solid line is a linear regression with slope significantly different from zero ($p < 0.0005$, regression F test; $r = 0.197$). **F.** Plot of the logarithm of the ratio of set- to movement-related activity. Conventions as in E. Regression slope is significantly different from zero ($p < 0.0002$, regression F test; $r = 0.212$). (From Johnson et al., 1996, reprinted with modification)

The analysis of the tangential organization of the region of the SPL including dorsal area 5 and MIP, and containing the cells of origin of the parietal projections to the frontal

cortex, reveals the existence of functional trends similar to those observed in the frontal lobe. Cells displaying directional activity during the movement (Fig 9 C) and positional (Fig 9 D) epoches of the task are distributed in a rather uniform fashion along the tangential domain of the dorsal bank of the IPS. Neurons displaying directional set-related activity, while observed in the more dorsal regions of the bank (Fig 9 B), are encountered more consistently as one moves ventrally in the cortex within the dorsal bank of the IPS (Fig 9 B). This is particularly true for directional signal-related activity which occurs after the presentation of the IS (Fig 9 A), and which is mostly segregated in the ventralmost part of MIP.
The relative activation of neurons within the dorsal bank of the IPS (Fig 9 E-F) show significant ($p < 0.0005$, regression F test) trends for the directional activity during the earlier epoches of the task to increase relatively as one moves ventrally (posteriorly) through the cortex.

5. Conclusions

5.1. VISUOMOTOR ACTIVITY IN THE FRONTAL LOBE

The objective of the physiological aspects of the present review was to describe the reaching-related portion of the dorsolateral frontal cortex as characterized by means of a commonly-used behavioral paradigm. This characterization involves identifying differing types of activity on the basis of their temporal correlation to the task. However, this task does allow us to make statements about the *relative* placement of the regions studied along the visual to motor continuum.

Neuronal activity which is temporally associated with the presentation of the visual stimulus in the absence of movement is commonly termed signal- or set-related. The degree to which these types of activity can be interpreted as "visually-related" activity is a matter of debate. Considerable controversy has been generated by reports of "visual" activity in the MI/PMd region [47, 62, 52]. Recent studies employing elaborate tasks to dissociate pure visual from visuomotor activity [1, 2, 63, 24] have concluded that even early MI and PMd activity is dependent on the motor significance of the stimulus and, thus, is not purely visual in nature. Thus, the signal- and set-related activities of the present study cannot be regarded as pure visual signals. In addition, we have referred to neuronal activity associated with the execution of the movement as movement-related. However, the fact that, in the task used, the movements were cued by a visual go-stimulus precludes us from ruling out a visuomotor nature of this activity as well.

Given that signal- and set- related activity is temporally dissociated from the execution of movement, it is evident that the visual aspect of this activity is weaker than it is for movement-related activity. Similarly, the dissociation of movement-related activity from the presentation of the visual instruction stimulus suggest a weaker visual driving of this type of activity. Thus, signal- and set-related activity most probably reflects a prevalence of visually-derived input.

5.2. THE SOURCES OF VISUAL INPUT TO THE REACHING-RELATED REGION OF THE FRONTAL LOBE. PARALLEL INPUTS AND INTRACORTICAL PROCESSING

Recent metabolic mapping studies using the 2-DG method have implicated the SPL in the processing of visual information for the control of movement. Savaki and colleagues [55] trained monkeys to perform a visual pattern discrimination task in which the correct responses were visually-guided reaching movements toward targets located in front of the animal. In intact monkeys, contralateral to the performing arm, they found an activation of the medial part of that portion of area 5 which lies in the medial wall of the IPS. This region approximately corresponds to deep area MIP and possibly part of VIP [20]. In monkeys in which the hemisphere contralateral to the performing arm had been surgically deprived of visual input, this same area was metabolically depressed. Savaki and colleagues concluded that this area was an "... important part of the circuit that normally mediates visually guided forelimb reaching ... ." The same pattern of activity, although to a lesser degree, was found for area MDP. These results extend earlier metabolic mapping studies [49]. In both of these studies, the metabolically-defined "visual" cortex includes the most posterior portion of the SPL and extends rostrally on both the medial wall of the hemisphere and within the deep part of the medial wall of the IPS. These areas partially correspond with the areas identified in the present study as projecting to the more rostral zones of the frontal lobe reaching representation. In our data, the labelled neurons in MIP and MDP appear to be situated at the dorsal edge of the metabolically-defined visually-related cortex.

Independent anatomical studies have shown that area PO [22, 30, 20] receives direct projections from a number of visual areas including V1, V2, V3, V4 and the medial temporal visual area (area MT) [20] and is reciprocally linked with MIP and MDP [20, 10]. Furthermore area 7m, located just ventrally to MDP on the medial surface of the hemisphere, receives a projections from ispilateral area PO [18]. The latter is therefore capable of supplying MIP, 7m, and probably MDP with visually-derived information, and therefore emerges as the main source of visual input to the SPL.

The dynamic properties of neurons in area PO [28, 29, 27] and the nature of the visual field representation [30] with the relative emphasis on the visual periphery suggests a role of this area in visuospatial processing [20]. Furthermore, PO is connected with the lateral intraparietal area area LIP, known to be involved in the analysis of visual space [5, 10, 8).

Thanks to its connections with visual area PO and to its projection to the arm region of the frontal lobe, areas MIP, MDP and 7m in the SPL emerge as the intermediate link in the cortical network for visual reaching.

Based on physiological data and network models, it has been proposed that neurons in the cortical regions involved in controlling reaching movements to visual targets perform their function by progressively matching the incoming patterns of visual and proprioceptive information to the appropriate motor command [13, 14]. Such a "progressive match" mechanism could be implemented by a network of neurons spanning multiple cortical regions and operating both at inter- and intra-areal levels. A multisynaptic pathway, along whith the constituent neuronal components, performing computations based on extrinsic afferent input and local input from surrounding neurons, would be well suited to implement such a mechanism [13, 14].

The direct comparison of the distributions of functional properties and anatomical connections in both frontal and parietal cortices reveals two main characteristics of these cortical regions: i) the existence along their tangential domain of trends of functional properties; ii) the existence of sets of parallel association connections linking cortical regions with similar functional properties. This suggests that the different inputs carrying the information necessary for reaching are largely segregated in nature and postulates the necessity of significant local interactions [see also 33] between assemblies of neurons sharing similar properties for the final shaping of the motor command. In this respect, the shift in the orientation of PDs of frontal cells during task epoches is of interest since it could be the expression of a shift in the origin of the coordinate system for reaching from a head to a shoulder-centered one, as suggested by psychophysical studies [58]. This hypothesis can be experimentally tested.

5.3. ORGANIZATION OF DIRECTIONAL PROPERTIES IN THE FRONTAL LOBE REACHING-RELATED REGION

For movements made in a given part of space, neurons in MI and PMd can be assigned a PD vector [31, 57, 15, 16, 17]. This vector represents the movement direction for which this cell would exhibit maximal activity. The conceptual similarity between PD vectors and the preferred orientations of neurons in the visual system [39, 40] suggests a possible organization of movement PDs within the motor cortices. There is some evidence that directionality is represented in columns which are perpendicular to the cortical surface [32]. A separate question is whether the distribution of PDs across the cortical surface is organized in some fashion. Since the orientations of PDs are distributed in 3 dimensional space, an orderly representation of these vectors in the cortex would necessarily require a more complicated organization than that of orientation columns in V1. However, an expected consequence of many possible organizational schemes is that, at some scale, neurons which are close to one another within the tangential extent of the cortex should have similar PDs.

The fact that our results fail to show any relationship between similarity of PD and distance between neurons at the larger of the 2 scales studied is consistent with the observation that PDs have a distributed, multiple representation across the cortical surface [32]. Early microstimulation studies implied that muscle representations within MI were segregated [19, 7]. Given such an organization, it could be expected that neurons within the representation for a particular muscle would possess the "preferred direction" of that muscle. Therefore, clustering within the cortex of preferred directions would be expected. Recent microstimulation studies, however, have found muscle and movement representations to be large, disjoint, and highly overlapping [41, 25]. At any one point in the motor cortex, neurons seem to have access to multiple muscles representations. As a result, neurons are not restricted by their output connectivity to any one particular preferred direction. The lack of relationship between similarity of PD and inter-neuronal distance are consistent with these recent microstimulation results.

Our data confirmed that the PDs of neurons located very close to one another tend to be similar. However, based on the results of earlier studies [32], we had expected to find a stronger relationship between similarity of preferred direction and distance between neurons recorded within the same microelectrode penetration. One interpretation for this discrepancy is that columns of neurons with similar preferred directions are, indeed, very narrow. Columns with a width on the order of 30-100 µm (such as the orientation columns of the primary visual cortex) would be difficult to detect unless microelectrode penetrations are made along tracks exactly parallel to the column. Furthermore, with such a narrow column, the probability of recording neurons of adjacent columns at the same microelectrode position would be increased. Any conclusive interpretation of the data shown here must therefore be tempered by these considerations.

Acknowledgements:
We are grateful to Oxford University Press for permission of reproducing some of the figures shown in this article. This research was supported by the Human Capital and Mobility Program of the Commission of the European Communities and by Ministry of University and Scientific and Technological Research of Italy.

References

1. Alexander GE, Crutcher MD (1990a) Neural representations of the target (goal) of visually guided arm movements in three motor areas of the monkey. *J Neurophysiol* **64**, 164-178.
2. Alexander GE, Crutcher MD (1990b) Preparation for movement: Neural representations of intended direction in three motor areas of the monkey. *J Neurophysiol* **64**,133-150.
3. Andersen RA, Essick GK, Siegel RM (1985) Encoding of spatial location by posterior parietal neurons. *Science* **230**, 456-458.
4. Andersen RA, Mountcastle VB (1983) The influence of the angle of gaze upon the excitability of the light-sensitive neurons of the posterior parietal cortex. *J Neurosci* **3**, 532-548.
5. Andersen RA, Asanuma C, Essick G, Siegel RM (1990a) Corticocortical connections of anatomically and physiologically defined subdivisions within the inferior parietal lobule. *J Comp Neurol* **296**, 65-113.
6. Andersen RA, Bracewell RM, Barash S, Gnadt JW, Fogassi L (1990b) Eye position effects on visual, memory and saccade-related activity in area LIP and 7a of macaque. *J Neurosci* **10**, 11761196.
7. Asanuma H, Rosen I (1972) Topographical organization of cortical efferent zones projecting to distal forelimb muscles in the monkey. *Exp Brain Res* **14**, 243-256.
8. Barash S, Bracewell RM, Fogassi L, Gnadt JW, Andersen RA (1991) Saccade-related activity in the lateral intraparietal area. 2. Spatial properties. *J Neurophysiol* **66**, 1109-1124.
9. Bauswein E, Fromm C (1992) Activity in the precentral motor areas after presentation of targets for delayed reaching movements varies with the initial arm position. *Eur J Neurosci* **4**, 1407-1410.
10. Blatt GJ, Andersen RA, Stoner GR (1990) Visual receptive field organization and cortico-cortical connections of the lateral intraparietal area (area LIP) in the macaque. *J Comp Neurol* **299**, 421-445.
11. Blouin J, Bard C, Teasdale N, Paillard J, Fleury M, Forget M, Lamarre Y (1993) Reference systems for coding spatial information in normal subjects and a deafferented patient. *Exp Brain Res* **93**, 324-331.
12. Boussaoud D (1995) Primate premotor cortex.: Modulation of preparatory neuronal activity by gaze angle. *J Neurophysiol* **73**,886-889.
13. Burnod Y, Grandguillaume P, Otto I, Johnson PB, Caminiti R (1992a) Reaching toward visual targets. II. Computational studies. In: Caminiti R, Johnson PB, Burnod Y (eds) *Control of arm movement in space.* Springer-Verlag, Berlin, pp 159-174.
14. Burnod Y, Granguillaume P, Otto I, Ferraina S, Johnson PB, Caminiti R (1992b) Visuomotor transformations underlying arm movements toward visual targets: a neural network model of cerebral cortical operations. *J Neurosci* **12**, 1435-1453.
15. Caminiti R, Johnson PB, Burnod Y, Galli C, Ferraina S (1990a) Shift of preferred directions of premotor cortical cells with arm movements performed across the workspace. *Exp Brain Res* **83**, 228-232.
16. Caminiti R, Johnson PB, Urbano A (1990b) Making arm movements within different parts of space: dynamic aspects in the primate motor cortex. *J Neurosci* **10**, 2039-2058.
17. Caminiti R, Johnson PB, Galli C, Ferraina S, Burnod Y (1991) Making arm movements in different parts of space: the premotor and motor cortical representation of a coordinate system for reaching to visual targets. *J Neurosci* **11**, 1182-1197.
18. Cavada C, Goldman-Rakic PS (1989) Posterior parietal cortex in rhesus monkey: I. Parcellation of areas based on distinctive limbic and sensory corticocortical connections. *J Comp Neurol* **287**, 393-421.
19. Chang HT, Ruch RC, Ward AA (1947) Topographic representation of muscles in the motor cortex of monkeys. *J Neurophysiol* **10**, 39-56.
20. Colby CL, Gattass R, Olson CR, Gross CG (1988) Topographical organization of cortical afferents to extrastriate visual area PO in the macacaque: A dual tracer study. *J Comp Neurol* **269**, 392-413.
21. Colby CL, Duhamel J-R (1991) Heterogeneity of extrastriate visual areas and multiple parietal areas in the macaque monkey. *Neuropsychologia* **29**, 517-537.

22. Covey E, Gattass R, Gross CG (1982) A new visual area in the parieto-occipital sulcus of the macaque. *Soc Neurosci Abs* **8**, 681.

23. Darling WG, Gilchrist L (1991) Is there a preferred coordinate system for perception of hand orientation in three-dimensional space? *Exp Brain Res* **85**, 405-416.

24. di Pellegrino G, Wise SP (1993) Visuospatial versus visuomotor activity in the premotor and prefrontal cortex of a primate. *J Neurosci* **13**, 1227-1243.

25. Donoghue JP, Leibovic S, Sanes JN (1992) Organization of the forelimb area in squirrel monkey motor cortex: representation of digit, wrist, and elbow muscles. *Exp Brain Res* **89**, 1-19.

26. Ferraina S and Bianchi L (1994) "Posterior parietal cortex: Functional properties of neurons in area 5 during an instructed-delay reaching task within different parts of space". *Exp Brain Res* **99**,175-178.

27. Galletti C, Battaglini PP (1989) Gaze-dependent visual neurons in area V3A of monkey prestriate cortex. *J Neurosci* **9**, 1112-1125.

28. Galletti C, Battaglini PP, Fattori P (1993) Parietal neurons encoding spatial locations in craniotopic coordinates. *Exp Brain Res* **96**, 221-229

29. Galletti C, Battaglini PP, Fattori P (1991) Functional properties of neurons in the anterior bank of the parieto-occipital sulcus of the macaque monkey. *Eur J Neurosci* **3**, 452-461.

30. Gattass R, Sousa APB, Covey E (1985) Cortical visual areas of the macaque: possible substrates for pattern recognition mechanisms. In: Chagas C, Gattass R, Gross C (eds) *Pattern Recognition Mechanisms*. Pontificiae Acadamiae Scientarium Scripta Varia, Rome, pp 1-20.

31. Georgopoulos AP, Kalaska JF, Caminiti R, Massey JT (1982) On the relations between the direction of two-dimensional arm movements and cell discharge in primate motor cortex. *J Neurosci* **2**, 1527-1537.

32. Georgopoulos AP, Kalaska JF, Crutcher MD, Caminiti R, Massey JT (1984) The representation of movement direction in the motor cortex: Single cell and population studies. In: Edelman GM, Gall WE, Cowan WM (eds) *Dynamic Aspects of Neocortical Function*. John Wiley & Sons, New York, pp 501-524.

33. Georgopoulos AP, Taira M, Lukashin A (1993) Cognitive neurophysiology of the motor cortex. *Science* **260**, 47-52.

34. Ghez C, Gordon J, Ghilardi MF, Christakos CN, Cooper SE (1990) Roles of proprioceptive input in the programming of arm trajectories. *Cold Spring Harbor Symp Quant Biol* **55**, 837-847.

35. Ghez C, Hening W, Gordon J (1991) Organization of voluntary movement. *Curr Opin Neurobiol* **1**, 664-671.

36. Gnadt JW, Bracewell RM, Andersen RA (1991) Sensorimotor transformation during eye movements to remembered visual targets. *Vision Res* **31**, 693-715.

37. Helms-Tillery SI, Flanders M, Soechting JF (1991) A coordinate system for the synthesis of visual and kinesthetic information. *J Neurosci* **11**, 770-778.

38. Hocherman S (1993) Proprioceptive guidance and motor planning of reaching movements to unseen targets. *Exp Brain Res* **95**, 349-358.

39. Hubel DH, Wiesel TN (1963) Shape and arrangement of columns in cat's striate cortex. *J Physiol (Lon)* **165**, 559-568.

40. Hubel DH, Wiesel TN (1974) Sequence regularity and geometry of orientation columns in the monkey striate cortex. *J Comp Neurol* **158**, 267-294.

41. Huntley GW, Jones EG (1991) Relationship of intrinsic connections to forelimb movement representations in monkey motor cortex: a correlative anatomic and physiological study. *J Neurophysiol* **66**, 390-413.

42. Johnson PB, Ferraina S, Caminiti R (1993) Cortical networks for visual reaching. Exp Brain Res **97**, 361-365.

43. Johnson PB, Ferraina S, Bianchi L, Caminiti R (1996) Cortical networks for visual reaching. Physiological and anatomical organization of frontal and parietal lobe arm regions. *Cereb Cortex* (*in press*)

44. Josin G (1988) Neural-space generalization of a topological transformation. *Biol Cybern* **59**, 283-290.

45. Karnath HO, Christ K, Hartje W (1993) Decrease of contralateral neglect by neck muscle vibration and spatial orientation of trunk midline. *Brain* **116**, 383-396.

46. Kawato M, Isobe M, Maeda Y, Suzuki R (1988) Coordinates transformation and learning control for visually-guided voluntary movement with iteration: A Newton-like method in a function space. *Biol Cybern* **59**, 161-177.

47. Kwan HC, Murphy JT, Wong YC (1987) Interactions between neurons in precentral cortical zones controlling different joints. *Brain Res* **400**, 259-269.

48. Lacquaniti F, Guigon E, Bianchi L, Ferraina S, Caminiti R (1995) Representing spatial information for limb movement: the role of area 5 in monkey. *Cereb Cortex* **5**, 391-409.

49. Macko KA, Mishkin M (1985) Metabolic mapping of higher-order visual areas in the monkey. In: Sokoloff L (ed) *Brain Imaging and Brain Function*. Raven Press, New York, pp 73-86.

50. Marsino T (1992) Brainstem control of orienting movements: intrinsic coordinate systems and underlying circuitry. *Brain Behav Evol* **40**, 98-111.

51. Raphan T, Dai M, Cohen B (1992) Spatial orientation of the vestibular system. *Ann NY Acad Sci* **656**, 140-157.

52. Riehle A (1991) Visually induced signal-locked neuronal activity changes in precentral motor areas of monkey: hierarchical progression of signal processing. *Brain Res* **540**, 131-137.

53. Roll R, Velay JL, Roll JP (1991) Eye and neck proprioceptive messages contribute to the spatial coding of retinal input in visually oriented activities. *Exp Brain Res* **85**, 432-444.

54. Sanes JN, Evarts EV (1984) Motor psychophysics. *Hum Neurobiol* **2**, 217-225.

55. Savaki HE, Kennedy C, Sokoloff L, Mishkin M (1993) Visually guided reaching with the forelimb contralateral to a "blind" hemisphere: a metabolic mapping study in monkeys. *J Neurosci* **13**, 2772-2789.

56. Schall JD, Morel A, Kaas JH (1993) Topography of supplemental eye field afferents to frontal eye field in macaque: implications for mapping between saccade coordinate systems. *Vis Neurosci* **10**, 385-393.

57. Schwartz AB, Kettner RE, Georgopoulos AP (1988) Primate motor cortex and free arm movements to visual targets in three-dimensional space. 1. Relations between single cell discharge and direction of movement. *J Neurosci* **8**, 2913-2927.

58. Soechting JF, Tillery SIH, Flanders M (1990) Trasformation from head to shoulder-centered representation of target direction in arm movements. J Cogn Neurosci **2**, 32-43.

59. Soechting JF, Flanders M (1992) Moving in three-dimensional space: frames of reference, vectors, and coordinate systems. *Annu Rev Neurosci* **15**, 167-191.

60. Taylor JL, McCloskey DI (1991) Illusions of head and visual target displacement induced by vibration of neck muscles. *Brain* **114**, 755-759.

61. Trotter Y, Celebrini S, Stricanne B, Thorpe S, Imbert M (1992) Modulation of neural stereoscopic processing in primate area V1 by the viewing distance. *Science* **257**, 1279-1281.

62. Wannier TMJ, Maier MA, Hepp-Reymond MC (1989) Responses of motor cortex neurons to visual stimulation in the alert monkey. *Neurosci Lett* **98**, 63-68.

63. Wise SP, di Pellegrino G, Boussaoud D (1992) Primate premotor cortex: dissociation of visuomotor from sensory signals. *J Neurophysiol* **68**, 969-972.

SPACE CODING IN INFERIOR PREMOTOR CORTEX (AREA F4): FACTS AND SPECULATIONS

L. Fogassi
V. Gallese
L. Fadiga
G. Rizzolatti
Istituto di Fisiologia Umana,
Università di Parma,
I-43100 Parma, Italy

1. Introduction

Area F4 is a premotor area which occupies the caudal part of inferior area 6 [see 32]. It's location is shown in Fig. 1. F4 controls head, face, and arm movements. As shown by Rizzolatti and coworkers, in this area most neurons are bimodal. They have tactile RFs, and corresponding visual RFs extending outward from the tactile fields into the space around the body [20, 22, 37, 38]. The visual field location is independent of eye position and does not change with gaze shifts [22]. Recently, by using a new behavioral paradigm in which during a fixation task monkeys were presented with moving stimuli driven by a robot arm, we studied quantitatively the type of space coding of F4 visual RFs, showing that most of them use a somatocentered frame of reference [16]. Some of these data will be briefly summarized here. New data on the somatosensory, visual and motor properties of F4 neurons will be also presented. The functional properties of F4

F. Lacquaniti and P. Viviani (eds.), Neural Bases of Motor Behaviour, 99–120.

neurons and their possible role in coding of space and in motor control will be discussed.

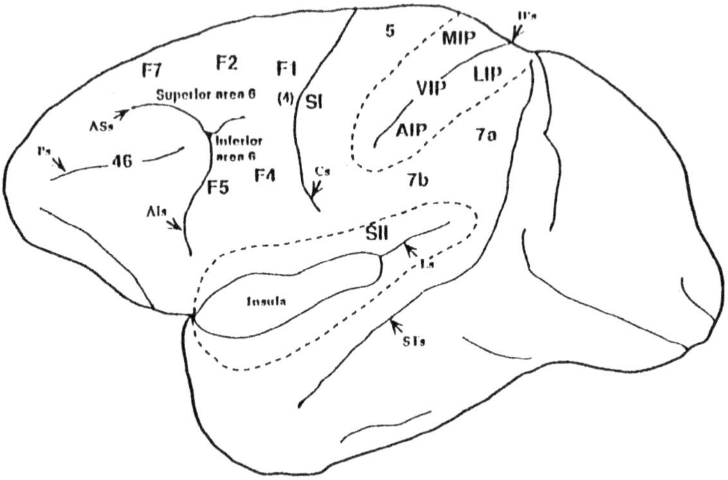

FIG.1 Lateral view of the monkey cerebral cortex showing the location of area F4. Abbreviations: AIs=inferior arcuate sulcus; ASs=superior arcuate sulcus; Cs=central sulcus; Ips=intraparietal sulcus; Ls=lateral sulcus; Ps=principal sulcus; STs=superior temporal sulcus.

2. Sensory properties of F4 neurons

We recorded from 539 F4 neurons in three hemispheres of two macaque monkeys. The majority of F4 neurons, as previously shown [20, 27, 37, 38], responded to sensory stimuli. The responsive neurons were subdivided into three categories: "somatosensory" neurons, "visual" neurons, and "bimodal", visual and somatosensory, neurons.

2.1. SOMATOSENSORY RESPONSE PROPERTIES

The somatosensory properties of F4 neurons were studied using touch of the skin, hair bending, light pressure of the tissue and slow and fast rotation of the joints. All testings were done with eyes open and closed. The somatosensory properties of "bimodal" and "somatosensory" neurons were indistinguishable and therefore will be described together. Out of 401 neurons that responded to somatosensory stimulation, 328 (81.8%) were activated by touch, 33 (8.2%) by pressure applied to the skin or passive movement of the joints, and 40 (10%) by both touch and joint rotation or deep pressure. The tactile receptive fields (RFs) of F4 neurons were typically large (see Fig. 2).

FIG.2 Example of a peripersonal RF of a bimodal "sagittally-directional" neuron. Shadowed area represents the tactile RF. Solid around the tactile field indicates the visual RF.

They were located on the face, neck, trunk, and arms. Most frequently they were contralateral to the recorded side (66%), some extended bilaterally (22%), and a few were strictly ipsilateral (12%). Table 1 summarizes the number and percentage of different RF locations.

TABLE 1. *Location of tactile RFs*

	Somatosensory	Bimodal	Total
Upper face	34 (29.8 %)	63 (24.8 %)	97 (26.4 %)
Lower face	43 (37.7 %)	112 (44.1 %)	155 (42.1 %)
Whole face	14 (12.3 %)	45 (17.7 %)	59 (16.0 %)
Trunk	6 (5.3 %)	9 (3.5 %)	15 (4.1 %)
Arm	0 (0.0 %)	7 (2.8 %)	7 (1.9 %)
Combination	17 (14.9 %)	18 (7.1 %)	35 (9.5 %)
Total	114 (100 %)	254 (100 %)	368 (100 %)

"Trunk" refers to RFs located on the neck, trunk, or both. "Combination" includes receptive fields covering two or more of the body parts listed in the table. The majority of the neurons falling into "Combination" category had RFs that included the face. Table 2 shows the location of responses to skin pressure or joint rotation of "somatosensory" and "bimodal" neurons.

TABLE 2. *Location of responses to proprioceptive and deep stimuli*

	Somatosensory	Bimodal	Total
Face	7 (18.9 %)	2 (5.6 %)	9 (12.3 %)
Trunk	16 (43.2 %)	13 (36.1 %)	29 (39.7 %)
Arm	12 (32.5 %)	16 (44.4 %)	28 (38.4 %)
Combination	2 (5.4 %)	5 (13.9 %)	7 (9.6 %)
Total	37 (100 %)	36 (100 %)	73 (100 %)

In contrast to tactile responses, proprioceptive and deep responses were mostly evoked by trunk and arm stimulation (39.7% and 38.4% respectively). Proprioceptive and deep responses were most frequently evoked by stimuli applied contralateral to the recorded side (60.5% of the neurons activated by proprioceptive stimuli), whilst only 9.3% of the neurons responded to ipsilateral stimuli, and 30.2% to both.

2.2 VISUAL RESPONSE PROPERTIES

Visual response properties were studied using 3D objects presented by hand at different positions and distances from the monkey. They were then moved toward and away from the monkey from different angles. Borders of the visual responding region (3D visual RF) were considered the external limits of that part of space whose crossing gave constant responses. Other types of visual stimulation (object rotation, movements of the experimenter's body, etc.) were also employed. According to the type of stimulation effective in activating them, visually responsive neurons were subdivided into five main classes (Table 3). The *first* and most represented class was formed by neurons with RFs located in the space around the monkey (peripersonal space) and responding best to stimuli moved along a sagittal plane ("Sagittally directional" neurons). All neurons but 5 were directionally selective. Almost all of them (n=249) preferred movements toward the monkey, two preferred movements away from it. The properties of this class of neurons will be described in more detail in the next paragraphs. The *second* class consisted of neurons which preferred movement directions along the tangential plane.

All these neurons were directionally selective, except two. Most neurons of this class had peripersonal RFs. The *third* class consisted of neurons that discharged phasically in response to an abrupt presentation of an object. One third of these neurons responded only if the object was presented in the peripersonal space. The *fourth* class comprised neurons that discharged tonically when an object was kept in the animal's peripersonal space. The *fifth* class was formed by a heterogeneous group of neurons. Some of them responded to rotation or jerky movements of 3D objects. Others were active in response to movements of the experimenter (e.g. leaning towards the monkey) or during monkey observation of its own hand. Nine neurons of this class preferred stimuli in the peripersonal space, whereas 11 fired also to stimulus presentation far from the animal. Finally, some neurons were difficult to characterize. They are listed in Table 3 as *"Others"*.

TABLE 3. *Classes of F4 visually responsive neurons*

	Unimodal	Bimodal	Total
First class	35 (53.0 %)	221 (84 %)	256 (77.8 %)
Second class	2 (3.0 %)	13 (4.9 %)	15 (4.6 %)
Third class	13 (19.8 %)	5 (1.9 %)	18 (5.5 %)
Fourth class	3 (4.5 %)	11 (4.2 %)	14 (4.3 %)
Fifth class	8 (12.1 %)	12 (4.6 %)	20 (6.0 %)
Others	5 (7.6 %)	1 (0.4 %)	6 (1.8 %)
Total	66 (100 %)	263 (100 %)	329 (100 %)

2.2.1 *Size and location of visual receptive fields*

The RFs of F4 neurons were usually large, typically extending for many degrees along both the horizontal and vertical plane. Confirming previous findings [20, 37, 38], visual RFs were typically located around the tactile field (see Fig. 2). Particular attention was paid to RF extension in depth. This property was studied in 94 "sagittally directional" neurons. All of them were classified as such by clinical testing and then confirmed using the quantitative testing (see below). According to their RF extension, they were subdivided into two broad categories: neurons with "peripersonal" fields and neurons with "far" fields. We defined as peripersonal fields those fields that starting from the animal's skin extended in depth up to 40 cm. Far fields were defined those fields that did not show a clear outer border. Table 4 shows the number of neurons belonging to each of these two categories. The table indicates also the number of fields whose outer border did not exceed 10 cm from the animal skin. Visual RFs were mostly located contralateral to the recorded side.

TABLE 4. *Sagittally directional neurons: visual RF depth extension*

		Unimodal	Bimodal	Total
Peripersonal	(< 10 cm)	1 (12.5 %)	24 (28 %)	25 (26.6 %)
Peripersonal	(10-40 cm)	3 (37.5 %)	59 (68.5 %)	62 (66.0 %)
Far	(> 40 cm)	4 (50.0 %)	3 (3.5 %)	7 (7.4 %)
Total		8 (100 %)	86 (100 %)	94 (100 %)

Out of 256 neurons studied for this property, 154 (60.2%) had RFs exclusively contralateral to the recorded side, 20 neurons (7.8%) had ipsilateral RFs, and 82 neurons (32.0%) had fields extending bilaterally. Table 5 shows the location of the visual RFs with respect to the animal's body. Most RFs were located around the face (68.3% of the studied neurons). Other RFs were around the trunk and arm (4.7% and 4.3% respectively). Neurons indicated as large field neurons (22.7%) had fields located around both face and trunk and/or arm.

TABLE 5. *Sagittally directional neurons: visual RF location*

	Unimodal	Bimodal	Total
Face	13 (37.1 %)	162 (73.3 %)	175 (68.3 %)
Trunk	3 (8.6 %)	9 (4.1 %)	12 (4.7 %)
Arm	1 (2.9 %)	10 (4.5 %)	11 (4.3 %)
Large field	18 (51.4 %)	40 (18.1 %)	58 (22.7 %)
Total	35 (100 %)	221 (100 %)	256 (100 %)

3. Coordinate systems of F4 neurons visual RFs

The visual properties of neurons of area F4 were studied quantitatevely by training monkeys to fixate a small light and by presenting moving stimuli driven mechanically by a robot arm (see Fig. 3). By changing the monkey's gaze location we were able to decide whether the field was "retinocentric" or somatocentered, whilst by the precise control of stimulus position we could delimit the extent in depth of the RFs. Only the main results of this study will be presented here. The full description of these properties is given elsewhere [see 17].

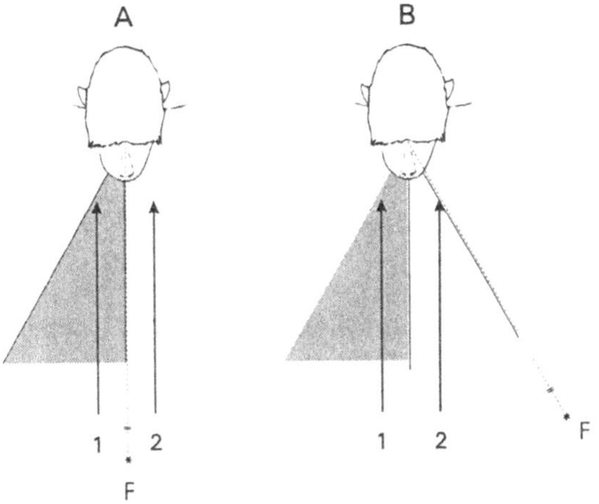

FIG.3 Schematic representation of the experimental procedure employed to study visual receptive fields (RFs). Two hypothetical RFs, one coded in retinotopic coordinates (space between the continuous lines), the other coded in somatocentered coordinates (shadowed area) are shown. In **A** the monkey fixates centrally. The two fields are in register. In **B** the animal fixates eccentrically (30° to the left). The retinotopic field follows the eyes, while the somatocentered field remains anchored to the head. In **1** the robot arm is moved inside the somatocentered receptive field, whereas in **2** is moved outside it. The robot arm started its trajectory at a distance of 50-70 cm from the monkey, 200 ms after the illumination of the fixation point, approached the monkey and reversed movement direction a few cm from it. The speed of the moving robot arm was 40 cm/s. Asterisk represents fixation point. The 3D trajectory of the moving stimulus with respect to the animal was reconstructed using a computerized movement recording system (ELITE System). The arrows indicate the trajectory of the robot arm.

3.1. SOMATOCENTERED RFs

In most studied neurons (90%), visual RFs were coded in somatocentered coordinates, whilst in few neurons (10%) visual RFs were retinocentric. An example of a somatocentered neuron is shown in Fig. 4. In A1, the monkey was looking straight ahead and a mechanically driven visual stimulus was moved toward it inside the RF. The onset of the response occurred when the stimulus was at about 25 cm from the monkey. In A2 the monkey was still looking straight ahead, but the stimulus was moved outside the medial border of the RF. There was no response. In B1 and B2 the gaze was deviated 30° to the left. The response was still present in B1, absent in B2. If the field would have been retinotopically organized a response should have occurred in B2. Note also that the responses in A1 and B1 were identical. This indicates that the angle formed by the gaze and stimulus trajectory did not affect the neuron's response. C and D show two control situations. C shows that the RF did not change position when the gaze was deviated 30° to the right. D, in which there was no moving stimulus, illustrates the background activity in the absence of RF stimulation. Another example of somatocentered visual neuron is shown in Figure 5. The examples presented above clearly indicate that in F4 there are neurons that code space in non-retinocentric coordinates. One may argue, however, that the responses of those neurons were not necessarily somatocentered, but could depend upon some visual cues present in the room where the monkey was tested. Although it is extremely unlikely that neurons with somatosensory and visual responses as those of F4 may code space in allocentric coordinates, nevertheless in a series of neurons we tested the allocentric hypothesis, by changing orientation of the monkey with respect to the walls of the recording room. None of the tested neurons was influenced by the changed environmental cues.

About 20% of somatocentered neurons showed a modulation of the response intensity and/or of the spontaneous activity with gaze deviation. Note that, although the imposed behavioral variable was the deviation of the eyes, eye deviation was constantly accompanied by a synergic, tonic increase of the activity of the ipsilateral neck muscles in the case of horizontal gaze deviations, or of the neck muscles controlling flexion or

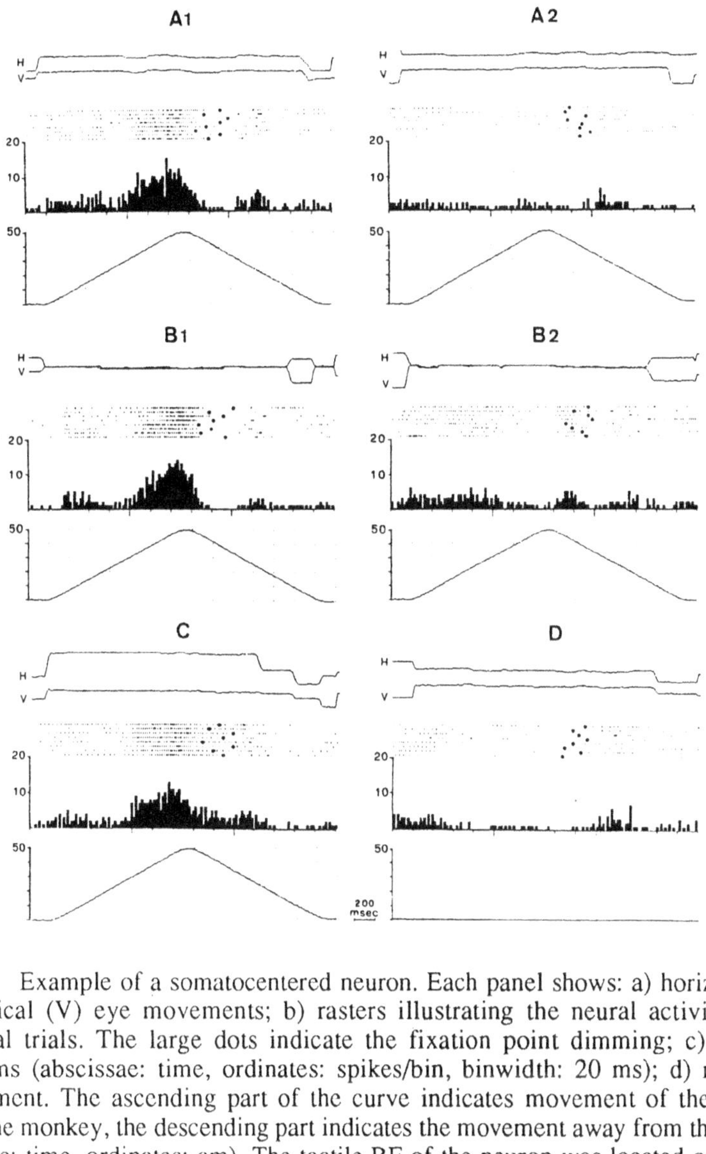

FIG.4 Example of a somatocentered neuron. Each panel shows: a) horizontal (H) and vertical (V) eye movements; b) rasters illustrating the neural activity during individual trials. The large dots indicate the fixation point dimming; c) response histograms (abscissae: time, ordinates: spikes/bin, binwidth: 20 ms); d) robot arm displacement. The ascending part of the curve indicates movement of the stimulus toward the monkey, the descending part indicates the movement away from the monkey (abscissae: time, ordinates: cm). The tactile RF of the neuron was located on the face contralateral to the recorded side. The visual RF was located around the tactile one (visual RF width: 70°; medial border of the field coincided with the 0° gaze axis. In **A**1, **B**1 and **C** the trajectory of the stimulus, moved inside the RF, was along a parasagittal plane 7 cm lateral to the head midline; the direction reversed when the stimulus was 4 cm from the orbital plane. In **A**2 and **B**2 the trajectory was identical to **A**1 but on the opposite side of the midline, outside the RF. **D** control trials. No moving stimulus was presented. Fixation point: **A**1, **A**2 and **D**=0°; **B**1 and **B**2=30° to the left; **C**=30° to the right. The response onset corresponded approximately to a distance of 25 cm from the orbital plane.

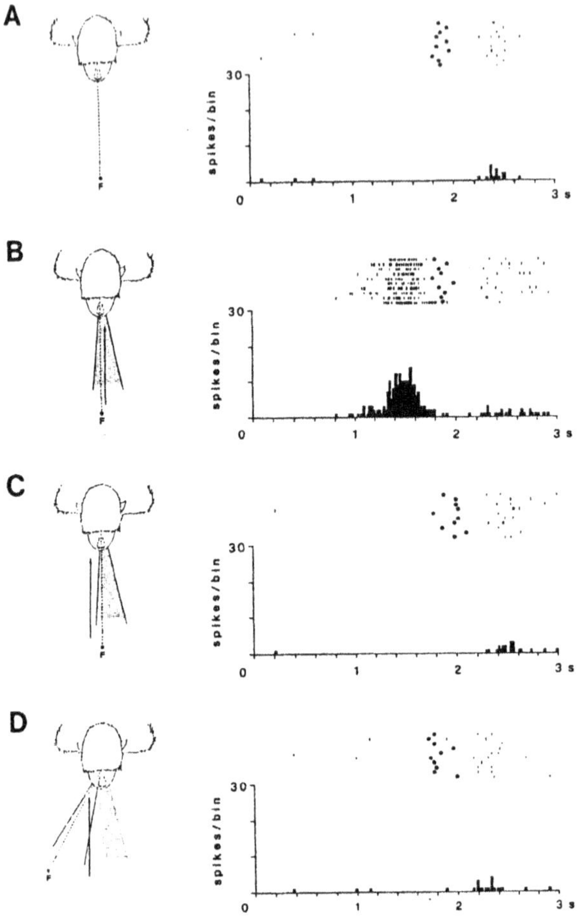

FIG.5 Somatocentered neuron. The tactile RF of the neuron was located on the face contralateral to the recorded side. The visual RF was located around the tactile one (visual RF horizontal width: 30°). The response onset corresponded approximately to a distance of 20 cm from the orbital plane. Other conventions as in Fig. 4.

extension of the head in the case of downward and upward gaze deviations, respectively [for similar observations see also 8, 30, 39, 42]. An example of a "modulated" neuron is shown in Fig. 6. The upper panels illustrate the modifications of the spontaneous activity, when the monkey fixated three different locations. When the gaze was deviated downward the spontaneous activity was high, whereas it was virtually absent when the gaze was deviated upward. The lower panels show the visual responses to moving stimuli with the same gaze locations. The strongest responses occurred in the condition in which the spontaneous activity was low. If one relates the RF location with neuron spontaneous and visually evoked discharge, the following pattern emerges: the spontaneous activity is maximal when the gaze is directed towards the visual RF,

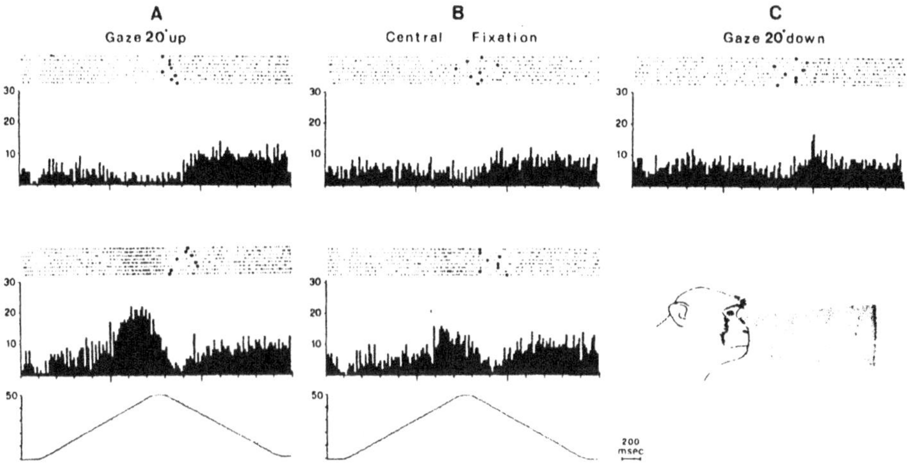

FIG.6 "Modulated" somatocentered neuron. The tactile RF was bilateral, located on the central and lower parts of the face. The visual RF was located around the tactile one. **Upper panels**: neuron's spontaneous activity during fixation of three different locations. Lower panels, **A** and **B**: neuron's activity when the stimulus was moved inside the visual RF during fixation of the same location presented in the corresponding upper panels. For other conventions, see Fig. 4.

whereas the response is maximal when the gaze is directed away from it. Since area F4 does not have connections with the cortical oculomotor centers [31], and its neurons are modulated by head, and not by eye positions, the "gaze dependent" modulation could be explained as determined by a concomitant neck muscles contraction. If this is true, modulated neurons could control the orientation of the head towards their visual RF.

When the eye-neck system is deviated away from the visual RF of the recorded neuron, its spontaneous activity is inhibited by the ongoing activity of neurons which control the contrasting neck positions. When the neuron's RF is stimulated, the response is minimal when the gaze-neck system is deviated toward the RF, because the "orienting error" is minimal and the expected head position almost coincides with that indicated by the neuron's RF. In contrast, the response is maximal when the gaze-neck system is away from the position required by the RF location. In this case the "orienting error" is large and requires a strong discharge to be corrected.

3.2. RETINOCENTRIC RFs

Neurons with retinocentric RFs represented 10% of the studied neurons, forming a rather homogeneous group. They were all bimodal, having their tactile RFs on the face, had visual RFs with the maximal response in the peripersonal space and preferred approaching visual stimuli. Tactile RFs were contralateral (70%) or bilateral (30%). The paucity of retinocentric neurons in area F4 is in agreement with the data by Graziano et al. [28]. However, Boussaoud et al. [8] reported that most cells in the "ventral premotor area" (inferior area 6) are "retinocentric" and modulated by gaze position. The different type of visual stimuli and the different behavioral requirements used in the study of these last authors could account for this discrepancy.

4. Motor properties

Out of 539 studied neurons, 299 (55%) fired in association with monkey's active movements. Of them 72 were purely motor, 82 were activated by somatosensory stimuli, 38 by visual stimuli, and 107 responded to both visual and somatosensory stimuli. The most represented movements were: neck and upper trunk movements (n=85, 28.4%), reaching, bringing to the mouth and other types of arm movements (n=73, 24.4%), upper, lower face and mouth movements (n=29, 9.7%). A large group of neurons (n=99, 33%) was not related exclusively to one of the above mentioned movements, but discharged in association with two of them. In this group particularly frequent were neurons that fired both during movements bringing the hand to the mouth and mouth opening (n=23, 7.7%) or arm reaching plus neck and trunk orienting movements (n=24, 8.0%). Motor properties were assessed also using intracortical microstimulation. Out of 305 stimulated sites, movements were evoked from 247 (81%). The movements most frequently elicited concerned: the neck and upper trunk (n=99, 40.1%), the face (n=52, 21.1%), the arms (n=36, 14.6%), and the mouth (n=17, 7.1%). Frequently, combined movements involving two body regions were observed (e.g. face plus neck or upper trunk, n=24, 9.7%; mouth plus neck or upper trunk, n=7, 2.8%).

5. Visual versus motor responses

Considering the large number of F4 neurons that discharge during active movements, one can argue that the responses to visual stimuli observed in F4 could actually be motor attempts made by the animal to avoid or to reach for the stimuli rather than real sensory responses. To rule out this possibility EMG activity was recorded from a series of muscles of the trunk, neck, face and shoulder during visual stimulation. We never observed any correlation between muscle activity and the presentation and approaching of visual stimuli to the monkey.

6. Somatocentered receptive fields and their role in motor control

Although the responses of F4 neurons to sensory stimuli are not a reflection of actual movements, the anatomical and functional organization of F4 strongly suggests that its sensory inputs are primarily used for movement organization. The crucial issue is to understand how they are used. We previously suggested that a convenient way to conceptualize the functional organization of F5, another premotor area [36], is to conceive it as a "vocabulary" of specific hand actions (e.g. precision grip, whole hand prehension), that can be retrieved either internally or by presentation of 3D objects congruent in size with the type of grip coded by the neurons. The present data could be interpreted in a similar way. Unlike F5, the actions controlled by F4 are movements of the head, movements of the arm directed toward the body and, possibly (see below), reaching movements directed away from the body. The function of sensory inputs in F4 would be that of selecting a location on the body or in the peripersonal space and to recruit neurons that control movements related to stimulus location. Let us examine this hypothesis by analyzing the different input-output couplings found in F4 neurons. Many neurons discharge during head movements and have tactile RFs on the face or on the head and visual RFs around the cutaneous one. By definition, a tactile RF indicates where a stimulus is located. Neurons controlling head movements have therefore the cutaneous information necessary for localizing the stimulus on the skin. The interesting point is that the same neuron that controls head movements on the basis of cutaneous information could do it also on the basis of visual information. The presence of a visual RF around the tactile one, and anchored to it, projects the cutaneous RF into space. This could allow the neuron to localize the stimulus even when the skin is not stimulated and to produce an appropriate movement in response to it. This function cannot be carried out by retinocentric neurons. The necessary condition, even for an apparently simple action as that of avoiding a stimulus coming toward the face, is to know its position relative to the head. This information is given to F4 by its neurons with somatocentered RFs. Another common input-output coupling present in F4 is that exemplified by neurons that have tactile RFs on the face and discharge in concomitance with monkey's active arm movements directed toward the tactile field. Also for these neurons when the stimulus touches the face its location is specified. In this case, however, the transformation of stimulus location into the appropriate movement is more complex than in the case of head movements, because, according to the hand starting positions, the arm trajectory to reach the same end point varies. Although the problem of trajectory formation is by no means solved [see 15, 29], it is since long time established that electrical or chemical stimulation of the skin can elicit limb movements that succeed in bringing the hand to the location of the stimulus such in the scratch reflex in higher vertebrates [40], or the wiping reflex in amphibians [34]. Particularly interesting is the observation that spinal frogs can reach the stimulus not only when it is located on the trunk but also on the contralateral hindlimb or ipsilateral forelimb, regardless of the limb positions [6, 18]. This indicates that the specification of a tactile location can be directly transformed into the end-point for the movement [see 7]. Independently of how this sensorimotor transformation is achieved, what is important here is that the same mechanism that brings the paw of the frog to its skin may account also for the localization of visual stimuli in the space around the monkey. If one considers F4 visual RFs as a three dimensional extension of the cutaneous fields, a stimulus located near to the animal would activate specific movements directing the arm towards it, as though it was a tactile stimulus, without requiring a complex transformation of visual coordinates into other types of coordinates. Only few neurons of our sample discharged in association with arm movements directed away from the body. This is most likely due to the fact that our behavioral paradigm favored the selection of neurons with stable visual responses near the face. The motor output of these neurons typically concerned axial movements or arm movements towards the face. The presence, however, of neurons discharging in association with reaching movements away from the animal, reported in previous experiments [20], indicates that this type of movements is coded in F4. Somatocentered RFs extending in depth for 30-40 cm could be the spatial frame of reference for these movements. In addition to somatocentered neurons with RFs anchored to the head or trunk, there are other somatocentered neurons with visual RFs anchored to the arm or hand [28, 38]. These neurons have tactile RFs on the hand and visual RFs located around the hand, or tactile RFs on the arm and visual receptive fields around the arm. When the arm moves, the visual RFs also move to the new arm location. It is possible that reaching movements directed away from the body can rely on a joint activity of neurons with somatocentered RFs anchored to the head or trunk such as those described in the present study, and neurons with somatocentered RFs anchored to the arm or hand. The somatocentered neurons with RFs anchored to the head and trunk will give the general frame on where the stimulus is located [21], while the "arm centered" neurons will provide arm on-line adjustments [see 27].

7. Space coding in F4 and in other areas

There are several coordinate systems in which visual space is coded in the cortex. In the primary visual cortex, and in other parieto-occipital areas, space is coded in retinocentric coordinates [see however 19]. Controversial is the issue of how space is coded in visual centers related to eye movements. According to Goldberg, Bruce and others a spatial map is not necessary for programming eye movements [10, 24, 25]. The spatial location of the target can be computed by mean of vector calculations taking into account its position and motor errors. In contrast, according to Andersen and his coworkers, neurons of areas LIP and 7a have visual RFs retinotopically organized, but their discharge is modulated by eye and head position [1, 5, 4, 3, 9]. Since the joint knowledge of the position of the stimulus on the retina and of the eyes in the orbit determines unequivocally where an object is located [see 43], it was proposed that neurons showing orbital effect mediate spatial vision. Orbital effect neurons would provide spatial information necessary for space perception, oculomotor control, as well as for other body parts movement organization [41]. The notion that the parietal oculomotor centers establish a general frame of reference for all behaviors requiring spatial vision is challenged by the data on the somatocentered visual RFs of F4. This RF organization and the practical absence of anatomical connections between the parietal oculomotor centers and the areas that control head, face and arm movements [2, 11, 12, 23, 31, 33] represent a strong evidence against the idea of a unique area responsible for different spatial behaviors. More recently somatocentered RFs were described in several other centers controlling body-movements. Graziano and Gross reported them in the putamen [26] and, subsequently [27], in area 7b of the parietal lobe as well as, confirming the above reviewed data, in inferior area 6 [27, 28]. Colby et al.[13, 14] found that many neurons in parietal area VIP are bimodal. Some of them have the characteristics typical of the somatocentered neurons. All together these data lend support to the view [35] that, in parallel to a circuit controlling eye movements, there is another circuit for different body part movements and that the two systems code space in a different way. An important aspect of somatocentered coding of space in F4 is the fact that in this area the stimulus location is coded explicitly at the single neuron level. This is particularly advantageous since it allows the parameter intensity of neuronal discharge to convey information about other features of the stimulus such as its movement direction or its velocity. This information can be used to program and guide in a more suitable way the various movement parameters such as movement onset, movement amplitude, etc. In conclusion, it appears that the way in which space is coded by different cortical areas radically differs according to the types of effectors they control.

REFERENCES

1. Andersen, R. A. (1989) Visual and eye movement functions of the posterior parietal cortex, *Ann. Rev. Neurosci.* **12**, 377-403.

2. Andersen, R. A., Asanuma, C., Essick, G., and Siegel, R. M. (1990 a) Corticocortical connections of anatomically and physiologically defined subdivisions within the inferior parietal lobule, *J. Comp. Neurol.* **296**, 65-113.

3. Andersen, R. A., Bracewell, R. M., Barash, S., Gnadt, J. W., and Fogassi, L. (1990 b) Eye position effects on visual, memory, and saccade-related activity in Areas LIP and 7a of macaque, *J. Neurosci.* **10**, 1176-1196.

4. Andersen, R. A., Essick, G. K., and Siegel, R. M. (1985) Encoding spatial location by posterior parietal neurons, *Science* **230**, 456-458.

5. Andersen, R. A. and Mountcastle, V. B. (1983) The influence of the angle of gaze upon the excitability of the light-sensitivity neurons of the posterior parietal cortex, *J. Neurosci.* **3**, 532-548.

6. Berkinblit, M. B., Feldman, A. G., and Fukson, O. I. (1986) Adaptability of innate motor patterns and motor control mechanisms, *Behav. Brain Sci.* **9**, 585-638.

7. Bizzi, E., Giszter, S. F., Loeb, E., Mussa-Ivaldi, F. and Saltiel, P. (1995) Modular organization of motor behavior in the frog's spinal cord, TINS **18**, 442-446.

8. Boussaoud, D., Barth, T. M., and Wise, S. P. (1993) Effects of gaze on apparent visual responses of frontal cortex neurons, *Exp. Brain Res.* **93**, 423-434.

9. Brotchie, P.R., Andersen, R.A., Snyder, L.H. and Goodman, S.J. (1995) Head position signals used by parietal neurons to encode locations of visual stimuli, Nature **375**, 232-235.

10. Bruce, C. J. (1988) Single neuron activity in the monkey's prefrontal cortex, in P. Rakic and W. Singer (eds.), *Neurobiology of neocortex*, Wiley, New York, pp. 297-329.

11. Cavada, C. and Goldman-Rakic, P. S. (1989) Posterior parietal cortex in rhesus monkey: II. Evidence for segregated corticocortical networks linking sensory and limbic areas with the frontal lobe, *J. Comp. Neurol.* **287**, 422-445.

12. Chavis, D. A. and Pandya, D. N. (1976) Further observations on cortico-frontal connections in the rhesus monkey, *Brain Res.* **117**, 369-386.

13. Colby, C. L., Duhamel, J.-R., and Goldberg, M. E. (1993) Ventral intraparietal area of the macaque: anatomic location and visual response properties, *J. Neurophysiol.* **69**, 902-914.

14. Duhamel, J.-R., Colby, C. L., and Goldberg, M. E. (1991) Congruent representations of visual and somatosensory space in single neurons of monkey ventral intraparietal cortex (area VIP), in J. Paillard (ed.), *Brain and space*, Oxford Univ. Press, Oxford, pp. 223-236.

15. Flanders, M., Helms Tillery, S. I., and Soechting, J. F. (1992) Early stages in a sensorimotor transformation, *Behav. Brain Sci.* **15**, 309-362.

16. Fogassi, L., Gallese, V., di Pellegrino, G., Fadiga, L., Gentilucci, M., Luppino, G., Matelli, M., Pedotti, A., and Rizzolatti, G. (1992) Space coding by premotor cortex, *Exp. Brain Res.* **89**, 686-690.

17. Fogassi, L., Gallese, V., Fadiga, L., Luppino G., Matelli, M. and Rizzolatti, G. (submitted) Coding of peripersonal space in inferior premotor cortex (area F4).

18. Fukson, O. I., Berkinblit, M. B., and Feldman, A. G. (1980) The spinal frog takes into account the scheme of its body during the wiping reflex, *Science* **209**, 1261-1263.

19. Galletti, C., Battaglini, P. P., and Fattori, P. (1993) Parietal neurons encoding spatial locations in craniotopic coordinates, *Exp. Brain Res.* **96**, 221-229.

20. Gentilucci, M., Fogassi, L., Luppino, G., Matelli, M., Camarda, R., and Rizzolatti, G. (1988) Functional organization of inferior area 6 in the macaque monkey: I. Somatotopy and the control of proximal movements, *Exp. Brain Res.* **71**, 475-490.

21. Gentilucci, M. and Rizzolatti, G. (1990) Cortical motor control of arm and hand movements, in M. A. Goodale (ed.), *Vision and action: the control of grasping*, Ablex, Norwood, NJ, pp. 147-162.

22. Gentilucci, M., Scandolara, C., Pigarev, I. N., and Rizzolatti, G. (1983) Visual responses in the postarcuate cortex (area 6) of the monkey that are independent of eye position, *Exp. Brain Res.* **50**, 464-468.

23. Godschalk, M., Lemon, R. N., Kuypers, H. G. J. M., and Ronday, H. K. (1984) Cortical afferents and efferents of monkey postarcuate area: an anatomical and electrophysiological study, *Exp. Brain Res.* **56**, 410-424.

24. Goldberg, M. E., Colby, C. L., and Duhamel, J.-R. (1990) The representation of visuomotor space in the parietal lobe of the monkey, *Cold Spring Harbor Symp Quant Biol* **55**, 729-739.

25. Goldberg, M.A. and Wurtz, R.H. (1991) Extraretinal influences on the visual control of eye movement, in D. R. Humprey and H.-J., Freund (eds) *Motor control: concepts and issues*, Wiley, Chichester, pp.163-179.

26. Graziano, M. S. A. and Gross, C. G. (1993) A bimodal map of space: somatosensory receptive fields in the macaque putamen with corresponding visual receptive fields, *Exp. Brain Res.* **97**, 96-109.

27. Graziano, M. S. A. and Gross, C. G. (1994) The representation of extrapersonal space: A possible role for bimodal visual-tactile neurons, in M. S. Gazzaniga (ed.), *The Cognitive Neurosciences*, MIT Press, Cambridge, Massachusetts, pp. 1021-1034.

28. Graziano, M. S. A., Yap, G. S., and Gross, C. G. (1994) Coding of visual space by premotor neurons, *Science* **266**, 1054-1057.

29. Kawato, M. (1993) Optimization and learning in neural networks for formation and control of coordinated movement, in D. E. Meyer and S. Kornblum (eds.), *Attention and performance, Vol. XIV,* The MIT Press, Cambridge, Massachusetts; London, England, pp. 821-849.

30. Lestienne, E., Vidal, P. P., and Berthoz, A. (1984) Gaze changing behaviour in head restrained monkey, *Exp. Brain Res.* **53**, 349-356.

31. Matelli, M., Camarda, R., Glickstein, M., and Rizzolatti, G. (1986) Afferent and efferent projections of the inferior area 6 in the macaque monkey, *J. Comp. Neurol.* **251**, 281-298.

32. Matelli, M., Luppino, G., and Rizzolatti, G. (1985) Patterns of cytochrome oxidase activity in the frontal agranular cortex of macaque monkey, *Behav. Brain Res.* **18**, 125-137.

33. Petrides, M. and Pandya, D. N. (1984) Projections to the frontal cortex from the posterior parietal region in the rhesus monkey, *J. Comp. Neurol.* **228**, 105-116.

34. Pflüger, E. (1853) Die sensorischen functionen des ruckenmarks bei wirbelttieren. Berlin.

35. Rizzolatti, G. and Camarda, R. M. (1987) Neural circuits for spatial attention and unilateral neglect, in M. Jeannerod (ed.), *Neurophysiological and neuropsychological aspects of spatial neglect*, North-Holland, Amsterdam, pp. 289-313.

36. Rizzolatti, G., Camarda, R., Fogassi, M., Gentilucci, M., Luppino, G., and Matelli, M. (1988) Functional organization of inferior area 6 in the macaque monkey: II. Area F5 and the control of distal movements, *Exp. Brain Res.* **71**, 491-507.

37. Rizzolatti, G., Scandolara, C., Matelli, M., and Gentilucci, M. (1981 a) Afferent properties of periarcuate neurons in macaque monkey. I. Somato-sensory responses, *Behav. Brain Res.* **2**, 125-146.

38. Rizzolatti, G., Scandolara, C., Matelli, M., and Gentilucci, M. (1981 b) Afferent properties of periarcuate neurons in macaque monkey. II. Visual responses, *Behav. Brain Res.* **2**, 147-163.

39. Roucoux, A., Vidal, P., Veraart, C., Crommelinck, M., and Berthoz, A. (1982) The relation of neck muscle activity to horizontal eye position in the alert cat. I. Head fixed, in A. Roucoux and M. Crommelinck (eds.), *Physiological and pathological aspects of eye movements*, Junk, The Hague, pp. 371-378.

40. Sherrington, C. S. (1906) The integrative action of the nervous system, Yale University Press, New Haven.

41. Stein, J. F. (1992) The representation of egocentric space in the posterior parietal cortex, *Behav. Brain Sci.* **15**, 691-700.

42. Vidal, P. P., Roucoux, A., and Berthoz, A. (1982) Horizontal eye-position related activity in neck muscles of the alert cat, *Exp. Brain Res.* **46**, 448-453.

43. Zipser, D. and Andersen, R. A. (1988) A back propagation programmed network that simulates response properties of a subset of posterior parietal neurons, *Nature* **335**, 679-684.

MOTOR DISTURBANCES AFTER CORTICAL LESIONS OF THE HUMAN BRAIN

H.-J. FREUND
Department of Neurology, Heinrich-Heine-University Duesseldorf
D - 40225 Duesseldorf

1. Introduction

Damage of different parts of the sensory-motor cortex in humans affects different aspects of motor function ranging from paresis as the most common and elementary motor dysfunction to the more complex motor disturbances such as motor neglect or the apraxias. Fig. 1 shows those areas where lesions are known to cause motor deficits. As illustrated, the 'motor deficit field' comprises not only the frontal motor areas, but also the whole parietal cortex.

The regional specificity of these areas is indicated by the typical deficits that can be seen after small lesions. Since many lesions are not restricted to particular areas and show a the wide scatter of lesion size and sites, the lesion approach should be confined to carefully mapped surgical excisions and small well defined lesions as shown by good CT or magnetic resonance tomography images (MRI). MRI even allows to distinguish purely cortical lesions from the more common combined cortical-subcortical lesion type. Only on the basis of this distinction is it possible to discentangle the effects of damage of local cortical circuitry from dysconnection syndromes due to the interruption of long transcortical fibre systems by the subcortical lesion component.

Another major factor influencing functional impairment is what Jackson called the 'lesion momentum'. Acute versus subacute or chronically developing brain damage has quite different functional implications. Whereas acute lesions cause deficits that are maximal initially and recover to some extent in most cases slowly developing tumours are frequently not associated with any deficit at all, except focal seizures. Also developmental lesions are much better compensated than acquired brain damage so that even major structural defects are surprisingly well compensated.

Not only lesion site but the functional organisation of the system that is damaged plays a major role for its liability to functional impairment. The term 'eloquent brain areas' designates those regions where acute acquired lesions cause major and persisting loss of function. Conversely damage of non eloquent brain areas usually shows rapid and complete recovery. This liability depends on two factors. First, the complexity of the level of organisation of the damaged part. Elementary aspects of sensation or motor control following damage of primary sensory or motor cortex seem to be less redundandly organized than more complex, widely distributed 'network' functions. Consequently, lesions of these nodal points cause severe, persisting deficits, whereas frontal lobectomies with removal of prefrontal cortex may have

F. Lacquaniti and P. Viviani (eds.), Neural Bases of Motor Behaviour, 121–129.

neglegeable sequelae. Second, bilateral versus unilateral type of organisation is another major factor determining functional outcome.

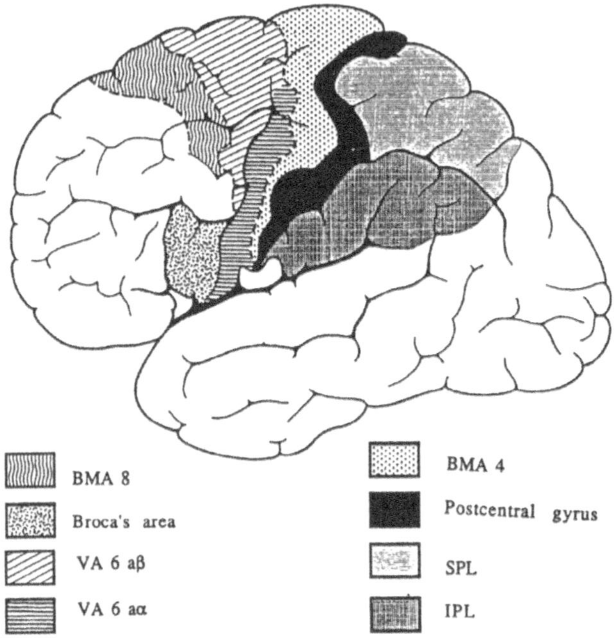

Figure 1. Lateral view of the human brain. Those areas where lesions are associated with distinct motor deficits are indicated by different symbols. BMA: Brodmann Area; VA: Vogt's Area; SPL: Superior Parietal Lobule; IPL: Inferior Parietal Lobule

It follows that there is an immense variability in the functional consequences of focal brain damage which depends on factors related to the temporal and spatial characteristics of the lesion and on the type of system organisation of the affected area. Only if these provisions of brain function are taken into account it is possible to use region specific functional deficits for the study of brain function and to distinguish local deficits from remote 'diaschisis' effects and from the impact on more global network functions such as attention.

The lesion approach in conjunction with the neuroimaging studies complements present view of the cortical organisation of motor control. It elucidates in particular the more complex aspects of motor functions such as the prominent role of parietal 'sensory' association cortex for the organisation of higher order motor behaviour. In the following, lesions of the frontal cortical motor fields will be discussed first, followed by the consideration of the motor deficits seen in parietal lobe damage.

2. Damage to the frontal cortical motor fields

The best known and most prominent motor dysfunction is the so called pyramidal syndrome following lesions of the precentral gyrus or its descending fibres. Acute destruction of this part is inevitably associated with contralateral weakness often associated with spasticity and increased tendon reflexes and muscle tone. The deficits are usually distally accentuated. Recovery is better in the proximal parts. What recovers least are the fractionated finger movements. This deficit mirrors the data obtained by electrical stimulation of motor cortex where low threshold stimuli elicit preferentially distal, twitch like motor responses from an area approximately coextensive with area 4 [1] and with imaging studies showing prominent precentral activations during skilled distal motor acts [2; 3]. Other facets of the motor disturbance in the pyramidal syndrome comprise weakness, the difficulty in the selection of certain muscles for action, disturbances of tactile placing, temporal deordering of muscle activation and loss of antagonistic inhibition. This pattern of deficits illustrates that the nature of the disturbance is purely executional.

3. Lesions of the supplementary motor area (SMA)

The SMA was first defined as a separate area by Penfield and Welsh [4] on the basis of stimulation studies. The major difficulty is that many reports on SMA lesions are describing the effects of lesions that are not confined to the SMA but in most cases comprise the cingulate gyrus and more prefrontal medial cortex, mostly including the underlying white matter. This is typically seen after infarcts in the territory of the anterior cerebral artery. The best examples of SMA lesions are surgical excisions [5; 6] of epileptic foci in medial frontal cortex, some of them however also including cingulate cortex. The typical deficits were a severe reduction of spontaneous motor activity most pronounced contralaterally and often accompanied by a reduction of speech and an emotional facial palsy. Disturbances of bimanual coordination and motor copying were also reported. These deficits receded during the first weeks. The only persistent deficit was a slowing of alternating serial movements of both hands. This reciprocal coordination disorder persisted whereas serial unimanual performances were unaffected. The transient nature of the initial disturbances is contrasted by the severe and persisting deficits seen after bilateral SMA damage. These are characterized by complete akinesia and mutism that may resolve to some extent in some patients. This dramatic difference between the uni- and bilateral cases is typical for the functional impairment that follows damage of bilaterally organized systems.

Electrical recordings of Bereitschaftspotentials followed by investigations of regional blood flow provided evidence that the SMA contributes to motor initiation and movement planning [7-10]. The Bereitschaftspotential as defined by Kornhuber and Deecke [7] is maximal over the SMA and showed an early negativity as early as one to two seconds before movement onset. It is seen between a conditioning stimulus and a subsequent movement indicating the readiness of the subject to move. Blood flow studies showed focal and significant increase of blood flow in the SMA also when the subject imagined a motor performance without executing it [2; 9; 11]. Again the brain recording and imaging studies provide a mirror image of the evidence obtained from the lesion data.

In the monkey, a pre-SMA area [12] or area F7 as designated by Luppino et al. [13] has been defined in front of the SMA. The stimulation of an area in front of the SMA in the human has elicited negative motor responses i.e. inhibition of ongoing movements or postures [14] whereas stimulation of the SMA proper evoked complex synergistic movements as if the contralateral hand were raised in preparation for its use in a complicated act while the gaze is directed by head and eye movement towards the hand [5; 15; 16].

Lesions of medial frontal cortex are sometimes associated with an impressive desinhibition of hand movements This so called alien hand syndrome is characterized by involuntary movements of the hand as if it would have its own will. Often, the hand has to be counteracted by the other hand in order to bring it under voluntary control. Compulsive grasping and groping are part of the syndrome. Whether the 'alien hand syndrome' results from lesions of the negative sensory motor area (NSMA; [14] and whether this NSMA corresponds to the pre-SMA as defined in the monkey or an area in front of it is not yet clear. Smaller lesions are required to settle this issue.

4. Dorsolateral premotor cortex

The parcellation of premotor cortex in the monkey will be discussed elsewhere in this book. The issue of homologies between the monkeys' and the human premotor cortex is still unsettled. Fulton originally designated the human premotor area as the frontal agranular cortex rostral to the precentral sulcus corresponding to the Vogts' area 6 ab or to the dorsal part of Brodman area 6. This area shows particular enlargement in the human where it is five times larger than area 4 as compared to a 1:1 relationship in the macaque [17]. As shown in fig. 1, a smaller ventral part of Brodman area 6 (Vogts' area 6 aa) lies on the precentral gyrus. When the human inferior part of precentral sulcus represents the homologue of the postarcuate sulcus in monkeys, area 6 aa covering the anterior part of the precentral gyrus would correspond to the monkey's postarcuate premotor area. The large part of PMC in front of precentral sulcus (area 6 ab) may correspond to the dorsal compartment of PMC as outlined by Wise [18] and by Humphrey and Tanji [19]. Because of the uncertainties of the role of the precentral part of area 6 our own position is to designate only the PMC anterior to the precentral sulcus (Vogts' area 6 ab) as premotor cortex [20] as originally proposed by Fulton [21]. This position is also taken, because even the smallest precentral lesion observed do not show any differential deficits other than causing the pyramidal syndrome irrespective of their more anterior or posterior location on the gyrus [22]. In the same vein, functional imaging has not provided evidence for activations across the precentral gyrus that would indicate differences between area 4 and area 6 aa.

Clinically, Foerster [1] described more than 40 patients in whom he had excised area 6 ab. The observed initial transient weakness was associated with a slowing and poverty of movements on the contralateral side which was proximally accentuated. Loss of the kinetic melody, disintegration of complex skilled movements and a disturbance of the kinetic structure of the motor acts have also been described. A special form of limb-kinetic apraxia was observed by Freund and Hummelsheim [20] in most patients with PMC damage. This was characterized by incoordination between the two arms or legs during such tasks as making a windmill movement with the arms or a pedalling movement with the legs. In contrast, distal bilateral movements like rotating the thumbs or performing a repertoire of bimanual coordination tasks were normal. These patients also showed deficits in sensory conditional motor learning, so that the patients were severely disturbed in the ability to learn the association between six hand postures demonstrated to them with particular sensory cues, whereas their ability to recognize an association between these sensory stimuli and different spatial locations was preserved [23]. Another deficit was a severe disturbance in rhythm reproduction from memory in the absence of difficulties to produce the rhythms under auditory pacing and without problems in rhythm discrimination [24].

Electrical stimulation of the human PMC elicited mainly proximal movements similar to those observed following SMA stimulation. Foerster [1] described flexion and extension synergies for the leg but a consistent pattern of arm movements with abduction and elevation of the arm thus resembling a gesture of reaching. Again, turning movements of trunk, head and eyes towards the elevated arm were usually associated. Stimulation with subdural electrodes extended these early studies [25]. Regional blood flow measurements showed increased flow particular when proximal movements were performed [3; 9]. Premotor activation was always bilateral.

Taken together, the results provide evidence for a bilateral organisation of premotor cortex and SMA. Consequently, clinical deficits are mild and mostly transient. Some dysfunctions are only disclosed when adequate test procedures are employed. The dysfunction affects both sides of the body. So far there is no evidence for a functional parcellation of the large premotor area. The few bilateral lesions with their dramatic effects on stance, gait and posture support the old concept of premotor cortex as a major extrapyramidal system that is bilaterally organised and subserves the cortical control of posture and locomotion. Broca's area is the well known premotor language area. The frontal eye field, which has been localized adjacent and anterior to the PMC (Brodmann area 8) on the basis of electrical stimulation [1; 15; 26] is now supposed to lie on the anterior part of the precentral gyrus as indicated by PET activation studies [27].

5. Parietal cortex

Lesions of different parts of the parietal cortex lead to distinctly different motor disturbances. Postcentral excisions [1] or lesions [28-30] cause motor deficits closely resembling those seen after deafferentation with prominent ataxia and dysmetria, poverty of movement and reluctance to move the affected extremity (motor neglect). These deficits are purely contralateral.

Lesions of the posterior parietal cortex cause severe impairment of motor behaviour - the apraxias. The apraxias are higher order motor disturbances of purposive movements that are not due to paresis, ataxia, aphasia or dementia. Different types of apraxia have been described. The classical parietal apraxias as defined by Liepman [31; 32] are (i) ideational apraxia characterized by a deficit of the conception of the movement so that the patient does not know what to do and (ii) ideo-kinetic (ideomotor) apraxia where the patient does not know how to perform a particular motor act, such as performing gestures. These apraxias are seen after lesions of the inferior parietal lobule (IPL), in particular of the area around the supramarginal gyrus. These 'conceptual' apraxias are supramodal, affect both sides of the body and are only seen after lesions of the language dominant hemisphere. On the basis of these apractic deficits Liepman postulated, that the left hemisphere is not only dominant for language but also for praxis. These apraxias are often associated with aphasia so that the apractic phenomena must be tested by imitating the movements demonstrated.

Lesions of the superior posterior parietal lobule (SPL) are associated with motor deficits that affect only the contralateral side and are restricted to motor acts performed under sensory guidance [33]. Usually only one modality is affected such as visuomotor apraxia seen after parieto-occipital lesion. This is the inability to direct the arm or eye properly towards a target. Tactile apraxia is the inability to explore and manipulate objects on the basis of tactile information. Active touch is therefore grossly disturbed or impossible. This disturbance can be partly compensated by vision. But perception gained from active touch differs from that obtained from passive touch or visual information. It can elaborate the unity, stability, plasticity and shape of phenomenal objects. The dysfunction is therefore associated with asterognosis, the inability to identify objects by active touch. The deficient object recognition and manipulation after SPL lesions illustrate the disturbance of both, the gnostic and practic functions thus representing a specific disturbance of sensory motor integration on the basis of a deranged interaction between the sensory and motor processes. These disturbances provide evidence that sensory association cortex is required for the elaboration of those motor acts designed to orient the respective receptor sheets to the target or to interact with objects. The situation is basically similar to aphasia, where Wernicke aphasics show pronounced speech dysfunctions in addition to their deficits in language comprehension. Accordingly, patients with left hemisphere strokes who are deaf mute and communicate by American Sign Language showed sign language aphasia which was actually an apraxia for the production of the spatial-temporal patterns of sign language, usually associated with difficulties to comprehend sign language as well [34].

Neuroimaging studies indeed showed parietal activation in the various parietal areas when complex manual performances were required as during visuomotor tasks [35] or active touch [2] or movement imagination [11]. In conlusion, there is a host of evidence emphasising the prominent role of the posterior parietal lobule for the conceptualization and elaboration of motor behaviour. Damage to those areas leads to severe and disabling motor disturbances.

6. Hemispherectomy and the problem of functional recovery

The clinician is sometimes confronted with the paradoxical situation that a small lacunar infarct of the internal capsule may render the patient completely hemiplegic, with little or no functional recovery, whereas the removal of a whole hemisphere may be followed by reasonable recovery allowing useful residual motor capacities. This can, however, only been observed when the removed hemisphere had developmental lesions followed by intractable epilepsy. Obviously not the size of the developmental lesion - it may be small - is the decisive variable for outcome, but the fact that the hemisphere was damaged early and later successively by the epileptic seizures [37]. We examined a patient who was hemispherectomized at age 18 and walked with only a barely perceptible limb. She could use the affected arm for many purposes and could open and close the fingers. Axial and proximal movements were almost normal [38] Such patients, in contrast to normal subjects, showed ipsilateral compound muscle action potentials induced by magnetic stimulation of the remaining hemisphere. that were of equal amplitude as the contralateral response over proximal muscles but showed reduced ipsilateral amplitudes over distal muscles [39]. This is suggestive that strengthening of ipsilateral corticospinal pathway may be a major factor underlying the excellent residual motor function in such cases.

That major recovery and functional reorganisation is not confined to developmental lesions was shown by the large scale plasticity of the human motor cortex described by Seitz et al. [36] in cases with tumours infiltrating the precentral gyrus. The patients had normal hand function on the contralateral side. Mapping regional blood flow changes related to voluntary finger movements in these patients showed activations solely outside the tumour. Compared with the unaffected side, the activations were shifted by 9-43 mm either along the mediolateral body representation of motor cortex or to premotor or parietal cortex. These data show that slowly developing lesions can induce large scale reorganisation in the adult nervous system.

7. Conclusions

Unilateral brain lesions produce motor deficits not only after damage of the frontal cortical motor fields, but also following parietal lobe damage. For preferentially unilateral organized systems such as primary motor cortex or anterior or superior posterior parietal cortex the deficits are purely contralateral and often severe and persistent. Unilateral damage of bilaterally organized systems such as the SMA or premotor cortex causes only mild and transient clinical deficits. Additional dysfunctions such as the disturbances of rhythm production or of motor learning can be disclosed by adequate testing. The conceptual apraxias (ideational and ideomotor) after damage after the inferior posterior parietal cortex produce bilateral disturbances of purposive motor behaviour illustrating that the language dominant hemisphere is also dominant for praxis. The apraxias therefore disclose another aspect of lateralized brain function and underline the significance of the posterior parietal lobule for higher order motor behaviour.

8. References

1. Foerster, O. (1936) Motorische Felder und Bahnen, in H. Bumke und O. Foerster (eds.) *Handbuch der Neurologie*, Springer-Verlag, Berlin, **6**, 1-357.
2. Roland P.E. (1985) Cortical organization of voluntary behavior in man. *Hum. Neurobiol.* **4**, 155-167.
3. Colebatch, J.G., Deiber, M.-P., Passingham, R.E., Friston, K.J., Frackowiak R.S.J. (1991) Regional cerebral blood flow during voluntary arm and hand movements in human subjects. *J. Neurophysiol.* **65**, 1392-1401.
4. Penfield, W., Welsh, K. (1949) The supplementary motor area in the cerebral cortex of man. *Trans. Am. Neurol. Assoc.* **74**, 179-184.
5. Talairach J., Bancaud, J. (1966). The supplementary motor area in man. *Int. J. Neurol.* **5**, 330-347.
6. Laplane, D., Talairach, J., Meininger, V. et al. (1977) Motor consequences of motor area ablations in man. *J. Neurol. Sci.* **31**, 229-238.
7. Kornhuber, D., Deecke, L. (1965) Hirnpotentialänderungen bei Willkürbewegungen und passiven Bewegungen des Menschen: Bereitschaftspotential und reafferente Potentiale. *Pfluegers Arch. Gesamte Physiol.* **284**, 1-17.
8. Orgogozo, J.M., Larsen, B., Roland, P.E. et al. (1979) Activation de l'aire motrice supplementaire au cours des mouvements voluntaires chez l'homme. *Rev. Neurol. Paris* **135**, 705-717.
9. Roland, P.E., Larsen, B., Lassen, N.A. et al. (1980) Supplementary motor area and other cortical areas in organization of voluntary movements in man. *J. Neurophysiol.* **43**, 118-136.
10. Kristeva, R., Keller, E., Deecke, L. et al. (1979) Cerebral potentials preceding unilateral and simultaneous bilateral finger movements. *Electroencephalogr. Clin. Neurophysiol.* **47**, 229-238.
11. Stephan, K.M., Fink, G.R., Passingham, R.E. et al. (1995) Functional anatomy of the mental representation of upper extremity movements in healthy subjects. *J. Neurophysiol.* **73**, 373-386.
12. Tanji, J. (1994) The supplementary motor area in the cerebral cortex. *Neurosci. Res.* **19**, 251-268.

13. Luppino, G., Matelli, M., Camarda, R., Gallese, V., Rizzolatti, G. (1991) Multiple representations of body movements in mesial area 6 and the adjacent cingulate cortex: an intracortical microstimulation study. *J. Comp. Neurol.* **311**, 463-482.
14. Lüders, H., Dinner, D.S., Morris, H., Wyllie, E., Comair, Y. (1994) Cortical electrical stimulations in humans: the negative motor area, in S. Fahm, H. Hallett, H. Lüders, C.D. Marsden (eds.) Negative Motor Phenomena. *Advances in Neurology*, Raven Press, New York.
15. Penfield, W., Rasmussen T. (1950) The cerebral cortex of man, Macmillan, New York.
16. Lim, S.H., Dinner, D.S., Pillay, P.K., Lüders, H., Morris, H.H., Klem, G., Wyllie, E., Awad, I.A. (1994) Functional anatomy of the human supplementary sensorimotor area: results of extraoperative electrical stimulation. *Electroencephal. and Clin. Neurophysiol.* **91**, 179-193.
17. Bailey, P.A., von Bonin, G. (1951) The Isocortex of Man. Urbana, Univ. of Illinois Press.
18. Wise, S.P. (1991) What are the specific functions of the different motor areas? in Humphrey, D.R., Freund, H.-J. (eds.) *Motor Control: Concepts and Issues*. Dahlem Konferenzen. Wiley, New York.
19. Humphrey, D.R., Tanji, J. (1991) What features of motor control are encoded in the neuronal discharge in different cortical motor areas? in Humphrey, D.R., Freund, H.-J. (eds.) *Motor Control: Concepts and Issues*. Dahlem Konferenzen. Wiley, New York.
20. Freund, H.-J., Hummelsheim, H. (1985) Lesions of premotor cortex in man. *Brain* **108**, 697-733.
21. Fulton, J.F. (1935) A note on the definition of the 'motor' and 'premotor' areas. *Brain* **58**, 311-316.
22. Freund, H.-J. (1987) Abnormalities of motor behaviour after cortical lesions in humans. In: F. Plum (ed): *Handbook of Physiology*. Section 1. The Nervous System. Vol. V, Higher Functions of the Brain. Part 2. American Physiology Society, Bethesda, pp 763-810.
23. Halsband, U., Freund H.-J. (1990) Premotor cortex and conditional motor learning. *Brain* **113**, 207-222.
24. Halsband, U., Ito, N., Tanji, J., Freund, H.-J. (1993) The role of premotor cortex and the supplementary motor area in the temporal control of movement in man. *Brain* **116**, 243-266.
25. Lüders, H., Lesser, R.P., Dinner, D.S. et al. (1987) Chronic intracranial recording and stimulation with subdural electrodes. in J. Jr. Engle (ed.) *Surgical treatment of the epilepsies*. Raven Press, New York, pp. 297-321.
26. Rasmussen, T., Penfield, W. (1948) Movement of head and eyes from stimulation of human frontal cortex. *Res. Publ. Assoc. Res. Nerv. Ment. Dis.* **27**, 346-362.
27. Petit, L., Orssaud, C., Tzourio, N. et al. (1993) PET study of voluntary saccadic eye movements in humans: Basal ganglia-thalamocortical system and cingulate cortex involvement. *J. Neurophysiol.* **69**, 1009-1017.
28. Déjérine, J. (1907) A propos de l'agnosie tactile. *Rev. Neurol. Paris* **15**, 781-784.
29. Head, H., Holmes, G. (1911-1912) Sensory disturbances from cerebral lesions. *Brain* **34**, 102-254.
30. Critchley, M. (1953) The parietal lobes. Hafner, New York.
31. Liepmann, H. (1900) Das Krankheitsbild der Apraxie ("motorische Asymbolie"), auf Grund eines Falles von einseitiger Apraxie. *Monatsschr Psychiatr Neurol* **8**, 182-197.
32. Liepmann, H. (1920) Apraxie, in *Brugsch's Ergebnisse der Gesamten Medizin*. Urban & Schwarzenberg, Berlin, pp. 518-543.

33. Pause, M., Kunesch, E., Binkofski, F., et al. (1989) Sensorimotor disturbances in patients with lesions of the parietal cortex. *Brain* **112**, 1599-1625.
34. Bellugi, U., Poizner, H., Klima, E.S. (1989) Language, modality and the brain. *Trends Neurosci.* **12**, 380-388.
35. Grafton ST, Mazziotta, JC, Woods RP, Phelps ME (1992) Human functional anatomy of visually guided finger movements. *Brain*, **115**, 565-587.
36. Seitz, R., Huang, Y., Knorr, U. et al. (1995) Large-scale plasticity of the human motor cortex. *NeuroReport* **6**, 742-744.
37. Gardener, W.J., Karnosh, L.J., McClure, C.C. et al. (1955) Residual function following hemispherectomy for tumour and for infantile hemiplegia. *Brain* **78**, 487-502.
38. Müller, F., Kunesch, E., Binkofski, F., Freund, H.-J. (1991) Residual sensorimotor functions in a patient after right-sided hemispherectomy. *Neuropsychologia* **29**, 125-145.
39. Benecke, R., Meyer, B.U., Freund, H.-J. (1991) Reorganisation of descending motor pathways in patients after hemispherectomy and severe hemispheric lesions demonstrated by magnetic brain stimulation. *Exp Brain Res* **83**, 419-426.

THE ROLE OF THE POSTERIOR PARIETAL CORTEX AND CEREBELLUM IN THE VISUAL GUIDANCE OF MOVEMENT

J.F.STEIN
University Laboratory of Physiology, Parks Rd.
Oxford, OXI 3PT (UK)

1. Introduction

A logical way to approach the problem of finding out how vision controls movement is first to describe how visual cortical areas connect to motor structures. There are two main outputs from the visual system which are often called the 'what' and 'where' pathways (Ungerleider & Mishkin, 1982; Mishkin, 1983)). The 'what' stream is fed by both the parvo and magnocellular components of the retinogeniculate pathway (Merrigan & Maunsell, 1993). It is thought to be specialised for the analysis of the shape and colour of objects in order to identify them, and it projects laterally into the inferotemporal cortex. So lesions there cause visual agnosia, the inability to recognise objects by sight.

In contrast the dorsomedial stream is dominated by the reticulogeniculate magnocells. It is responsible for signalling the timing of visual events, hence the motion of visual targets; and it projects dorsally to the posterior parietal cortex (PPC). Its main function is probably to guide movements under visual control; so to call it the 'where' stream is really somewhat misleading (Stein & Glickstein, 1992, Goodale & Milner, 1992). Lesions here cause neglect, the inability to attend to the side contralateral to the lesion, and targets are mislocalised and missed when aimed for. So the PPC is where we should start our analysis of the visual guidance of movement. However we should not be too categorical about dividing visual processing into two separate what and where pathways. This dichotomy is a considerable oversimplification, because there are large and functionally important cross connections between the pathways; and also much feedback from higher visual areas to the lower areas that supply both streams. Nevertheless so long as it is remembered that it is a simplification, the what and where categorisation captures features that are important for understanding the visual guidance of movement.

F. Lacquaniti and P. Viviani (eds.), Neural Bases of Motor Behaviour, 131–151.

2. Posterior Parietal Cortex

In monkeys the posterior parietal cortex consists of Brodman's areas 5 and 7, the superior and inferior parietal lobules respectively. These may be subdivided into areas 5a,5b,7a,7b; also the intraparietal sulcus contains further areas: lateral interparietal (LIP), ventral interparietal (VIP) and medial intraparietal (MIP). In humans areas 5 and 7 are both situated in the superior parietal lobule, whereas the inferior parietal lobule consists of two new areas, area 39 (the angular gyrus) and 40 (supra-marginal gyrus). The posterior parietal cortex lacks prominent layer 4 granule cells, but all layers are clearly visible; so it is known as eulaminate or homotypical. This is characteristic of multimodal sensorimotor association areas.

2.1 SUPERIOR PARIETAL LOBULE

In monkeys the superior parietal lobule contains cells which are dominated by somatosensory input. They have complex receptive fields which put together 'matching' cutaneous, proprioceptive and motor signals; in other words they respond to the kind of stimulation of the muscles and skin that would occur during a normal movement, and their discharge is much greater during active movement than when the limb is moved passively. For example, when the biceps contracts the elbow flexes, so that the skin over the biceps touches that over the forearm flexors. In the superior parietal lobule there are neurones which discharge when these particular conditions co-occur: simultaneous contraction of the biceps and stimulation of these two areas of skin (Sakata, 1973). But their discharge is much reduced if the signals are the result of the experimenter moving the limb passively, rather than the monkey moving it actively.

So the superior parietal lobule is probably involved in the somaesthetic control of movements, particularly those of the hand; hence lesions in this area impair animals' ability to make accurate palpatory movements (Stein, 1978). Even though it is not paralysed the limb becomes clumsy and the animal is loath to use it. In humans lesions here lead to astereognosis, the inability to identify objects by palpation alone, together with clumsiness and neglect of the contralateral side of the body. Sometimes the patient fails to remember even to dress the contralateral limbs (dressing apraxia); and sometimes the contralateral side of the body is denied altogether (anosognosia). Lesions on the left side are particularly associated with apraxia, the inability to produce sequences of movements on demand even though the same sequence can often be produced automatically. For example if a patient is asked to scratch his ear he can't do it but if his ear is tickled with a feather he does it so automatically, In humans lesions of the right superior lobule are particularly associated with impaired visual guidance of movement (Freund, this volume).

2.2 INFERIOR PARIETAL LOBULE

But it is the inferior parietal lobule in monkeys which is dominated by visual input. The cells have very large receptive fields; and they are not retinotopically organised. Most cover at least a quarter of the contralateral visual field, usually extending at least 10 degrees into the ipsilateral field. By itself such a receptive field would not be much use for localising objects. But it has been argued that the combined output of a network of many such cells could do so (Zipser & Andersen, 1989; Pouget & Sejanowski, 1993).

In order to locate targets accurately independently of the direction in which the eyes happen to be pointing retinotopic coordinates have to be converted to an egocentric reference frame, by associating retinotopic locus with signals indicating the position and movements of the eyes, head and body. Neurons in 7a and LIP receive inputs related to the position of the eyes in the orbit and also to the position of the head. Their response to a visual stimulus delivered at the same retinal location is therefore greater for one particular position of gaze and less for others (Andersen et al, 1985). This variation can often be fitted by a sloping plane. But its slope is quite shallow; hence the strength of eye position signals is only large at the extremes of gaze.

Zipser and Andersen (1989) showed that the combination of retinal and eye position signals found in the responses of PPC neurones could be generated by a neural network trained by back propagation to combine the typically gaussian retinotopic receptive fields found in the primary visual cortex with a scalar representing eye position in order to transform the coordinate system for target location from retinotopic to egocentric, independent of eye or head position. However eye position signals only become available after an eye movement; and Duhamel, Goldberg and coworkers (1992) have shown that the receptive field of each PPC neuron is actually remapped to a new location just before each eye movement. Corollary discharge indicating the vector of the upcoming eye movement is added to the retinal vector of the position of the receptive field of the PPC neurone, so that each time the eyes move the coordinates of the receptive fields remain constant with respect to the fovea. It seems likely that this remapping of PPC receptive fields is the usual mechanism for allowing for small eye movements; whereas eye position signals may be more important for larger eye movements which approach the limits of the orbit. In normal life less than 10% of saccades are over 10 degrees in size however.

The other important feature of the receptive fields of inferior parietal lobule neurons is that their discharge is much enhanced when the animal makes an eye or limb movement towards a target within their receptive field, or even if the animal merely directs attention to such a target without making any overt movement (Bushnell & Goldberg, 1981).

In sum the results of recording from PPC neurones in awake trained monkeys suggest that they are involved a) in coordinate transformations and b) attending to objects. These conclusions are supported by the results of lesions in both monkeys and man. Inactivation of the inferior parietal lobule in monkeys leads to inability to direct either the contralateral or ipsilateral limb accurately into contralateral space, together with a tendency to neglect contralateral targets (Stein, 1978).

In humans contralateral neglect is the most prominent symptom of PPC lesions (Heilman & Valenstein, 1993). The coordinate system of the space that is neglected is obviously important. But no simple geometric description captures all its features. Usually the patient fails to notice details on the contralateral side of space in everyday life, copying drawings or drawing from memory. Often it is also I object-centred'; the patient neglects the contralateral side of objects, whether they straddle the midline, or if they are situated wholly in the contralateral half of space, or even wholly on the ipsilateral side of the patient. Thus the PPC must be provided with a clear concept of what is a single objects as opposed to multiple separate ones, probably by its connections with the inferotemporal cortex. Such patients are often unable to adapt the orientation of their hands to the orientation of an object to be grasped, even though perceptually they can describe its orientation correctly. The converse of this is that some patients with partial inferotemporal cortex lesions can orientate their hands correctly to pick up differently orientated objects; but they cannot report their orientation correctly, because they do not perceive them accurately (Goodale et al. 1991).

3. PPC Space map

We tend to accept, without really thinking about it, that our brains probably contain a multimodal topographic map of 'real' space. After all our consciousness presents us with a picture of the world in which vision, audition, somaesthesia and our actions are seatnlessly combined to give us a unitary view of the world. So it is often assumed that a topographic representation of the world is held in the PPC. Generating such a map would present great problems however. It would require that the distortions introduced by the nonlinear magnification factors found in the primary sensory maps be removed; that the movements that we make with our sensory surfaces all the time be allowed for; and that all the sensory maps be converted into a common coordinate system. In other words the PPC would have to build up a neural representation with a 1:1 topographical correspondence to the physical world outside.

However conscious introspection probably misleads us here. There is no evidence that such a topographic real space map exists in the PPC or anywhere else in the brain (Stein 1992). Moreover if such a map really did exist 'someone' would be required to inspect the map. The problem of how the map was used would merely be shifted one stage further into the nervous system; and, like Descartes, we would still be left wondering whether there really is a little person in the pineal gland to look at it.

The striking effect of PPC lesions is not on sensory perception however, but on active exploration of the world; they impair the ability to direct attention and to guide eye, limb and body movements towards a target. Probably therefore the PPC contains not a map, but distributed neural networks which are used for routing sensory inputs through to guide movement. There seem to be separate circuits for whatever kind of motor activity is required, and they are selected according to the direction of attention. LIP seems to be specialised for aiming saccades, VIP for the direction of the head, 7a for limb movements and so on (Rizzalotti, this volume). Perhaps we should liken the PPC to a telephone exchange rather than to a map!

4. Hemispheric Specialisation

The PPC and the temporal lobe just below it are the two cerebral regions which show the greatest asymmetry in humans; the right inferior parietal lobule and the left planum temporale are up to 20 times larger than their counterparts on the other side (Geschwind & Galaburda, 1987). Hence right parietal lesions cause neglect twice as often as left parietal lesions do; and left sided lesions cause speech and comprehension problems twenty times more commonly than right sided lesions do (Springer & Deutsch, 1992; Corballis, 1991).

5. Acquired Dyslexia

A situation in which both these specialisations manifest their importance is in acquired developmental dyslexia Specific acquired dyslexia occurs when lesions of the angular gyrus, usually on the left, also affect the fibres linking the two angular gyri through the corpus callosum (Hinshelwood, 1896). Under these circumstances the visual input from both the left and right hemifields is disconnected from the lan uage areas in the left temporal cortex; so the patient cannot read even though other aspects of language may be entirely unaffected (Springer & Deutsch, 1992). Actually lesions of the right PPC often cause reading problems also, but this is often attributed to neglect.

There are two basic processing pathways which are used for reading (Behrman & Moser, 1990). For groups of letters and words that can be recognised by sight the visual system feeds directly into the semantic lexicon, which holds the meaning of words, so that the meaning of the word can be retrieved at sight. This is the direct 'visual' route. However when learning to read unfamiliar words which cannot be recognised on sight, a 'phonological' strategy must be employed. The letters are translated into phonemes, thence the sound of the word is recovered, hence its meaning.

Both acquired and developmental dyslexics therefore exhibit two different kinds of syndrome depending on whether the visual/semantic or phonological route for reading is most damaged. The classic symptom of acquired 'deep' dyslexia is the semantic error, e.g. dog is read cat. There is no visual or phonological similarity between the word and the patient's error; instead their meanings are semantically related (Marshall & Newcombe, 1966). Patients with deep dyslexia also tend to make visual errors in which visual detail is lost (e.g. child is read chill, grill is read girl; and words with irregular spelling are found particularly difficult (e.g. wood, yacht, dough). So these patients probably have lesions which have damaged the direct visual route to meaning; but their phonological route is less affected. Therefore they can read unfamiliar regular words with little difficulty e.g. non words such as flub and smig.

In acquired 'surface' dyslexia the phonological route is impaired; so that patients can read familiar words on sight, but not unfamiliar words which require the phonological strategy. Thus the hallmark of surface dyslexia is the inability to read non words, such as the examples given above (Marshall & Newcombe, 1966).

6. Developmental Dyslexia

Developmental dyslexia is extremely common; 5 to 10% of 8-10 year old children have unexpected difficulties learning to read which cannot be accounted for by low.intelligence or psychiatric, cultural or sociological problems (Rutter & Yule, 197). Most have a strong family history and 4 times as many boys as girls are affected. These facts suggest that these reading difficulties are the result of congenital neurological impainnent in these children. This conclusion has recently been given much stronger backing by postmortem and MRI studies which show that developmental dyslexics have absent or reversed asymmetry of the speech areas in the planum temporale; and that this is associated with abnormal development of the brain with multiple ectopias and microgyria in the cerebral cortex, particularly in the parietotemporal association areas (Galaburda et al. 1985).

Semantic errors are much less common in the developmental equivalent of deep dyslexia than in acquired deep dyslexia. This is probably because, unlike adults who were fluent readers, these children never acquired the fluency of the visual route that would lead them to make semantic errors if it became damaged. Instead they make more visual than phonological errors and they find reading non words easier than irregular words. Their visual route is probably impaired because they suffer perceptual instability when trying to read small print as we will discuss later.

The developmental equivalent of surface dyslexia is phonological dyslexia. These children make mainly phonological errors. They have difficulty in splitting the continuous sounds which make up the spoken word into the separate phonemes that are required for coding it alphabetically. So they have particular difficulty reading non words; and they are relatively good at reading familiar irregular words.

However neither deep nor surface dyslexia are particularly common among developmental dyslexics in their pure forms; most developmental dyslexics have both phonological and visual problems (Stein 1991, Eden et al. 1993).

7. The Magnocellular Theory of Developmental Dyslexia

It has recently been shown that magnocells in dyslexics' brains are mildly abnormal (Livingstone et al. 1992). The special functions of the magnocellular system have been most studied in the visual system (Merrigan & Maunsell, 1993). So the psychophysical tests that have been developed to study the magnocellular system in humans have been employed in dyslexics. The magnocellular system is thought to provide the visual system with high contrast sensitivity at low spatial and high temporal frequencies. Bill Lovegrove and colleagues have been amassing evidence since the early 80s that dyslexics have mildly impaired low spatial and high temporal frequency contrast sensitivity functions (Lovegrove & Williams, 1993); and we have been able to confirm his results. Thus in dyslexics as a group the flicker sensitivity, flicker VEPs and motion sensitivity are impaired compared with controls matched for age or younger normal readers matched for reading age (Mason et al. 1992; Comelissen et al. 1994). However this very mild impairment (no more than 6 decibels) cannot directly cause their poor reading by itself. Some intermediate process must intervene.

7.1 EYE MOVEMENTS AND THE MAGNOCELLULAR SYSTEM

The magnocellular system dominates the control of eye movements. Hence a mild deficiency of timing in the visual system resulting from dyslexics' impaired magnocellular system may translate into instability of ocularinotor control. In particular unstable fixation could cause difficulties for a child who is attempting to fixate small letters in order to learn them (Stein & Fowler, 1993). In fact many dyslexic children do complain that small letters appear to move around when they are trying to fixate them. So they find that they have problems determining which way round or which order they should be in. This is a particular difficulty when the eyes are converged for inspecting small print at the reading distance because the two eyes can wobble independently and therefore confuse the child about which letter ought to be seen where (Eden et al. 1994). Using a simple clinical test for binocular instability we have been able to show that children with unstable binocular control tended to make more visual than phonological errors, when compared with children who had stable control. Often their instability can be helped by the simple expedient of occluding the left eye for reading because this simplifies their binocular control problem. We found that this treatment caused visual dyslexics to make less visual errors; and over 6 months of occlusion their reading improved at twice the rate that the children who received no treatment did (Stein & Fowler, 1985).

The hypothesis that dyslexic children's unstable binocular fixation causes them to mislocate and missequence letters implies that they should be unable to locate any small visual target, not just letters. We therefore tested the children in a dot localisation task in which the child was shown a target dot for some seconds (long enough for any wobble to manifest itself), and then a test dot was flashed either to the right or left of it. The child was asked to point whether the test flash appeared to move right or left. As expected the dyslexics made many more errors than normal readers (Riddell et al. 1990).

Interestingly we found that most of the dyslexics' errors were made on the left, whereas the normal children were slightly better on the left. Subsequently we have found many more visual tests in which dyslexics with unstable binocular control tend to make more errors on the left side (Eden et al. 1993). When asked to draw clocks they tend to squash all the figures onto the right hand side and neglect the left. They tend to neglect more targets on the left in cancellation tasks. They reproduce the Rey figure making more mistakes on the left. Their ability to judge the orientation of lines is worse on the left. This tendency to make more errors on the left is reminiscent of right PPC patients with left sided neglect. They suggest that these developmental dyslexics' problems may result from impaired development of the right PPC.

Most dyslexics suffer from both visual and phonological problems however. Their phonological impairments may be associated with impaired temporal processing in the auditory domain. Paula Tallal and colleagues (1993) have shown that the voice onset time in dyslexics is slower than in nonnals; and we have confirmed this in a different way (Cornelissen et al. 1994) We have also shown that their pure tone discrimination, their sensitivity to frequency modulation and to binaural phase differences are all significantly worse than norinals (McAnally et al, 1994). These impairments are all consistent with the hypothesis that impaired development of magnocells in the auditory system cause dyslexics to have impaired auditory temporal processing, in a similar way to how their vsual magnocellular deficit leads to impaired visual temporal processing. It is surely highly significant therefore that Galaburda and colleagues have recently reported that magnocells in the auditory thalamic relay nucleus in dyslexic brains are also abnormal (Galaburda et al 1994)

Dyslexics are often reported to be particularly clumsy; and recently it has been confirmed that they have mild impairments on the motor side as well. If asked to stand on a beam and perform mental arithmetic at the same time they tend to fall off (Fawcett & Nicolson, 1992). Likewise if they are asked to stand on one leg whilst blindfolded they tend to fall over. Balance employs some of the largest and most rapidly conducting fibres in the body passing between the vestibular system, cerebellum and spinal cord. Again therefore these motor symptoms are consistent with the hypothesis that magnocells in dyslexics fail to develop quite normally.

We can therefore speculate that developmental dyslexia is a congenital consequence of impaired development of magnocellular systems which leads to all the perceptual and motor problems outlined here; and, wnong other things, causes their impaired reading, spelling and writing performance. Since reading etc., has only been common for the last hundred years perhaps one should not treat dyslexia as a 'disease', but rather as an individual difference which probably carries compensatory advantages, as evidenced by the great men who were probably dyslexic (Alexander the Great, Churchill, Einstein, Leonardo de Vinci etc.).

8. Visuomotor Pathways

The prestriate and posterior parietal cortical visual areas project to motor structures via four main pathways, namely the corticocortical route via the superior longitudinal fasciculus, and three subcortical routes via the basal ganglia, superior colliculus and cerebellum (Stein & Glickstein, 1992). The corticocortical pathway projects symmetrically around the central sulcus, so that area 18 and 19 project to prefrontal cortex, area 7 projects to area 8, and area 5 projects to area 6.

Note that there are no direct projections from visual cortical areas to the motor cortex. The corticocortical pathway is often assumed to be the most important route by which vision guides movement. However if in monkeys the superior longitudinal fasciculus is completely divided by cutting deeply through the base of the central sulcus right down to the ventricle, they can still pick up morsels of food from a moving turntable with no great difficulty (Myers et al. 1968). Hence although the corticocortical pathway may be important for the initiation of visually guided movements it is by no means essential for them. In contrast, if the visual fibres projecting to the cerebellum are lesioned the visual guidance of eye and limb movements is decimated (Holmes, 1939; Stein, 1978).

The visuomotor pathway via the basal ganglia is probably not very important for the visual guidance of movement either. It is well known that in patients with basal ganglia disease movements are often considerably improved if they are given strong visual cues; for example chalk marks on the floor help them to walk (Glickstein & Stein, 1992). Jonathan Miller filmed a patient with Parkinson's disease who, stimulated by the sight of an approaching ball, could easily catch it; but in striking contrast he simply could not throw it back again. The return throw required him to generate movements spontaneously without strong visual guidance. It is this production of internally generated movements independent of sensory guidance which the basal ganglia seem to be specialised for. Their visual input is probably for helping to select what movement to make, what movement programme to choose, rather than to guide movements (Stein 1986, Passingham, 1987).

The visual guidance of movements seems to be a prime function of the cerebellum. Cerebellar lesions profoundly impair visually controlled movements. All the components of the visual dorsomedial, magnocellular, 'where' stream, including its prestriate relays and the posterior parietal cortex, give large projections to the pontine nuclei which relay to the cerebellar cortex (Glickstein et al. 1980). Pontine neurones therefore respond best to the rate and direction of movement of targets or to optic flow; but they are not interested in the orientation of lines, colour or the shape of objects. Indeed they respond better to a pattern made up of random dots moving in their preferred direction than they do to an orientated grating (Baker et al. 1976). Thus pontine neurons deliver to the cerebellar cortex signals about both the motion of targets and of the optic flow created by the observer's own movement (Thier et al. 1988) . Lesions in the pontine nuclei therefore have a particularly devastating effect on pursuit eye and limb movements.

9. The Cerebellum

Amazingly the cerebellum contains nearly half the neurones in the whole CNS. This is because it boasts 10^{11} granule cells. It is the largest motor structure in the brain. If its intricate pleated surface were unfolded its area would approach that of the cerebral cortex, being over 3 metres in length, although only 10 cm wide (Braitenberg, 1988). The cerebellar hemispheres have grown larger in humans relative to other animals than even the cerebral cortex; and they receive a larger amount of input from the cerebral cortex than any other structure. The inputs, internal circuitry and outputs of the cerebellum are known in very great detail. Yet despite all this information its function is frustratingly far from being understood.

Part of the reason is that the cerebellum is not on any of the direct pathways which link sensory and motor structures. So lesioning it does not completely eliminate any function. It is detached from the brainstem, lying over the pons and medulla, below the overhanging occipital lobe, separated from it by a deep fold of dura mater, the tentorium. It is linked to the brainstem by fibres running in the superior, middle and inferior cerebellar peduncles on each side. The cerebellum therefore fortns side pathways which operate in parallel with the direct links between sensorimotor centres. Its most likely function is to adjust the strength of other circuits in order to make them automatic and optimise their actions in different circumstances. It has been called the brain's autopilot, a metasystem whose function is to calibrate sensorimotor circuits in the rest of the brain (Allen & Tsukahara, 1974; Stein & Glickstein, 1992).

Deep within the substance of the cerebellum are the three cerebellar nuclei on each side. These provide almost its only output. The most medial, fastigial, nucleus receives fibres from the Purkinje cells of the most medial part of the cortex, the vermis. This medial compartment helps to control posture and whole limb movements by its connections with axial and limb girdle muscles via the vestibular nuclei and the reticular formation. The anterior and posterior interpositus nuclei serve the intermediate, paraverinal, part of the cerebellum. They project indirectly to ipsilateral limb muscles via the red nucleus. The most lateral cerebellar nucleus is the dentate which receives from the lateral cerebellar cortex including the dorsal parafiocculus. It projects mainly to the contralateral premotor and motor cortices via the ventrolateral thalarnic nucleus. The flocculus and nodulus lie below the rest of the cerebellum; this is the only region of the cerebellar cortex which receives primary sensory afferent fibres directly from the periphery, from the vestibular apparatus; and it sends its Purkinje cell axons out of the cerebellum altogether, back to the vestibular nuclei, rather than going via a cerebellar nuclear relay. Hence these vestibular nuclear neurones may be considered displaced cerebellar ones.

9.1 INPUTS

The largest input to the cerebellar cortex comes via the mossy fibres; the most important source of these are the pontine nuclei. There are over 20 million neurones on each side of the pons. These are supplied by virtually every efferent fibre in the cerebral peduncles. Over

95% of these are the axons of layer 5 corticofugal pyramidal cells which terminate in the pontine nuclei. The remainder are corticobulbar and corticospinal fibres which also give collaterals to the pontine nuclei. Thus the cerebellum receives information about the output of virtually the whole cerebral cortex, except for the areas lying at its outer margins; there are very few projections from the striate, cingulate, prefrontal or lateral parts of the temporal lobe, to the cerebellum.

The axons of the pontine neurones project to the cerebellar cortex as mossy fibres. For example visual pontine axons project as mossy fibres all over the contralateral cerebellar cortex, but particularly to the dorsal paraflocculus and uvula (Glickstein et al. 1980). The dorsal parafiocculus is not part of the vestibular flocular nodular complex. It receives no vestibular input, but is instead a ventrolateral extension of the cerebellar hemispheres. The uvula is the lowest folium of the vennis, located just above the nodulus.

Each mossy fibre splits into about 16 terminals, known as rosettes; and each of these makes contact with around 30 granule cell dendrites in structures known as glomeruli. Thus each mossy fibre contacts around 500 granule cells; but each granule cell has on average 4 dendrites, so it receives from 4 different mossy fibres. Golgi cells inhibit the mossy fibre glomeruli. Each golgi cell receives over 200 direct mossy fibre inputs and also another 5000 inputs from parallel fibres; and it inhibits a block of over 6000 granule cells. So the golgi cell probably acts to stabilise the discharge rate of a set of parallel fibres, whenever either its mossy or parallel fibres are discharging.

The way one mossy fibre diverges to supply 500 granule cells suggests that this may be an expansion recoding system in which one input fibre is distributed into a much larger number of association cells. If the strength of the synapses between parallel fibres and the dendrites of Purkinje cells is adjustable, then each Purkinje cell could serve as a classifier of which set of mossy fibres were active at any particular time.

9.2 CLIMBING FIBRES

Climbing fibres (CF) are a unique and characteristic feature of the cerebellar cortex. The 1: 1 relationship between a climbing fibre and the Purkinie cell that it supplies contrasts greatly with the enormous divergence and convergence in the relationship between mossy fibres and Purkinje cells. All climbing fibres derive from the inferior olive. Unlike for mossy fibres there is a fairly strict point to point topographical relationship between the inferior olive and the cerebellar cortex. Nevertheless each olivary axon divides deep in the cerebellar white matter to form up to 10 branches which project climbing fibres to Purkinje cells situated in separate sagittal strips. These supply around 500 synaptic terminals to each Purkinje cell. This ensures a high 'safety factor', so that whenever its CF discharges a complex spike always occurs in the Purkinje cell. This consists of a large sodium depolarisation followed by a plateau on which are superimposed a number of spikelets. These are the result of the activation of a high threshold calcium conductance if the cell is sufficiently depolarised (Llinas & Sugimori, 1980).

Simple spikes are elicited when a sufficient proportion of the 200,000 parallel fibres that pass through the dendritic tree of a Purkinje cell (perhaps less than 0.5% of them) discharge synchronously within 1 msec of each other. David Marr first suggested that the particular set of parallel fibres to which a Purkinje cell will respond preferentially in its role as pattern analyser, may be imprinted on it by the activity of climbing fibres (Blomfield & Marr,

1970), and Albus (1971) showed how long term depression of parallel fibre synapses by climbing fibres may mediate such imprinting.

Ito and colleagues confirmed long tertn depression by showing that if parallel and climbing fibres are excited simultaneously, that particular set of parallel fibre synapses is depressed whereas others are not (Ito, 1989). The influx of calcium that occurs as a consequence of the climbing fibre discharge probably activates a guanalate cyclase in Purkinje cells which downregulates the activity of the postsynaptic glutainate receptors that happened to be active at that time. In addition nitric oxide may be released from neighbouring basket and glial cells and this may downregulate the release of glutamate from neighbouring presynaptic terminals (Garthwaite, 1988). Balancing this effect, if pre and postsynaptic activity only depolarises the cell slightly, exciting only simple spikes without activating the high threshold calcium conductance, then long term potentiation, rather than depression, of that synapse may occur (Crepel & Jaillard, 1991). These mechanisms for synaptic plasticity explain how the weights of the synaptic inputs to Purkinje cells can be modified according to experience, and thus enable the cerebellar cortex to act as a pattern classifier.

Climbing fibres respond differentially to sensory inputs that are a consequence of the subject's own movements (reafference) compared with those that are a consequence of events in the outside world (exafference). So a tap on the sole of the foot causes inferior olive cells to fire, but the contact of the sole of the foot with the ground during locomotion does not (Gellman et al. 1985). If the sole of the foot is stimulated unexpectedly during the swing-phase of locomotion when such stimulation does not normally occur climbing fibres will respond, but less reliably than if the animal were not moving at all (Armstrong 1974).

The dorsal accessory olive receives visual information from the accessory optic tract in the pretectal region. Cells here respond to motion of optic texture across the retina, optic flow (Graf et al 1988); but their responses seem to be stronger if the optic flow is caused by motion in the outside world, rather than movements of the animal's own eyes. Another important point to note is that climbing fibres discharge on average only about once per second. So a single climbing fibre cannot provide quantitative information about, for example, motor error, as some people claim. Instead climbing fibres probably signal the timing of events in the outside world and help to distinguish these from the sensory reafference which is a consequence of your own movement. This is an important distinction to make because events in the outside world cannot usually be predicted; whereas reafferent signals indicate how a movement went off, and can therefore be compared with how it was predicted to go. In other words reafferance provides the error information which is necessary to improve the programming of movement, and so it is important to have some means of distinguishing it from exafference about stimuli impinging from the outside world.

9.3 VESTIBULAR OCULAR REFLEX

An important feature of the cerebellum is that it is separate from the rest of the brain; it does not lie on any of the direct pathways linking sensory with motor centres. For example the basic circuit which mediates the vestibulocular reflex is a three neurone arc (Precht, 1978). The semicircular canals project to the vestibular nuclei; vestibular axons then run in the medial longitudinal fasciculus to supply the motor neurones which control the external ocular muscles. Thus if the right horizontal canal is stimulated by rightwards movements of the head, after a latency of only 10 msecs the left lateral rectus and the right medial rectus muscles contract, causing the eyes to move to the left and thus to compensate for the head movement. So the gaze remains pointing straight ahead. The cerebellum has no part to play in this direct pathway. However it receives mossy fibre input from all three sites: the labyrinth and the vestibular and ocularmotor nuclei; and it projects inhibitory synapses back to them. Thus it forms a side pathway in parallel with the basic reflex circuitry.

If the flocculus is lesioned, the vestibulocular reflex still occurs, but it loses its accuracy (Carpenter, 1972; Takemori & Cohen, 1974); the amount of eye movement elicited is no longer strictly equal and opposite to the amount of head movement. The main function of the flocculus therefore seems to be to adjust the parameters of the VOR to calibrate it properly for different conditions. And this seems to be the general rule of the cerebellum; it operates in parallel with other circuits in order to calibrate them to optimise their performance in different circumstances. Hence it is often called a metasystem or supervisor (Allen & Tsukahara, 1974; McKay & Murphy, 1979).

The basic circuitry of the VOR provides no feedback pathway. The semicircular canals feed straight forward through to the eyes. But if the system is poorly calibrated the can retina signal that objects being fixated are slipping off the fovea. This information is sent to the cerebellum which adjusts the strength of VOR synapses until they stop doing so. So the relationship between the number of impulses emitted from the semicircular canals following a head movement, and the number of impulses required of the eye muscle motor neurons in order to achieve an exactly compensating eye movement, cannot be fixed and immutable. It must be able to change over time to compensate for growth of the head and eyes, changing optical factors, wearing spectacles etc. So the vestibulocular reflex is continuously being recalibrated by the flocculus. Hence if the flocculus is damaged the VOR cannot adapt to new conditions.

If you put on a pair of reversing spectacles the VOR becomes maladapted; each time you move your head to the right and the eyes move left the reversing optics cause the gaze to swing rightwards. This is the main reason why wearing such spectacles is initially such an unpleasant experience, often causing people to throw up. However Gonshor & Melville-Jones (1973) persuaded volunteers to continue wearing reversing spectacles for several days, and he was able to show that the vestibulocular reflex eventually reversed. So now when the head moved to the right the eyes did so also; hence after the spectacles reversed the line of sight, the gaze now kept pointing straight ahead.

When the VOR is miscalibrated just after putting on the reversing spectacles targets, instead of being stabilised, progressively slip further and further off the retina. The quantitative size of this slip is signalled to the flocculus by mossy fibres; and the fact that this is unwanted slip rather than optic flow caused by your own eyes moving is probably signalled by climbing fibre discharges. The flocculus then adjusts the parameters of the VOR pathway to reverse its direction of action, and thus avoid retinal slip. That is why if the flocculus is damaged the reflex cannot adapt to the new situation.

Ito and Miles recorded from floccular neurones before and after such adaptation, and they showed that the discharge of these neurones alters as would be expected if they play a crucial part in this kind of adaption. Ito found in rabbits that before adaptation there were 2 populations of floccular Purkinje cells, which fired in phase or out of phase with the head movement. The out of phase neurones were in the majority. Therefore they tended to inhibit ipsilateral eye movements and to favour those made in the opposite direction to the head, as is normally required of the VOR. However after the VOR was reversed after adaptation to reversing prisms the proportion of inphase neurones increased dramatically. These tend to cause the eyes to move in the same direction as the head as they depress the eye muscles which move the eyes in the opposite direction to the head. So Ito suggested that long tenn 'plastic' changes take place in the flocculus to reverse the normal proportions of in and out of phase P cells; hence they exert continuous control over the vestibular ocular reflex to keep it reversed (Ito, 1984).

In Ito's scheme the only long term plastic changes would take place in the cerebellar cortex. However Miles et al. (1980) came to a rather different conclusion from studying adaptation of the VOR in the monkey. They found that floccular neurones appear to discharge in relation to the direction of the gaze of the animal, so their modulation did not correlate with either head or eye movements alone. Hence both before and after adaptation when the gaze was stationary, neurones in the monkey flocculus did not change their discharge when the head was moved. They only did so when retinal slip was occurring, which was the signal that indicated that the VOR needed recalibrating. So Miles proposed that long tenn modification of synaptic strength must occur both in the flocculus and in the direct vestibulocular reflex pathway. Species and experimental differences probably explain their different conclusions. However the 'bottom line' is the same; the flocculus is essential for recalibrating the VOR to adapt to changed optical conditions.

9.4 CONDITIONING THE NICTITATING MEMBRANE REFLEX

Another much studied example of adaptation of a feedforward reflex is conditioning the nictitating membrane reflex (NMR). The nictitating membrane is the third eyelid that rabbits possess which moves across when the eye retracts for protection if the cornea or surrounding skin is stimulated, If a stimulus to the cornea (the normal unconditioned stimulus, US) is preceded by a flash of light (the conditioning stimulus, CS) for 100 or so pairings the flash of light alone will begin to evoke the eyelid response; ie. the reflex has become conditioned. Such conditioning only occurs if the flash of light precedes the comeal stimulus. In other words the conditioning stimulus must predict the future delivery of the unconditioned stimulus. This conditioning survives ablation of the whole neocortex, including the hippocampus (Oakeley & Russel, 1977). However if a small region of the anterior cerebellar cortex, paravermal lobule VI, is removed on one side the conditioned reflex on that side disappears (Thompson et al. 1988; Yeo et al. 1990). But the normal unconditioned response of the nictitating membrane to the corneal stimulation remains unchanged; and also the reflex can still be conditioned on the other side.

The region of the cerebellar cortex which is required for conditioning the NMR is determined by the topography of the climbing fibre input. The principal olive together with the medial and dorsal accessory olivary nuclei have a strict topographical relationship with the cerebellar cortex. Thus the region of the inferior olive which receives from the ophthalmic division of the Vth nerve which serves the skin of the cornea, projects climbing fibres to this crucial area of the cerebellar cortex (Yeo et al. 1990). Thus the normal unconditioned stimulus for the reflex specifies the area of cortex which mediates conditioning. During conditioning the climbing fibres which project to paravermal lobule VI are excited, and probably these reduce the efficacy with which concurrent visually drive parallel fibres excite those particular Purkinje cells by long-term-depression (LTD).

Therefore these Purkinje cells no longer inhibit their target neurones in the interpositus nucleus and so visual input alone can elicit the reflex.

9.4.1 Calibration as Prediction. Conditioning the nictitating membrane reflex is a simple experimental paradigm in which a stimulus comes to predict that a particular motor response is going to be required. Soon the predictive stimulus alone is enough to provoke the motor response. In fact the calibration of all feedforward processes can be seen as predictive in this way. In order to work properly feedforward controllers have to be loaded with a programme which is properly calibrated for the task in hand. This calibration can only be correct if it can draw on a precise prediction of what the sensory consequences of the execution of the programme will be.

The next example of how the cerebellum calibrates feedforward processes is the way in which visual motion information is used to predict where a target will be at the end of a movement. When tracking a moving target whose trajectory is not entirely predictable we usually resort to making a series of intermittent steps rather than a smooth continuous movement. The advantage of this strategy is that it combines the virtues of both feedforward and feedback control. Each intermittent movement consists of a negative feedback component equal in size to the current distance between target and track, the current error, and also a second component which is calculated to allow for the predicted continued movement of the target during the time the next movement will take. The speed with which the target is moving is multiplied by the expected duration of the movement in order to programme it to be the right size to catch up with the target. If both components are computed accurately the next movement lands precisely on target (Miall et al. 1988). Neurones in the dorsal paraflocculus which is functionally part of the lateral hemispheres rather than of the vestibular flocculus, receive motion signals relayed via the pontine nuclei from the posterior parietal and prestriate visual areas discussed earlier (Marple Horvat & Stein, 199). If the nuclear target of the dorsal parafiocculus, the ventral dentate nucleus, is inactivated by local cooling or anaesthetic in animals trained to track moving targets their tracking becomes more intermittent and jerky and much less accurate. Before inactivation both current error and target velocity are good predictors of the size of the movement; i.e. they are used to compute the size that the movement should be. During inactivation however, the amplitude of each movement is no longer adjusted to compensate for the speed with which the target is moving. In other words the velocity of the target is no longer used to predict where it will have reached by the end of the next movement; and so the size of the next movement is no longer scaled appropriately (Miall et al. 1987)

9.4.2 Visuomotor Adaptation. The most direct way to show that the cerebellum is concerned with calibrating visuomotor feedforward control processes is to test its function during adaptive learning since this is when recalibration is most required. In the same way that the cerebellum is essential for adapting the vestibulocular reflex to reversing prisms, it is now known to be required for adapting reaching movements for the deviation of gaze that occurs when prisms are placed in front of the eyes. When subjects view through such prisms the visual axes are deviated, and so they initially misaim. After a few trials however, the visuomotor system adapts to the new situation and the subject's aim regains its accuracy. If the prisms are now removed misaiming occurs in the opposite direction for a few trials until readaption occurs (Harris, 1965). But if the lateral cerebellum including the dorsal parafloculus is removed in animals (Baizer & Glickstein, 1974) or damaged in humans, such adaptation no longer occurs. Recently several PET studies have been published which confirm that the cerebellum is highly active during adaptive motor learning in humans also (e.g. Friston et al. 1992)

Similar adaptation can be established in animals trained to track a moving target with a lever if the wnplitude of the lever movement required to track the target, i.e. the gain of the system, is altered. As might be expected monkeys' adaption to this gain change is impaired if the cerebellar dentate nucleus is inactivated (Kerr et al. 1993). During adaptation to changes in loading of the lever simple spike discharges corresponding to the inaccurate movements increase; then as adaptation occurs they decrease again as accuracy is regained. This change is accompanied by increased climbing fibre activity (Gilbert and Thach, 1977), but it has proved difficult to show, as one would like, that this increase occurs in the same cells whose simple spike activity decreased as the new movements were learnt. (Ojakangas & Ebner, 1992); and so there are still some who disbelieve that the climbing fibres have any role to play in motor learning (Llinas, 1977). Nevertheless recent results do suggest that complex spike activity may help to retrain simple spikes to fire only when most appropriate.

All these examples suggest that the cerebellum is probably involved in calibrating feedforward processes, whether reflex or voluntary. The climbing fibres probably report when something happens in the outside world, and the simple spike activity occurring at the same time provides the quantitative error information necessary to correct the movement at the time, and to be incorporated into an improved program subsequently. So the fundamental processing operation of the cerebellar cortex for which everybody has been searching may well be to combine climbing and parallel fibre discharges to set the level of activity of Plurkinje cells in order to calibrate movements precisely.

9.4.3 Internal Models. To further understand how the cerebellum may calibrate feedforward control processes we really need to develop computer simulations of them because the simple ideas, such as negative feedback or feedforward, which we use implicitly when designing experiments, do not really capture the complexity of most motor control situations. A useful way of conceptualising the control problems involved is to consider the cerebellar cortex as laying down 'internal models' of the movements to be performed.

Feedforward control requires a programme to be assembled and calibrated before it can be used. A convenient way of generating such a programme is to set up an internal model of the required movement which can then be played out when required. This internal model can be seen as a predicting process. It represents the movement before it actually happens; in order to do so it has to predict its outcome in terms of the expected sensory consequences of the movement.

Two rather different kinds of simulation of these internal representations have been developed by control theorists. An 'inverse model' generates the sequence of signals which will cause a controlled object to produce a desired movement. It is said to be the inverse of the controlled object because if it is correctly formulated, on being operated on by the controlled object, it will produce precisely the required motor output. Such inverse filters are popular with control engineers because they can be designed with mathematical precision to perform to any desired specification of accuracy (Kawato et al. 1992). But it is very difficult to design them in such a way that they are able to adapt easily to changing conditions; yet such plasticity is an essential feature of neurological control systems. Inverse filters are only useful when they are perfectly calibrated. Unless they are completely accurate, they can actually be very dangerous. So just when they are most required for adaptating to new conditions they cannot be used, until they have been completely recalibrated.

The second kind of internal representation suggested for programming movements is known as a 'forward' model. Here the model is a neural replica of the required movement, rather than its inverse. This means that its output is a prediction of the outcome of the desired movement. Since only our senses can tell us how a movement went off the output of the model must be expressed in the form of a prediction of the sensory consequences of the movement. We saw earlier how such predictions may contribute to programming movements correctly. But the output of such a forward model cannot be used directly to control the limbs, because it predicts the sensory reafference which will be returned as a consequence of the movement rather than specifying the signals required to actually drive the muscles to make it. Nevertheless if a forward model is put in an internal feedback loop its output can then be compared with the signal commanding the desired movement. This comparison then yields the inverse signal which can be used for actually contracting the muscles. There are many advantages to this approach and the idea that the cerebellum is a forward predictor of this type is very appealing (Miall et al. 1993).

First, because forward models provide a prediction of the sensory consequences of a movement, comparison with feedback from the appropriate sense organs indicates whether or not a movement went off as expected. Unlike an inverse filter therefore, even in the early stages of adapting to a new situation a forward model will always provide some sort of signal which can be used to control movement, even though it may still be inaccurate. Reafferent signals from the moving limb report how the movement actually went off. In the Smith Predictor version of this idea a delayed output from the model is compared with these feedback signals and necessary corrections computed. These may then be used not only to correct the current movement, but also to update and adapt the internal model for changed conditions.

Second, forward models allow you to imagine and rehearse movements without actually having to make them. Much cognitive activity involves such mental rehearsal of possible actions; moreover it is probable that mental rehearsal can actually improve the skill with which movements are later performed. Recent PET measurements of regional cerebral blood flow have shown that the cerebellum is indeed activated during mental rehearsal (Decety et al. 1990). Thus the hypothesis that the cerebellum lays down forward models for controlling movements gives rise to clear predictions of the kinds of signals that we can expect it to generate, and these are being energetically pursued.

10. Conclusions

Clearly the posterior parietal cortex and cerebellum work closely together in the control of movement. The posterior parietal cortex is a multimodal sensorimotor association area which mediates the direction of attention. Since one has to attend to a target in order to locate it or to direct a voluntary eye limb or body movement towards it, the PPC plays a crucial part in the sensory guidance of movement; and so it is not surprising to find that it provides a large projection to the cerebellum. The location information comes in the form of the vector which would land the eye or the limb on the target; it is not represented by coordinates in a 'real space' topographic reference frame. The cerebellum then converts this vector into a programme which will automatically activate the appropriate muscles by the correct amount at the right time and in the right sequence to bring the limb precisely on to the target. It probably achieves this by generating a forward model of the limb for predicting the outcome of any chosen set of muscle contractions, so that their sensory consequences can be compared with the required vector, and corrections computed by the cerebellum before the actual movement ever takes place. These predictions are then checked against the actual outcome and any required adjustments of the model incorporated, so that the skill of the movement is automatically optimised. In other words the posterior parietal cortex can be described as a vector generator and the cerebellum as a sensorimotor predictor.

Acknowledgements.
The research on which these ideas are based was supported by the Wellcome Trust and the Medical Research Council, UK.

11. References

Albus JS (1971) Cerebellar learning. Math.Biosci., 10, 25-61.

Allen GI & Tsukahara M (1974) Cerebrocerebellar interactions. Physiol.Review., 54, 957-1006.

Andersen RA, Essig GK & Siegel RM (1985) Encoding of spatial location. Science, 230, 456-458.

Armstrong, DM (1974) Functional significance of the inferior olive. Physiol.Rev., 54, 358-417.

Baizer JS & M Glickstein (1974) Role of cerebellum in prism adaptation. J. Physiol.(Lond), 36, 34P.

Baker, J, A Gibson, M Glickstein & J Stein (1976) Visual cells in the pontine nuclei of the cat. J.Physiol.(Lond)., 255, 414-433.

Behrinan M & Moser MC (1990) A connectionist account of neglect dyslexia. J.Cognitive Neurosci., 2: 96-122.

Blomfield M & Marr D (1970) Cerebellar learning. Nature, 227, 1224-1226.

Braitenberg V (1988) The cerebellum and the physics of movement, in Glickstein, Yeo & Stein, see below.

Bushnell MC & Goldberg ME (1981) Visual attention. J.Neurophysiol., 46, 775.

Carpenter, RHS (1972). Cerebellectomy and the transfer function of the vestibulo-ocular reflex in the decerebrate cat. Proc.R.Soc.Lond.B., 181, 353-374.

Corballis MC (1991) The lopsided ape. Oxford University Press.

Comelissen PL, Richardson AR, Mason A, Fowler MS & Stein JF (1994) Contrast sensitivity and coherent motion detection measured at photopic luminance levels in dyslexics and controls. Vision Research, 35, 1483-1494.

Comelissen PL, Hansen PC, Bradley L & Stein JF (1994) Analysis of perceptual confusions between 9 sets of consonant-vowel sounds in normal and dyslexic adults. Cognition (in press).

Crepel FC & Jaillard D (1991) Long tenn changes in synaptic strength in cerebellar P cells. J.Physiol. 432, 123-141.

Decety JH, Sjoholm S, Ryding E, Sternberg D & Ingvar D (1990) Cerebellum participates in mental activity. Brain Res., 53.5 313.

Duhamel P & Goldberg M (1992) Updating of the representation of visual space by intended eye movements. Science, 255 , 90-93.

Eden GF, Stein JF, Wood HM & Wood FB (1994) Differences in eye movements and reading problems in dyslexic and normal children. Vision Research, 34, 1345-1358.

Eden GF, Stein JF & Wood FB (1995) Temporal and spatial visual processing in reading disabled and normal children. Cortex (In press)

Eden GF, Stein JF & Wood FB (1993) Visuospatial ability and language processing in reading disabled and normal children. In: 'Facets of Dyslexia and its Remediation', ed. S.F.Wright & R.Groner, Elsevier.

Fawcett AJ & Nicolson RI (1992) Automatisation deficits in balance for dyslexic children. Perceptual and Motor Skills, 75, 507-529.

Friston K, Passingham R, Frackowiack R & Brooks D (1992) PET study of motor learning. Proc. Royal Soc.Biol., 248, 223-235.

Galaburda AM, Sherrnan GF, Rosen GD, Aboitz F & Geschwind N (1985) Developmental dyslexia: Four consecutive cases with cortical anomalies. Annals of Neurology, 18, 222-233.

Galaburda A (1992) Magnocellular defect in developmental dyslexia. Proc.NY Acad.Sci., 682, 70-83.

Galaburda A, Rosen G & Shennan M (1994) Impaired magnocellular organisation in the medial geniculate nucleus of dyslexics Proc.N.Y.Acad.Sci., 91, 8010-8013.

Garthwaite J (1988) Excitation of NMDA receptor releases Nitric Oxide. Nature, 336, 385-387.

Gellman R, Gibson AR & Houk J (1985) Inferior olivary neurones as event- markers. J.Neurophysiol., 54, 40-52.

Geschwind, N & Galaburda, AM (1987) 'Cerebral Lateralisation: Biological Mechanisms, Associations and Pathology, MIT Press, Cambridge, Mass. & London, UK.

Gilbert C & Thach T (1977) Purkinje cell activity during motor learning. Brain Res., 128, 309.

Glickstein MG (1992) The cerebellum & motor learning. Current opinion in Neurobiology, 2, 802-806

Glickstein MG, JL Cohen, B Dixon, A Gibson, M Hollins, E La Bossiere and F Robinson (1980) Corticopontine visual projections in macaque monkeys. J.Comp.Neurol., 190, 209-229.

Glickstein MG & JF Stein (1991) Paradoxical movements in Parkinson's disease. TINS, 14, 480-482.

Glickstein MG, Yeo C & Stein JF (1988) The Cerebellum and Neuronal Plasticity. Plenum Press.

Gonshor A & G Melvill-Jones (1973) Changes of human vestibulo-ocular response induced by vision-reversal during head rotation. J.Physiol.(Lond), 234, 102-103.

Goodale MA, Milner AD, Jakobson LS & Carey DP (1991) A neurological dissociation between perceiving objects and grasping them. Nature, 349, 154-156.

Goodale MA & Milner AD (1992) Separate visual pathways for perception and action. TINS, l5: 20-25.

Graf W, J Simpson & CS Leonard (1988) Spatial organization of visual messages of the rabbit's cerebellar flocculus. 11. Complex and simple spike responses of Purkinje cells. J.Neurophysiol., 6.0 2091-2121.

Harris CS (1965) Perceptual adaptation to inverted, reversed and displaced vision. Psychol.Rev., 72, 419-444.

Heilman K & Valenstein EC (1993) 'Clinical Neuropsychology', Oxford University Press, Oxford, UK

Hinshelwood C (1896) A case of dyslexia, a peculiar form of word blindness. Lancet 2, 1451-54.

Holmes G (1939) The cerebellum of man. Brain, 62, 1-30.

Ito M (1984) The Cerebellum and Neural Control. Raven Press.

Ito M (1989) Long-term depression of Purkinje cell responses. Ann.Rev.Neurosci., 12, 5-102.

Kawato M & Gomi H (1992) Neural network model for VOR/OKR learning. TINS, 15, 445-452.

Kerr GK, Miall RC & Stein JF (1993) Visuomotor adaptation during inactivation of the cerebellar nuclei. Human Movement Science, 1.2 71-84.

Livingstone MS, Rosen GD, Drislane FW & Galaburda AM (1991) Physiological and anatomical evidence for a magnocellular defect in developmental dyslexia. Proc.Nati.Acad.Sci., 88, 7943-7947.

Llinas R (1977) The Inferior Olive - its role in motor learning. Science, 190, 1230.

Llinas R & Sugimori K (1980) Ionic currents in Purkinje cells. J.Physiol., 305, 197-220.

Lovegrove WJ & Williams MC (1993) Visual temporal processing deficits in specific reading disability. p 311-330 in Willows DM ed. 'Visual Processes in Reading and Reading Disabilities'. L. Erlbaum.

McAnally K (1994) Reduced auditory temporal resolution in dyslexic subjects. 23rd Rodin Academy International Conference, Malta.

McKay WA & Murphy JT (1979) Cerebellar modulation of reflex gain. Progress in Neurobiol., 13 361-392.

Marshall JC & Newcombe F (1966) Syntactic and semantic errors in paralexia. Neuropsychologia, 4, 169-176.

Miall RC, DJ Weir & JF Stein (1987) Visuo-motor tracking during reversible inactivation of the cerebellum. Exp.Brain Res., 65, 455-464.

Miall RC, DJ Weir & JF Stein (1988) Planning of movement parameters in a visual tracking task. Behav.Brain Res., 27, 1-118.

Miall RC, Weir DJ, Wolpert DM & Stein JF (1993) Is the cerebellum a Smith Predictor? J.Motor Behav., 25 , 203-216.

Miles FA, DJ Braitman & BM Dow (1980) Long-term adaptive changes in primate vestibulo-ocular reflex. Electrophysiological observations in flocculus of adapted monkeys. J.Neurophysiol., 43, 1477-1493.

Mishkin M (1983) Object vision and spatial vision. TINS, 61 414.

Myers RE, RW Sperry & N.M. McCurdy (1968) Neural mechanisms in visual guidance of limb movements. Arch.Neurol., 7, 195-202.

Oakley DA & IS Russell (1977) Subcortical storage of Pavlovian conditioning in the rabbit. Physiol.Behav., 18, 931-937.

Ojakangas CL & Ebner TJ (1992) Simple and Complex spike activity in the cerebellum during visuomotor adaptation in monkeys. J Neurophysiol., 6.8 2222-2236.

Passingham R (1987) Two cortical systems for directing movement. CIBA Symposium 132, p.151-164.

Pouget J & Sejnowski T (1993) Egocentric representation in early vision. J.Cog. NSci. 12, 675-687.

Precht W (1978) Neuronal Operations in the Vestibular System. Berlin, Heidelberg, Springer-Verlag.

Riddell P, Fowler MS & Stein JF (1990) Spatial discrimination in children with poor vergence control. Perceptual & Motor Skills, 70, 707-718.

Rutter M & Yule W (1975) The Concept of Specific Reading Retardation. J. Child Psychol. 1.6 181197.

Sakata H, Takaoka Y, Kawarasaki A & Shibutani H (1973) Somatosensory properties of neurones in area 5 of the rhesus monkey. Brain Res., 64, 85-102.

Springer SP & Deutsch G (1992) Left Brain, Right Brain. Freeman.

Stein JF (1978) Effects of cooling PPC in monkeys. In: 'Active Touch', ed.G.Gordon, pp.79-90. Pergamon Press.

Stein JF (1986) Role of cerebellum in visual guidance of movement. Nature, 32.3 217-221.

Stein JF (1991) (ed.) Vision and Visual Dyslexia; Vol.13 of Encyclopedia of Vision and Visual Dysfunction. Macmillan Press, London.

Stein JF (1992) The representation of egocentric space in the posterior parietal cortex. Behav. Brain Sci., 15, 691-700.

Stein JF & Fowler MS (1985) Effect of monocular occlusion on visuomotor perception and reading in dyslexic children. Lancet, July, 69-73.

Stein JF & Fowler MS (1993) Unstable binocular control in children with specific reading retardation. J.Res. in Reading, 16, 30-45.

Stein JF & Glickstein M (1992) The role of the cerebellum in the visual guidance of movement. Physiol.Rev., 72, 967-1018.

Stein JF, Riddell P & Fowler MS (1988) Disordered vergence eye movement control in dyslexic children. Brit.J. Ophthalmol., 72, 162-166.

Takemori S & Cohen B (1974) Loss of visual suppression of VOR after floccular lesions in monkeys. Brain Res., 72, 203-212.

Tallal P (1993) Temporal information processing in the Nervous System. Ann. N.Y.Acad.Sci., 68.2 27-48.

Thier P, W Kochler & VW Buttner (1988) Neuronal activity in the dorsolateral pontine nuclei of the alert monkey modified by visual stimuli and eye movements. Exp.Brain Res., 70, 496-512.

Thompson RF (1988) Cerebellar role in conditioning nictitating membrane reflex. TINS, 11, 152-156.

Ungerleider LG & Mishkin M (1982) Two cortical visual systems. In: 'The Analysis of Visual Behavior, ed. Ingle DJ, Goodale MA & Mansfield RJW. MIT Press.

Yeo C & Glickstein MG (1990) Conditioning the nictitating membrane reflex. Exp. Br. Res., 60, 87-96.

Zipser D & Andersen RA (1988) A back-propagation programmed network that simulates response properties of a subset of posterior parietal neurons. Nature (Lond.), 331, 679-684.

DO EXECUTED, IMAGINED AND SUPPRESSED SACCADIC EYE MOVEMENTS SHARE THE SAME NEURONAL MECHANISMS IN HEALTHY HUMAN ?

L. PETIT †•, C. ORSSAUD †•, N. TZOURIO •,
B. MAZOYER •, A. BERTHOZ †
† *Labortoire de Physiologie de la Perception et de l'Action, CNRS-Collège de France, Paris, France.*
• *Groupe d'Imagerie Neurofonctionnelle, Serv. Hosp. F. Joliot, DRM-CEA, Orsay and EA 1555, Univ. Paris 7, France.*

Introduction

Visual exploration is a fundamental element of the repertoire of motor behaviours and saccadic eye movement are an important element of this repertoire. That is due to the physiological particulars of the human visual system. Indeed, the highly heterogeneous structure of the retina does not allow for a homogeneous and simultaneous perception of the totality of the visual field. Eye saccades are shifts of gaze produced by the oculomotor system. Our knowledge regarding the cortical and sub-cortical structures involved in the control and execution of saccadic eye movements has been mainly provided by stimulation and electrophysiological studies in the monkey and in the cat [1]. The larger part of such data resulting from anatomical and electro-physiological studies concerned the role of different cortical and sub-cortical structures in the programming, the control and the execution of visually guided horizontal saccades [1,2]. Recent positron emission tomographic (PET) studies made in healthy humans were also

F. Lacquaniti and P. Viviani (eds.), Neural Bases of Motor Behaviour, 153–164.

about visually guided horizontal saccades [3-7]. These studies have confirmed the involvement of a network composed of the primary visual areas involved in the perception of the visual stimulus, of the frontal eye fields (FEF), the supplementary eye fields (SEF), the superior parietal cortex and the cerebellum, in the execution of this type of saccades. The data presented in this chapter address the question of the cortical and sub-cortical mechanisms underlying the control of executed, imagined and suppressed self-paced horizontal saccades in total darkness [8-10]. Given the intricacy of ocular fixation, eye movements and visual attentional orienting, our findings will also be discussed as regard to their role in the theory of hierachical gating [11,12].

Material and Methods

Normalized regional cerebral blood flow (NrCBF) was measured using PET and ^{15}O-labelled water in twelve healthy right-handed young male volunteers during three different protocols. Vertical and horizontal electrooculograms (EOG) were recorded to check the execution of the tasks and a black chamber was placed around the PET tomograph and the subject. These experimental paradigms were approved by the Ethical Committee of the French Atomic Energy Commission, and all subjects gave informed written consent.

SELF-PACED SACCADES

NrCBF measurements, three at rest and three during self-paced saccades (each rest condition was followed by one fixation condition) was measured in 5 right-handed healthy volunteers with a constant 15 min. delay between two measurements. During the rest condition no indication was given except to relax, to keep eyes opened and not to make eye movements. In the fixation condition, subjects were asked to fixate a red LED that appeared at the primary central eye position 45 s before labelled water injection. Thirty seconds later, the central fixation point was turned off and the subjects were asked to hold their gaze at the recall LED position until the end of data acquisition, i.e. for 95 s. Each examination was completed in total darkness.

IMAGINED SACCADES

NrCBF was measured in 5 right-handed normal volunteers during two different conditions. In the first one (control), subjects had to hold their gaze on a central red LED which was present throughout the whole experiment. Participants were instructed to relax, to keep their eyes open and to avoid movements of eyes or of the body. In the second condition (saccade execution), subjects were asked to execute self-paced voluntary horizontal saccades, but the central red LED was kept lit. In the third condition (saccade imagination), subjects had to imagine that they performed the same horizontal saccadic eye movement as in the second condition while the central LED helped them to avoid actual eye movements. This series of three tasks was repeated twice in the same order with a constant time delay of 15 min. between two PET measurements.

FIXATION OF AN IMAGINED CENTRAL POINT.

NrCBF measurements, two at rest and four during self-paced saccades (each rest condition was followed by two saccade conditions) was measured in 4 right-handed healthy volunteers with a constant 15 min. delay between two measurements. During the rest condition no indication was given except to relax, to keep eyes opened and not to make eye movements. In the saccade condition, subjects were asked to execute self-paced voluntary horizontal saccades at maximal amplitude, the task being started 45 seconds before the injection and pursued during all data acquisition duration (2 min.). Each examination was completed in total darkness, without any cue.

DATA ANALYSIS.

For each protocol, rCBF was measured using i.v. bolus injections of 80 mCi of ^{15}O-labelled water, either with the TTV03 time-of-flight PET camera (CEA-LETI, F) [13], or with the ECAT 953B/31 (Siemens, D) PET camera [14]. CBF maps were produced using a single 80 second duration scan normalized by the whole brain number of counts and corrected for head attenuation. For each subject, sets of 3 mm contiguous axial and sagittal slices were also acquired on a 0.5 T magnetic resonance imager (MRMAX, General Electric, USA). Our data analysis method has been detailed elsewhere [15]: it combines functional (PET) and anatomical (MRI) data in order to test task specific significant CBF increases in volumes of interest (VOIs) having anatomical boundaries. VOIs definition is based on a parcellation of each subject brain in a set of 100 anatomical structures. Using a volumetric reconstruction of MRI data (Voxtool, General Electric, Buc, F), major gyri and sulci are first identified on the hemisphere surfaces and marked onto the axial MRI slices. ROIs corresponding to the intersections of the structures with the MRI slices are then manually drawn on each axial MRI slice. After alignment of the PET and MRI volumes [16], NrCBF is eventually computed in each VOI by averaging the NrCBF of its various pieces defined on the MRI slices. Values obtained during each reference condition are then subtracted to each respective saccadic condition. The mean value over the N pairs of such measurements was tested to zero using a two-tailed Student's t test.

Results

The various NrCBF changes are presented in Table 1. Figure 1 and 2 illustrate in different subjects the cortical activations observed during both self-paced and imagined saccade tasks. Each of our protocol have shown the involvement of a common cortical network composed of the supplementary motor area (or supplementary eye fields, SEF), of the median cingulate gyrus and, bilaterally, of the precentral gyrus. The precentral activation, that consists of both motor and premotor contributions, was localized at the level of the frontal eye fields (FEF), i.e. 42 mm above the bicommissural plane. In addition, sub-cortical actiavtions including lenticular nucleus and thalamus as well as the cerebellar vermis was observed during the execution of self-paced saccades while such NrCBF increases failed to reach significance during imagined saccade.

TABLE 1. Activation during the different oculomotor tasks are compared to their respective control condition. Values are averaged NrCBF variations expressed in percentage of whole brain counts.

		Self-paced saccades (N=8)	Imagined saccades (N=10)	Imagined fixation (N=13)
Anatomical VOI		mean ± s.d.	mean ± s.d.	mean ± s.d.
Suppl. Motor Area		5.2 ± 2.3 ***	7.3 ± 4.1 ***	4.0 ± 2.6 ***
Median Cingulum		4.8 ± 3.2 **	2.7 ± 3.4 *	2.2 ± 2.9 *
Precentral Gyrus	L	6.8 ± 2.2 ***	4.3 ± 2.7 **	0.9 ± 1.4 *
	R	4.7 ± 2.8 **	3.6 ± 2.2 **	1.9 ± 3.4
Lenticular Nucleus	L	5.1 ± 2.4 ***	1.8 ± 3.4	-0.6 ± 3.1
	R	5.8 ± 4.3 *	4.5 ± 6.7	-0.6 ± 4.1
Thalamus	L	2.1 ± 3.3	0.4 ± 3.1	-4.2 ± 4.2
	R	2.7 ± 3.1 *	1.5 ± 3.9	-0.7 ± 3.6
Cerebellar Vermis		3.2 ± 3.4 *	2.0 ± 3.9	-1.8 ± 4.0

*: $p<0.05$; **: $p<0.005$; ***: $p<0.0005$; Student's paired t-tests.
L: left, R: right; s.d.: standard deviation.

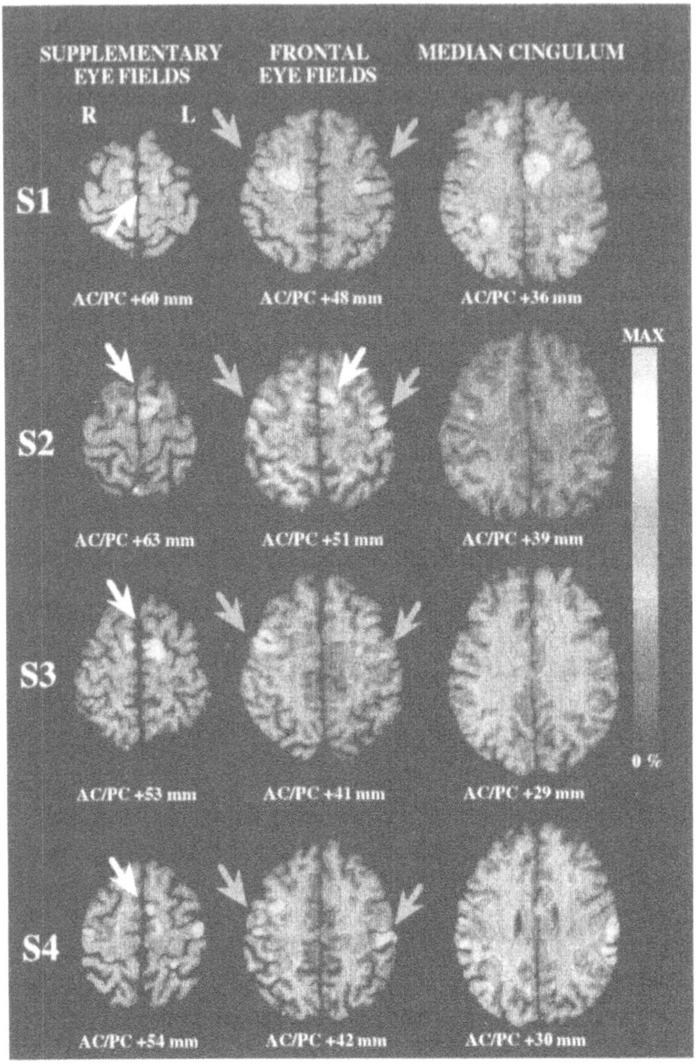

Figure 1. NrCBF PET activations observed in four subjects at three different brain levels during self-paced voluntary horizontal saccadic eye movements as compared to the rest control condition and superimposed on each corresponding MRI slices. R: right, L: left. AC-PC: Anterior and posterior commissure plane.

Color scale related to CBF increase (%) in pixel counts normalized by whole brain.

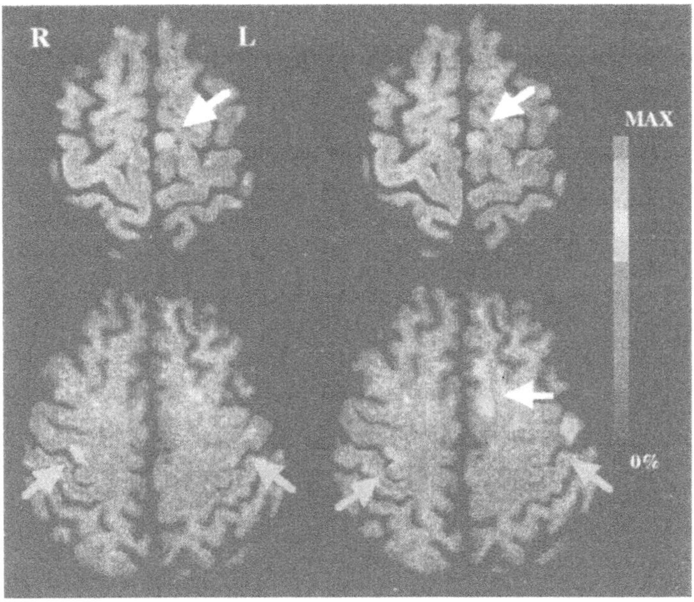

Figure 2. Single subject NrCBF activation maps obtained by subtracting averaged NrCBF maps during the control task from the averaged NrCBF maps during self-paced saccades (left) or imagined saccades (right). Median cingulate, and both bilateral precentral gyrus (FEF) and supplementary motor area (SEF) activations are superimposed on the coresponding axial MRI slices at two different brain levels.

Discussion.

THE ROLE OF INHIBITION IN THE GATING OF EXECUTED, IMAGINED AND SUPPRESSED SACCADES.

A common fronto-cingulate network composed of the frontal (FEF) and supplementary (SEF) eye fields, and of the median cingulate gyrus was selectively activated during the execution of self-paced horizontal saccades [8] as well as during imagined saccades [9]

and during fixation of an imagined central point [10]. To such a cortical network, one should add a consistent activation of sub-cortical structures (lenticular nucleus, thalamus, cerebellar vermis) during the excution of self-paced saccades. Similar sub-cortical regions were slightly activated during the imagined saccades task.
An important role is assigned to the ganglio-thalomo-cortical loop in the elaboration mechanisms of motor and oculomotor strategies [17,18]. It is interesting to note that the network of regions activated during the execution of horizontal saccades in total darkness includes both the cortical structures of the oculomotor loops (FEF,SEF) and the sub-cortical ones of the motor loop (lenticular nucleus) [17]. Such loops have been worked out mainly on the basis of electro-physiological and anatomical studies in monkeys and it is not possible to confirm an accurate equivalence with similar loops in human. Thus, PET technique offers us a means of checking and extending these data from animal to human, even when, because of the PET spatial temporal resolution, it is not possible to state positively that the different areas observed belong exactly to such loops.
The imagined saccades task have shown that part of the stuctures involved in the preparation to the execution of horizontal saccades and for their elimination could be isolated [9]. Our findings were consistent with those of Decety et al on imagined skeleton movements [19], and sustained the idea that motor imagery has to be considered as the manifestation of an internal representation of the planning and the execution of movements, whether ocular or skeletal. In addition, asking to the subjects to hold their gaze at the recall central position in total darkness also led to the activation of the similar fronto-cingulate network that during the executed or imagined saccades task performance. Such a result may reveal a part of the cortical composition of the foveal fixation system commanding a tonic inhibition on the saccadic system. Therefore, these results indicate that it is possible to observe the activation of an inhibitory system through the study of variations of the regional cerebral blood flow.

Taken together, our results are consistent with the hierarchical gating theory of executed, imagined or intended saccadic eye movements, recently proposed by Berthoz [11,12]. In that case, eye saccades are produced by neuronal mechanisms which involve several parallel loops which area arranged in cascade in such a way that excitatory processes are inhibited at several levels. This allows an internal simulation of eye movements even in the absence of overt saccades. More particularly, the collicular reticular premotor network responsible for the stimulation of extra-ocular neurones is under the control of the superior colliculus fixation neurones and of the omnipause neurones in the brainstem, the inhibition lifting of which authorizes the descending signals, which until then are only intentions of movements, to become ocular orientation movements. This first group itself is under the inhibitory control of the substantia nigra pars reticulata and of cortical structures such as the frontal and supplementary eye fields. A saccade can start only when the SNr neurones are in turn inhibited by the striatal neurones, thus causing a disinhibition of the collicular and the omnipause neurones.

Therefore, the present fronto-cingulate network may be viewed as the involvement of either the saccade-related neurons and/or the fixation-related ones located in each of the three cortical areas. For example, the imagined saccades task may involve both fixation-related neurons activation (for blocking the saccade execution) and saccade-related neurons activation (for cognitive control and timing of covert saccades), whereas the imagined fixation task may only activate the fixation-related neurons which explains the reduced fronto-cingulate activation observed in the present results during the latter task (Table 1).

THE PART OF THE VISUAL ATTENTIONAL SYSTEM IN THE HIERARCHICAL GATING MODEL

It was noticed that the orientation system of visual attention locates itself implicitly to the observation of cerebral structures engaged in the control and the production of horizontal saccades. In fact, though it is possible to shift one's attention without moving one's eyes, each eye movement is accompanied by a shifting of attention. In that case, visual attention appears as an important mechanism that can govern at the same time visual perception and eye movements. Neurophysiological studies confirm the existence of strong connections between foveal fixation, eye movements and attentional disorders. More particularly, patients suffering from the Balint's syndrome offer an impairment of ocular fixing combined with disorders of the peripheral visual attention and a quasi-absence of ocular movements (review in [20].

Each task studied in this chapter includes an attentional element. The imagined fixation condition can be considered as being the first part of the attentional process described by Posner as attention orientation, i.e. alignment of attention with a sensory entry or a semantic structure [21]. The fronto-cingulate network activated in each of our studies could be part of the fronto-parieto-cingulate network described by Mesulam regarding controlled attention [22]. In such a network, the frontal element, including FEF and SEF, would supply the distribution map of orientation and exploration movements where eye movements and fixation periods follow one another. It also fits with the theorical scheme offered by Rizzolatti regarding premotor attention [23].The latter considers that attention activation is a consequence of the activity of "pragmatic" maps of orientation and exploration movements. In the case of human, as well as other primates, because of the existence of a fovea, the central role of visual-spatial attention is thus filled by the pragmatical cards codifying the oculomotor space. Thus, covert orienting occurs when the programme for the saccadic eye movement has been completed but such a movement is internally blocked.

In conclusion, the present results provide a consistent anatomo-functional substrate both to the premotor theory of spatial attention and to the hierarchical gating theory of saccadic eye movements, which therefore be viewed as a common model to mediate the close interactions between ocular fixation, eye movements and directed visual attention.

References

1. Wurtz, R.H. and Goldberg, M.E. (1989) The neurobiology of saccadic eye movements, Elsevier, Amsterdam.

2. Leigh, R.J. and Zee, D.S. (1991) The neurobiology of eye movements, F.A. Davis Company, Philadelphia.

3. Melamed, E. and Larsen, B. (1979) Cortical activation pattern during saccadic eye movements in humans: Localization by focal cerebral blood flow increases, Ann Neurol 5, 79-88.

4. Fox, P.T., Fox, J.M., Raichle, M.E., and Burde, R.M. (1985) The role of cerebral cortex in the generation of voluntary saccades: A positron emission tomographic study, J Neurophysiol 54, 348-369.

5. Fox, P.T., Raichle, M.E., and Thach, W.T. (1985) Functional mapping of the human cerebellum with positron emission tomography, Proc Natl Acad Sci USA 82, 7462-7466.

6. Paus, T., Petrides, M., Evans, A.C., and Meyer, E. (1993) Role of the human anterior cingulate cortex in the control of oculomotor, manual, and speech responses: A positron emission tomography study, J Neurophysiol 70, 453-469.

7. Anderson, T.J., Jenkins, I.H., Brooks, D.J., Hawken, M.B., Frackowiak, R.S.J., and Kennard, C. (1994) Cortical control of saccades and fixation in man: A PET study, Brain 117, 1073-1084.

8. Petit, L., Orssaud, C., Tzourio, N., Salamon, G., Mazoyer, B., and Berthoz, A. (1993) PET study of voluntary saccadic eye movements in humans: Basal ganglia-thalamocortical system and cingulate cortex involvement, J Neurophysiol 69, 1009-1017.

9. Lang, W., Petit, L., Höllinger, P., Pietrzyk, U., Tzourio, N., Mazoyer, B., and Berthoz, A. (1994) A positron emission tomography study of oculomotor imagery, NeuroReport 5, 921-924.

10. Petit, L., Tzourio, N., Orssaud, C., Pietrzyk, U., Berthoz, A., and Mazoyer, B. (1995) Functional neuroanatomy of the human visual fixation system, Eur J Neurosci 7, 169-174.

11. Berthoz, A. (1995) Neural basis of decision in perception and in the control of movement. in Damasio, A., (ed.). The neurobiology of decision, Springer, Amsterdam, in press.

12. Berthoz, A. (1995) The role of inhibition in the hierarchical gating of executed and imagined movements, Cogn Brain Res, in press.

13. Mazoyer, B., Trebossen, R., Shoukroun, C., Verrey, B., Syrota, A., Vacher, J., Lemasson, P., Monnet, O., Bouvier, A., and Lecomte, J.L. (1990) Physical characteristics of TTV03, a new high spatial resolution time-of-flight positron tomograph, IEEE Trans Nucl Sci 37, 778-782.

14. Mazoyer, B., Trebossen, R., Deutch, R., Casey, M., and Blohm, K. (1991) Physical characteristics of the ECAT 953B/31: A new high resolution brain positron tomograph, IEEE Trans Med Imaging 10, 499-504.

15. Mazoyer, B.M., Tzourio, N., Frak, V., Syrota, A., Murayama, N., Levrier, O., Salamon, G., Dehaene, S., Cohen, L., and Mehler, J. (1993) The cortical representation of speech, J Cogn Neurosci 5, 467-479.

16. Pietrzyk, U., Herholz, K., and Heiss, W.D. (1990) Three-dimensional alignment of functional and morphological tomograms, J Comput Assist Tomogr 14, 51-59.

17. Alexander, G.E., Delong, M.R., and Strick, P.L. (1986) Parallel organization of functionally segregated circuits linking basal ganglia and cortex, Annu Rev Neurosci 9, 357-381.

18. Parent, A. and Hazrati, L-N. (1995) Functional anatomy of the basal ganglia. I. The cortico-basal ganglia-thalamo-cortical loop, Brain Res Rev 20, 91-127.

19. Decety, J. (1995) Do imagined and executed actions share the same neural substrate? in Berthoz, A., Deecke, L., and Lang, W., (eds.). Mental representation of movement. Cognitive Brain Research, in press.

20. Colby, C.L. (1991) The neuroanatomy and neurophysiology of attention, J Child Neurol 6, S88-S116.

21. Posner, M.I. (1980) Orienting of attention, Q J Exp Psychol 32, 3-25.

22. Mesulam, M.M. (1990) Large-scale neurocognitive networks and distributed processing for attention, language, and memory, Ann Neurol 28, 597-613.

23. Sheliga, B.M., Riggio, L., and Rizzolatti, G. (1994) Orienting of attention and eye movements, Exp Brain Res 98, 507-522.

THE ROLE OF THE SUPERIOR COLLICULUS IN FAST AND SLOW GAZE ORIENTATION

A. ROUCOUX, M. MISSAL AND M. CROMMELINCK
Lab. of Neurophysiology
Université Catholique de Louvain
NEFY 5449, UCL
1200 Bruxelles
Belgium

AND

P. LEFEVRE
Centre for systems engineering and applied mechanics
CESAME, Université Catholique de Louvain
4, av. G Lemaître
1348 Louvain-la-Neuve
Belgium

1. Introduction

The orientation of the line of sight (gaze) is necessary for the exploration of the visual environment. Classically, eye movements are divided into two broad categories: movements that orient gaze, and movements that stabilise gaze. Stabilisation of gaze is accomplished by the combined actions of the vestibulo-ocular and the optokinetic reflexes. Orientation is performed by saccades, smooth pursuit and vergence. In each of these categories, eye movements can be subdivided into fast (typically quick phases or saccades) or slow (typically slow phases or smooth pursuit) movements. Neurophysiologically, the control of fast and slow eye movements is considered to be performed by different structures.

The orientation of gaze is usually not done by an eye saccade alone, but rather by a combination of a saccade and a head rotation. The saccade being shorter than the head movement, a vestibular slow phase, that compensates the final portion of the head mevement, follows the saccade.

F. Lacquaniti and P. Viviani (eds.), Neural Bases of Motor Behaviour, 165–183.

In addition, the vestibular slow phase is not always completely "turned off" during the saccade or near its end and thus adds itself to the saccadic movement. In orienting movements, the interaction of slow and fast types of eye movements is thus rather complex. Interactions between slow and fast eye movements also occur during pursuit and vestibular or optokinetic nystagmus. Vergence movements are also intimately mixed with saccades. The existing categorisations of gaze movements thus tend to obliterate their complexity, especially when normal everyday-life situations are concerned. They also seem to suggest that these different categories of movements are elaborated and controlled by separate structures. The aim of this paper is to demonstrate that one of these structures, the Superior Colliculus (SC), considered unanimously as controlling exclusively fast orienting gaze movements, is also taking part in the control of slow movements.

2. The Superior Colliculus

The SC is a multilayered structure receiving convergent inputs from visual, auditory and somatosensory modalities. It also contains neurons involved in the generation of movements of eyes and head. There are two principal groups of layers of neural tissue in the SC: the superficial and deep layers.

2.1. THE SUPERFICIAL LAYERS

Ganglion cells arising from the periphery of the retina send collateral branches directly to the superficial layer of the SC. Most inputs to the SC from the central region of the retina reach the superficial layers after having been relayed through the lateral geniculate nucleus and visual cortex. There is also some direct input from the central retina and the fovea. The organisation of visual afferents is retinotopic, with the fovea occupying the anterolateral third and with the periphery extending posteromedially. The representation of the visual field is not uniform. In the monkey, the central 10 deg of the visual field take up more than 30 % of the SC; the projection of the rest of the visual field is compressed into the remainder of the structure. The representation of the central visual field is derived primarily from the contralateral eye.

All neurons recorded from the superficial layers are visually responsive. Only Y and W ganglion cells (more concerned with movement) project massively to the SC; X cells (more concerned with analysis of form) project mainly to the visual cortex and only in small proportion to the SC. Broadly speaking, the receptive fields of visually responsive neurons in the SC are larger and less precisely tuned for stimulus features than the receptive fields of cells in the lateral geniculate nucleus or striate cortex. The activity of collicular neurons appears to be more concerned with the location of the stimulus than with stimulus features. The diameter of the excitatory region of the receptive fields of the collicular cells is about 1 deg for cells serving the foveal region and as large as 10-30 deg for cells serving the retinal periphery. Neurons with receptive fields in the upper visual field are located medially; those with receptive fields in the lower visual field are found laterally. Units with receptive fields near the centre of the visual field reside anteriorly; those responsive to peripheral stimuli are located posteriorly.

2.2. THE DEEP LAYERS

The deep layers receive inputs from several sensory modalities, from motor areas, and from other areas which are neither purely sensory nor purely motor. Cortical inputs to the deep layers originate in the frontal eye fields, prefrontal cortex, parietal cortex, temporal lobe and occipital lobe.

The SC exerts its most direct control over orienting movements of the eyes, head and pinnae through the descending pathways. These fibres leave the deep layers in two major tracts: the crossed predorsal bundle (tectospinal tract) and the ipsilateral tectopontine-tectobulbar tract. The predorsal bundle sends axon collaterals to the mesencephalic and pontine reticular formation. These axons (in the cat) continue caudally, terminating in regions implicated in head movement (Grantyn and Grantyn, 1982). The tectopontine-bulbar tract sends axons into visual areas (parabigeminal nucleus, dorso-lateral pontine gray), oculomotor areas (reticular formation) and areas involved in pinnae movement in the cat (paralemniscal region of the lateral pontine tegmentum). Among these regions is the area where the reticulospinal fibres originate. Projections from the deep layers are also sent to deep layers of the opposite SC. In the deep layers, some visually responsive neurons (visuomotor cells) also discharge in association with saccadic eye movements. Other cells respond to somatosensory and/or auditory stimuli. The visual receptive fields of deep layer cells are larger than the receptive fields of superficial neurons for corresponding regions of the visual field. The convergence of inputs from many brain areas representing a variety of sensory modalities is a striking feature of the organisation of the deep division of the SC.

Electrical stimulation of deep layers in alert monkeys evokes conjugate, contraversive saccades; it does not produce smooth pursuit, vergence or nystagmic movements. The latency of saccades is approximately 20-30 ms. The direction and amplitude of evoked saccades are a function of the site of stimulation in the SC. Within limits, the trajectory of the saccade is independent of stimulation parameters as well as the location of the eye in the orbit, at the time the stimulus is delivered. Several authors observed the correspondence between the motor map and the overlying sensory map formed by the retinotopic projections to the superficial layers of the SC.

In cats, results differ in several ways (Guitton et al., 1980; Roucoux et al., 1980; McIlwain, 1986). The amplitude of the saccades produced by collicular stimulation in cats is much more dependent upon the parameters of stimulation (current intensity, duration of stimulation train, etc.) (Paré et al., 1994). There is an influence of initial eye position upon the direction and amplitude of the evoked movements. This influence is qualitatively and quantitavely different depending upon whether the stimulation site is located in the anterior two-thirds, the posterior one-third, or the most caudal zone of the SC.

In the cat, Grantyn and Berthoz (1985) and Munoz and Guitton (1985) recorded from identified tecto-reticulo-spinal neurons (TRSNs). These neurons are visually responsive, have large visual receptive fields, and generate a sustained discharge that seems to encode gaze motor error (the amplitude and direction of the difference between the current gaze position and the location of the visual target. This is also the case when the head is free to move. Berthoz et al. (1986) observed that some TRSNs generate presaccadic bursts with instantaneous frequency profiles closely correlated with the profile of saccadic eye velocity. More recently, two types of TRSNs have been described in the cat (Guitton and Munoz, 1991). TRSN's in the rostral part of the SC, near the representation of the fovea, discharge tonically when the animal fixates a visual or memorised target. This type of neuron has also been found in the monkey (Munoz and Wurtz, 1993). TRSN's in the caudal part of the SC, in the cat, (Berthoz et al., 1986; Munoz et al., 1991), have a visual response followed by a sustained activity leading to a phasic burst preceding the gaze saccade. In the monkey, similar neurons (buildup neurons) have been described (Munoz and Wurtz, 1995).

2.3. ROLE IN GAZE CONTROL

The pattern of afferent and efferent connections of the SC is consistent with the proposal that this structure is involved in transforming signals from several sensory systems into motor commands for initiating and guiding orienting responses of the eyes, head, and pinnae. Classically, the SC has been thought to provide the initial 'trigger' to release a gaze saccade, with its desired size coded in the location of activity on the SC map. It is now clear that such saccade release is also associated with the silencing of so-called 'fixation' cells in the foveal zone of the SC map (Munoz et al., 1991). But neither of these facts relate to SC activity during a saccade. First, Berthoz et al.(1986) and Rohrer et al.(1987) found that SC cells coded a desired eye velocity signal. More recently, experimental results show evidence for the coding of dynamic gaze error inside the intermediate layers of the SC (Waitzman et al., 1991; Munoz et al., 1991; Munoz and Wurtz, 1995): that is, efferent SC cell activities are modulated with decaying eye velocity or gaze error during saccades, whether the head is fixed or free. So, the SC plays a key role in both initiating saccades and in controlling them during the whole period of their progress.

3. Slow correcting movements

A poorly studied phenomenon that influences the trajectory of gaze shift appears at the end of the saccade: the eye may "drift" at a low velocity during a variable period of time. These movements are called "glissades" (Weber and Daroff, 1972) or "postsaccadic drifts" (Kapoula et al., 1986). Collewijn and coworkers (1988a and 1988b) showed that postsaccadic drifts of human subjects were a combination of version and vergence components. The version component was target directed.

In the cat, the end of the saccadic eye displacement, or the intersaccadic period between a principal saccade and the following corrective saccade, often shows some long lasting, slow movements. Most of the saccadic eye shifts of cats end abruptly but some of them are followed by a large displacement different from the preceding high velocity movement. As some of these movements are obviously driven by a system reducing the residual error at the end of the saccade, as illustrated below, we shall refer to these eye shifts as: "slow correcting movements" (SCM's).

3.1. DESCRIPTION OF SLOW CORRECTING EYE MOVEMENTS.

Slow correcting movements were detected as slow eye displacements following a saccade during which the eyes moved at a low, approximately constant velocity (Missal et al., 1993). We define "final SCM's" as slow movements whose end-time coincided with the end of the gaze displacement. Intermediate SCM's are defined as slow movements during the period between the end of a saccade and the initiation of a corrective one, or a reacceleration of the eye. During the intersaccadic period between two saccades following each other with a short latency, the eye covered sometimes an important angular distance by a slow movement. SCM's were not frequent (between 2 and 20% of the saccades). The slow displacements of the eyes studied were always target-directed, and usually followed an hypometric saccade. Slow velocity correcting eye movements occurred in the studied cats on the vertical and the horizontal component independently or together. Figure 1 shows an intermediate SCM on the vertical component, followed by a corrective saccade. The velocity (30 deg/s) of the SCM was well above the noise level and the change in vertical eye position reached 4 deg.

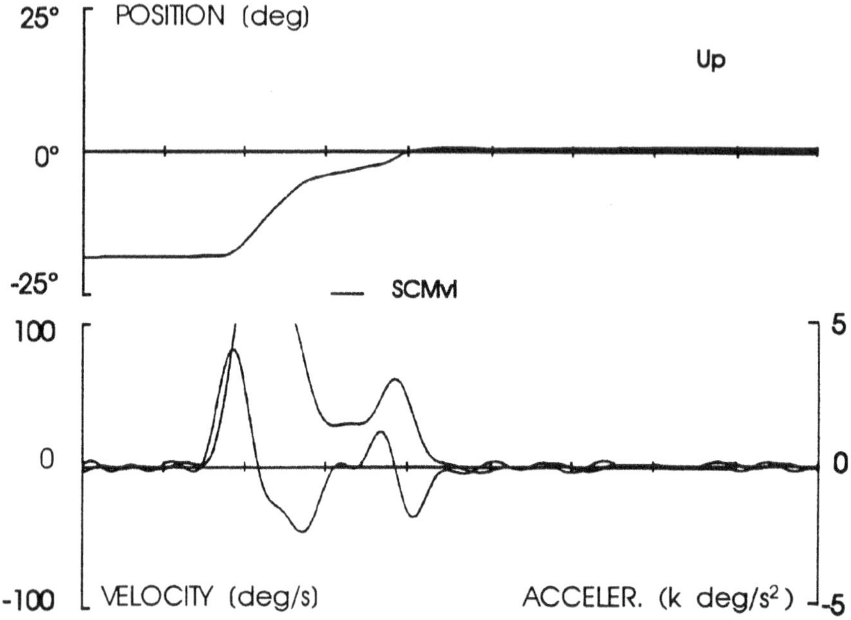

Figure 1. Intermediate slow correcting eye movement on the vertical component (SCMvi). The target is at zero on the vertical axis. Upper trace: position. Lower trace: vertical component of velocity and acceleration. The horizontal black bar under the position trace indicates the duration of the slow movement. Time scale: one tick every 100 msec.

Figure 2 shows an example of a saccadic eye movement followed by a SCM; eye position is projected on a fronto-parallel plane (XY plot). The target was at 5 deg on the vertical axis (black square). The last part of the movement (marked by an arrow) was achieved entirely by a slow velocity movement (part of the trajectory where the points are closely spaced). At the onset of the SCM, the direction of the movement was modified. This clearly illustrates the corrective role of these slow movements. Similar movements were observed on the vertical and horizontal components for all the cats studied.

A tight correlation between amplitude of the slow movement and duration or velocity was found. The duration reached values as high as 350 ms to cover an angular distance of only 4 deg. For a comparison, a saccade of the same amplitude (4 deg) lasts around 50 ms. The mean velocity of terminal SCM's varied from 7 deg/s up to 19 deg/s. The mean velocity of

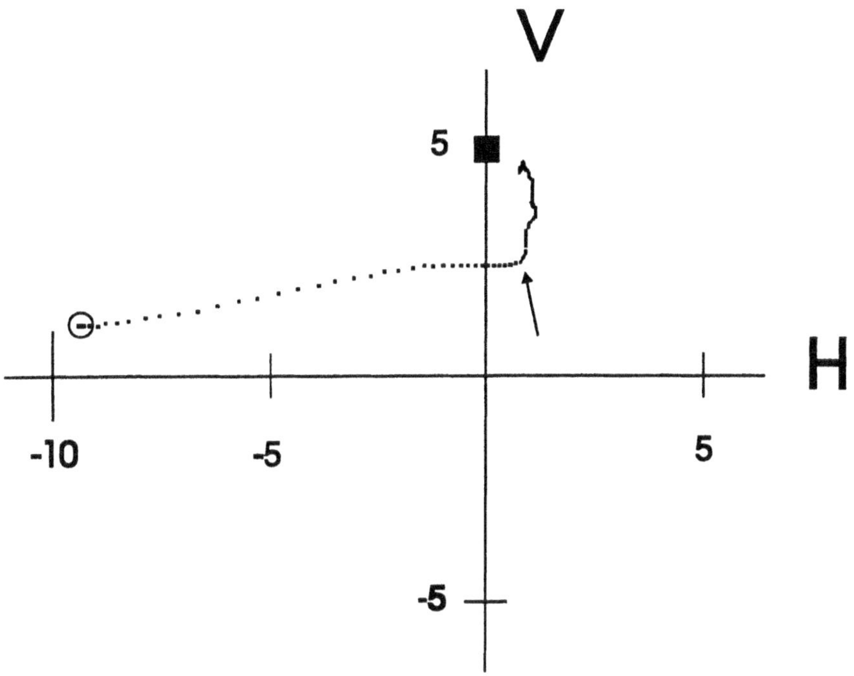

Figure 2. XY plot of a saccade with a terminal vertical slow correction. The cat made a saccade from the upper left quadrant to a target 5 deg on the vertical axis. Dots are successive eye positions every 2 ms projected onto a fronto-parallel plane. The slow movement corresponds to the part of the trajectory where the dots are more closely spaced, pointed by an arrow. Note that the direction of the movement is corrected during the slow movement. Abbreviations: V, vertical; H, horizontal.

intermediate SCM's varied between 5 deg/s and 38 deg/s but their duration was lower and varied only between 50 and 180 ms. Also during vertical gaze shifts terminal slow movements had lower velocity and longer duration than intermediate ones. The mean velocity tended to saturate for large amplitude SCM's. If SCM's play a corrective role, their amplitude should be related to the amplitude of the residual error when the rapid saccadic phase of eye displacement is over. This is the case (Missal et al., 1993).

4. Smooth eye movements evoked by SC electrical stimulation.

Even though the SC is usually considered as implicated exclusively in the control of saccadic eye movements, a few observations suggest that it may not be the case. Electrical stimulation of the deeper layers of the cat's SC sometimes evokes smooth eye movements in the head-fixed cat, as has been briefly mentioned (McIlwain 1990). When long trains of stimulation are applied, smooth eye movements occur between "staircase" saccades and following single saccades. The exact nature of these smooth movements and the possible role played by the SC in their control have not been thoroughly investigated.

The idea that the SC could also be involved in the control of smooth eye movements is not new. In the monkey, Schiller and Koerner (1971) have shown the existence of a premotor activity in the deeper layers of the SC, correlated with smooth pursuit eye movements. They concluded that the SC, and more particularly the units related to eye movements, could be involved in smooth pursuit. It is shown here that postsaccadic smooth eye movements can be evoked by electrical stimulation of the SC.

4.1. DESCRIPTION OF EVOKED SMOOTH EYE MOVEMENTS.

Microstimulation of the deep layers of SC was performed on cats trained to fixate spots of light. For stimulation periods shorter than 150 ms, only saccadic eye movements were observed. Smooth eye movements (SEM's) appeared in all cats studied when the stimulation period exceeded the duration of the latency period plus the saccadic period (around 150 ms). SEM's were present in both horizontal and vertical components and they were contra-laterally directed at least when the anterior part of the SC was stimulated. During the postsaccadic smooth movement, both eyes moved conjugately.

Figure 3 shows a smooth movement after an oblique saccade. The gaze shift was composed of a saccade and a SEM. The smooth movement started at the end of the saccade. The mean velocity of SEM was 12.5 deg/s in the horizontal component and 39.7deg/s in the vertical. For longer stimulation trains, "staircases" of saccades interspersed by SEM's were evoked. Intersaccadic SEM's were faster than those occurring between the last evoked saccade and the end of the stimulation (final SEM's). In fact, the mean velocity of the SEM was correlated with the amplitude of the preceding saccade. The direction of SEM's was close to the one of the preceding saccade except when horizontal and vertical SEM velocities were largely different or if the eye was already in a very eccentric vertical or horizontal position.

Mean velocity of SEM's increased with their amplitude and, moreover, seemingly depended also on the amplitude of the preceding saccade. If a staircase of saccades was evoked, the duration of the interspersed SEM's were limited to the durations of intersaccadic intervals, a few tens of milliseconds. But, when stimulation evoked only a single saccade, the duration of SEM's could last as long as the stimulation period beyond the end of the

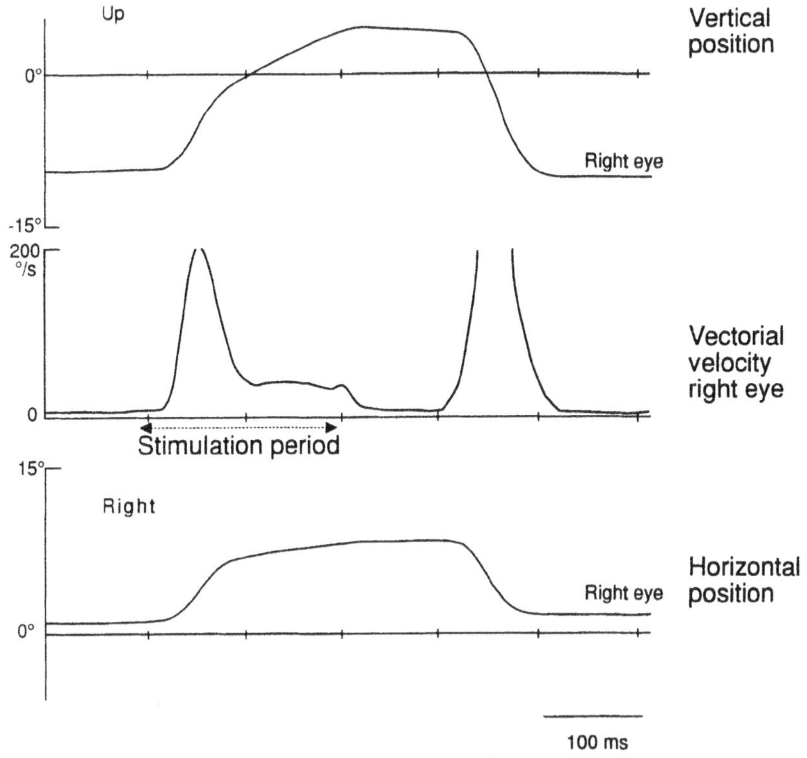

Figure 3. Head-fixed gaze shift evoked by electrical stimulation of the left SC. Upper trace, vertical position; middle trace, vectorial velocity; lower trace, horizontal position. Beginning of the trace synchronised on target extinction. Duration of stimulation: 200 ms at 600 Hz (current 1.5 times the threshold). The latency of the saccade (34 ms) plus the saccadic duration (68 ms) was 102 ms.

saccade, provided that the limits of the oculomotor range or the goal were not attained. Thus, the total duration of gaze displacement, comprising saccades and SEM's was correlated with the duration of stimulation.

Moving the electrode within the SC modified SEM characteristics. For a given stimulation site, the direction of the SEM's was identical among the different stimulation trials, but their mean velocity was modulated by the parameters of the current. Figure 4 shows the vectorial amplitude/mean velocity relationship for final SEM's obtained at two different sites with the same set of stimulation parameters. Stimulation in the rostral site (open squares) produced saccades of 5.31.6 deg (n=35). Stimulation in the caudal site (filled squares) produced saccades of 18.96.1 deg (n=29). Mean vectorial velocity of SEM's evoked at both sites increased with amplitude. But, SEM's evoked at the caudal site were larger and faster than those following the rostral site stimulation.

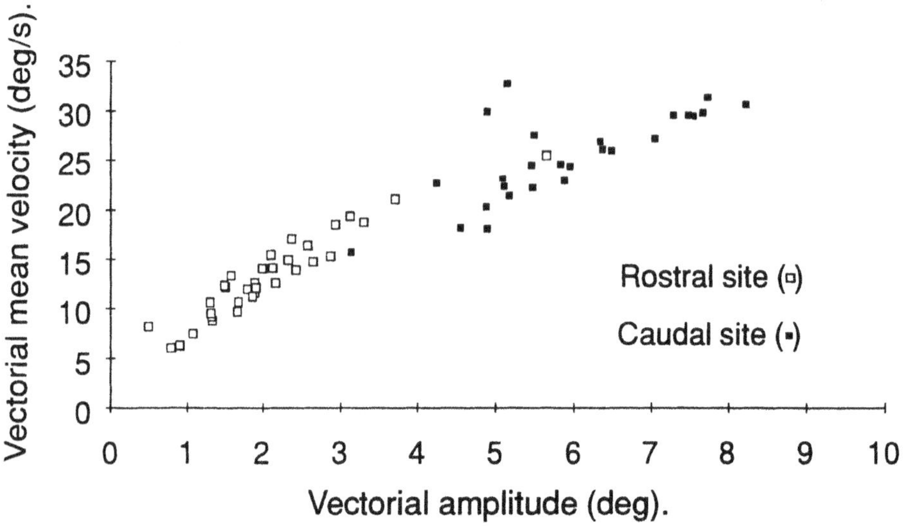

Figure 4. Relationship between SEM vectorial amplitude and vectorial mean velocity for a rostral (open squares) and a caudal (filled squares) site of stimulation for final SEM's. All stimulation parameters set at the same value (200 Hz, 400 ms, current 1.5 times the threshold). Vectorial mean velocity is higher for SEM's evoked from the caudal site where large amplitude saccades are coded.

If saccades were evoked from the rostral part of the SC, where coded amplitudes are small ("retinotopic" saccades; Guitton et al. 1980), the smooth eye movements were in the same direction as the saccade. SEM's following saccades evoked from the caudal SC and directed toward a restricted region in the orbit ("goal directed" saccades) brought the eye closer to the centre of this region. The velocity of SEM's decreased as the eye approached the centre of the goal region. Eventually, velocity dropped to zero during the stimulation period because the eye had reached the goal. In this situation, SEM's could also be restricted to one component only if, on the other component, the saccade brought the eye already near the centre of the goal region.

5. Modelling the interaction between fast and slow gaze orienting movements

Most of the existing models of the saccadic system consider that saccades are always followed by pure fixation, for stationary targets. However, experimental data (Missal et al., 1993; Olivier et al., 1993 in the cat; Weber and Daroff, 1972 in man) clearly show that gaze saccades are often followed (or even preceded) by slow gaze displacements. These peri-saccadic "drifts" are usually thought to be due to a mismatch between the pulse and the step components in the saccadic command.

These conjugate peri-saccadic drifts might have a corrective role, and the Superior Colliculus (SC) could be involved in this process, since it is a key site for saccadic control. This hypothesis was recently proposed on theoretical (Lefèvre et al., 1994) and experimental grounds (Missal et al., 1993; Olivier et al., 1993). In this section, a modelling approach is used to illustrate this hypothesis.

5.1. THE CLASSICAL VIEW

Most models of the saccadic system rely on one unique pathway to generate eye saccades, the so-called burst generator. All these models are extensions of the model proposed by Robinson and colleagues (Zee et al., 1976), which is based on a local feedback loop evaluating residual gaze error. The error is defined by comparing desired eye position with an internal copy of instantaneous eye position, and is then fed into the saccadic burst generator, to generate desired eye velocity. When the error signal reaches near-zero levels, the saccade ends crisply and the eye stops its movement, fixating the new target. More recent models of the saccadic system place the Superior Colliculus inside this feedback loop, but they all rely on one unique pathway relating instantaneous gaze error to eye velocity: the saccadic burst generator.

In these models, the saccadic system is usually supposed to produce saccades of appropriate size. If saccades stop too early (undershoot), the only possibility to acquire the target is then to trigger a second, corrective saccade. In other words, the residual gaze error signal can only influence the eye trajectory through the burst pathway (saccade). Within this context, slow peri-saccadic gaze displacements are classically attributed to a peripheral cause, i.e. a pulse-step mismatch. However, the clear corrective function of SCM's is not compatible with the explanation given above (see Figures 1 and 2).

5.2. THE NOVEL HYPOTHESIS

We recently proposed a model of gaze control (Lefèvre and Galiana, 1992) that can provide a hypothesis for the control of slow corrective gaze displacements. This model is based on gaze velocity feedback to the Superior Colliculus and proposes an original distributed parameter system to model collicular function. Figure 5 shows a simplified version of the original model

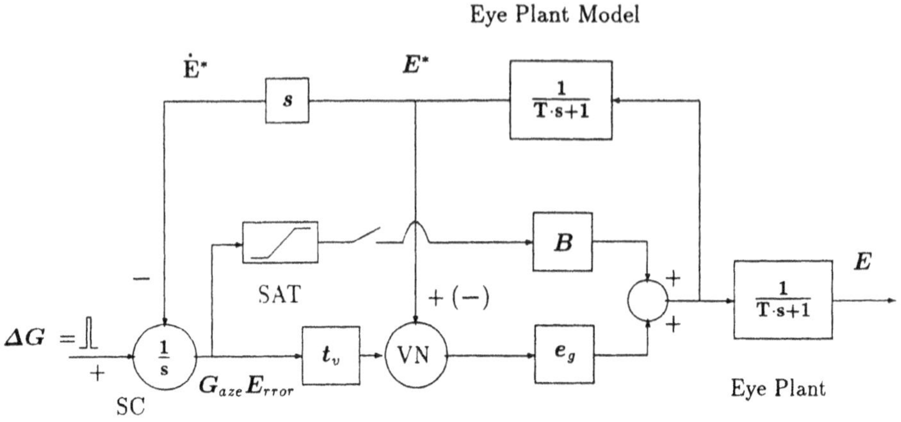

Figure 5. Lefèvre et al. (1994) model of the saccadic system. The switch in the model (output of SAT) is closed only during saccadic mode: this is the *fast* pathway; t_v, e_g and B are static gains; T is the eye plant time constant. ΔG is the desired gaze displacement, E is the eye position and E^* is an internal copy of E. Parameters were adapted to fit cat responses: $B = 2$, $t_v = 0.5$, $T = 200$ ms, $e_g = 0.5$ (saccadic mode), $e_g = 0.98$ (slow phase mode), SAT = gain of 4 with output saturation ± 20. *Laplace* notation is used. From Lefèvre et al.(1994).

(Lefèvre et al., 1994), for head fixed movements. In this model the integral action is performed by a combination of an eye plant internal model with a positive feedback loop; there are two pathways from gaze command (ΔG) to eye movement (E) in figure 5, a *fast* route (classical, through saccadic bursters), and a *slow* pathway, active during both saccades and slow phases (fixation or SCM's).

In this model, there is a natural way to explain peri-saccadic slow gaze displacements: through the slow pathway, linking gaze command (ΔG) with eye displacement (E) when the saccade is stopped (saccadic burst gate open). If one supposes that peri-saccadic SCM's are controlled by this *slow* pathway, dynamics of saccades are expected to be correlated to dynamics of SCM's: that is, faster saccades of equal size should be followed by faster SCM's for equal post-saccadic errors.

5.3. SIMULATION RESULTS

Simulations are based on the model proposed in Figure 5. Model parameters are chosen to fit cat responses, in order to compare these results with the experimental data of Missal et al.(1993) and Olivier et al.(1993). Figures 6 and 7 present two simulation examples, in the case of a rather slow final SCM (figure 6) and of a rather fast intermediate SCM combined with a corrective saccade and with a small terminal SCM (figure 7).

Figure 6 presents an eye saccade of 20 degrees, followed by a large SCM, for an initial 30 deg gaze error. The saccadic mode was switched off 10 deg away from the target. At the time of phase transition, eye velocity drops from 240 deg/s to 25 deg/s instantaneously, even though the residual error (the driving signal) does not change instantaneously.

Figure 7 shows simulation results related to a faster SCM, which is followed by a corrective saccade. In this simulation, the saccadic mode was arbitrarily interrupted during a period starting 30 ms after saccade onset and lasting 100ms. The gain of the slow pathway from the SC to the brainstem (T_v) was increased from 0.5 to 2, in order to generate faster SCM's. At the time of saccade interruption, eye velocity drops from 330 deg/s to 45 deg/s instantaneously. The same abrupt transition in eye velocity occurs when the corrective saccade starts and finishes.

This model is capable of reproducing both intermediate and final SCM's of variable dynamics, relying on physiologically plausible mechanisms: the gain T_v is proportional to population cells recruitment at the collicular output. These results are compatible with SEM cat data: an increase in stimulus intensity or frequency results in a higher SEM velocity.

6. Discussion

The data presented above show that the end of saccadic eye movements sometimes reveal the presence of another control system, inducing eye SCM's, that is qualitatively different from the saccadic rapid phase which is present at the beginning of the gaze shift. The slow movements induced have a corrective function. The data also suggest that the SC might be implicated in the generation of these SCM's. The model presented proposes a plausible mechanism by which the SC controls SCM's.

6.1. ARE THE EVOKED SMOOTH EYE MOVEMENTS COMPARABLE TO SLOW CORRECTING EYE MOVEMENTS?

It has been shown here that slow post-saccadic movements in the cat, observed when orienting towards a visual target, play a corrective role. Thanks to these slow movements, the animal compensates the residual error, i.e.

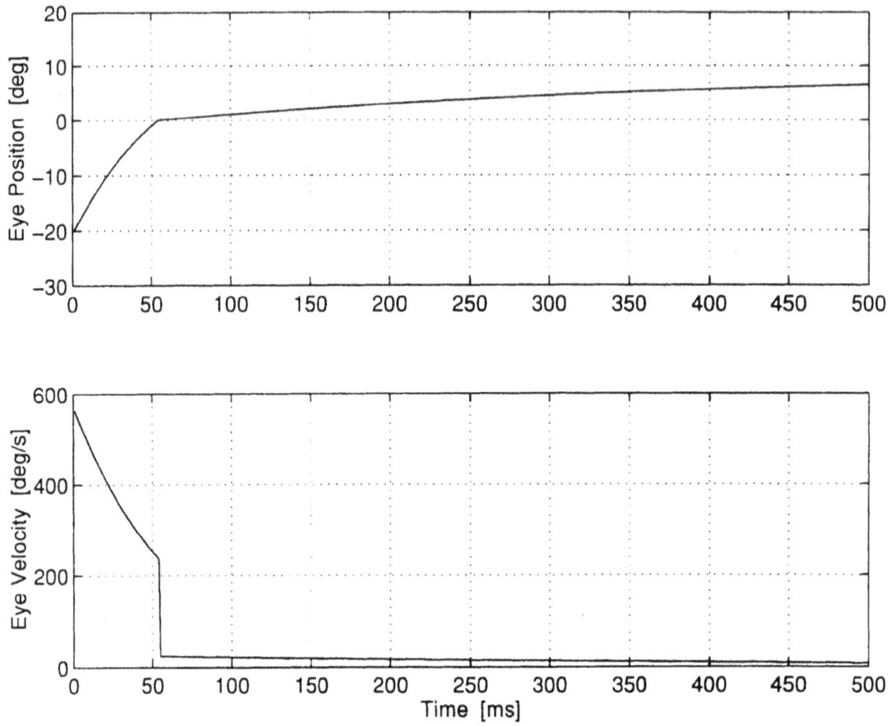

Figure 6. Simulation result obtained with the model presented in figure 5. The initial eye position is -20 deg and the desired gaze shift is 30 deg. The saccade is artificially stopped when residual gaze error is still large (10 deg). The system enters then in slow phase mode (see text), and the remaining gaze error drives the SCM. Top: position trace. Bottom: velocity trace.

the remaining retinal error. SCM's are observed in 2 to 20 % of eye saccades. These movements may be related to smooth pursuit. In this case, the stimulus triggering smooth pursuit would not be the velocity of the target, but rather its position in relation with the target. This plausible hypothesis has been put forward to explain smooth movements observed in man when a position error is maintained artificially.

SCM's of head-fixed trained cats are characterized by a period of low and approximately constant velocity of the eye after the saccade (Missal et al. 1993; Olivier et al. 1993). Following a saccade, there is often a residual error, i.e. a remaining angular distance between the actual position of the eye and the position of the target. Hypometric saccades are a particular case of the more general motor strategy consisting of undershooting the target,

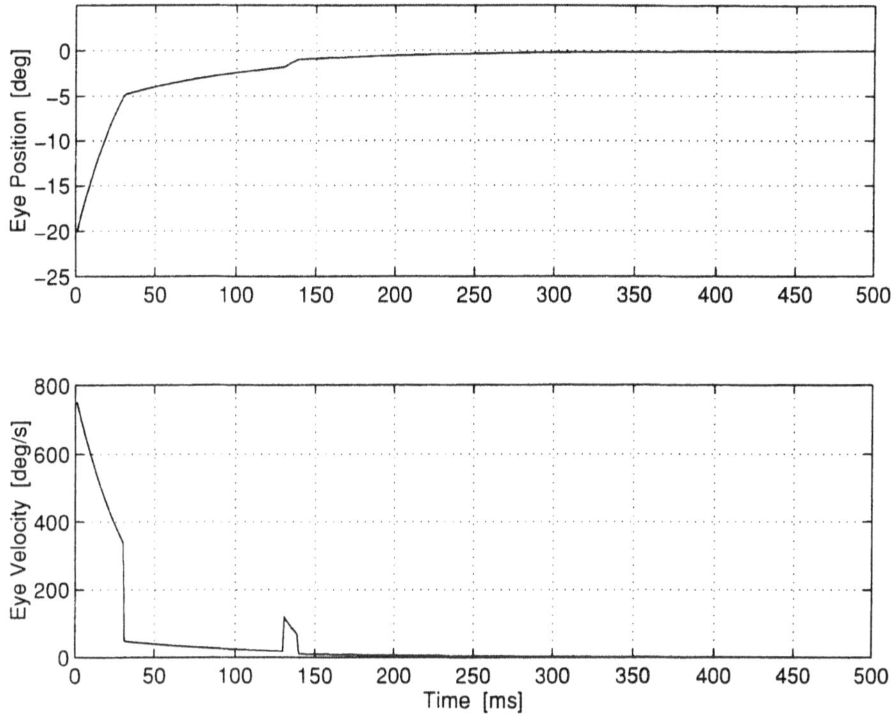

Figure 7. Simulation result obtained with the model presented in figure 5. The initial eye position is -20 deg and the desired gaze shift is 20 deg. The saccade is artificially interrupted during a period starting 30 ms after saccade onset and lasting 100 ms. During this 100 ms period, the system enters in slow phase mode, and the remaining gaze error drives the SCM, which is then followed by a corrective saccade. Top: position trace. Bottom: velocity trace. In this simulation, parameter t_v is set to 2, in order to reproduce an intermediate SCM with faster dynamics.

whatever the motor system involved in the action (Jeannerod 1988). It has been shown (Missal et al. 1993) that the amplitude of postsaccadic SCM's was adapted to reduce the residual error in amplitude and direction. Mean velocities and amplitudes of SCM's and SEM's are in the same range.

Between saccades, SCM's are faster than when the eye is near the target because the residual error is larger (Lefèvre et al.,1994). In the case of SEM's, their velocity depends on the position of the electrode: more caudal sites corresponding to larger errors elicit faster SEM's (Figure 4). Thus, it seems plausible to postulate that SEM's evoked by deep SC stimulation are analogous to SCM's occurring at the end of visually triggered gaze shifts.

6.2. DOES GAZE ORIENTATION RELY ON TWO PARALLEL PATHWAYS: A SLOW AND A FAST ONE?

An attractive hypothesis would be to consider the orienting system of the cat as composed of only two subsystems (a slow and a rapid one) acting together to reduce the position and velocity errors. This idea has been proposed by Lefèvre and Galiana (1992) in a model of the gaze system implying a control loop through the SC. Model simulations produce eye drifts whose time courses are controlled by dynamic error. It is assumed that SC output can follow two pathways: a fast saccadic one and a slow phase one. Both systems are active during an orienting movement. If there is still an error signal on the SC motor map when the saccadic generator is under omnipause neurons inhibition, the signal going through the slow phase pathway, reduces the residual error by a slow correction. A similar hypothesis has been proposed by Steinman et al. (1990) to account for their results obtained with human subjects gazing in realistic conditions of stimulation. Their hypothesis also suggests that the different subsystems controlling gaze movements could be reduced to only two independent ones: a slow (smooth) one and a fast (saccadic) one. This hypothesis is still debated and, of course, remains to be further tested experimentally.

6.3. DO SLOW AND FAST ORIENTING MOVEMENTS SHARE THE SAME REURONAL MACHINERY?

Olivier et al. (1993) recently demonstrated that TRSN's neurons sometimes have a prolonged postsaccadic discharge whose presence increases the probability of observing slow postsaccadic eye movements. It has also been shown that signals originating in the SC can be conveyed to ocular motoneurons without passing through the premotor structures controlling saccadic eye movements (Grantyn and Grantyn 1982). There is a direct connection between the deep SC and ocular motoneurons, through the TRSNs, even if the efficiency of this connection is probably low (Olivier et al. 1993). It is believed that stimulation of the deeper layers of the SC activates these pathways. However, even if they are activated during the total duration of the gaze shift, their motor effect is probably very weak during the saccade, but they might play a role in controlling slow correcting movements.

6.4. MIGHT THE SC PLAY A ROLE IN SMOOTH PURSUIT CONTROL?

So far, little is known about the pathways that could be involved in controlling smooth pursuit in the cat. At the sensory level, it is known that the visual system of the cat is able to detect fast moving spots of light up to very high speeds. Ganglion cells of the retina within the area centralis respond with actions potentials for spots moving through their receptive field with velocities ranging from a few degrees per second up to 200 deg/s for transient cells. In the lateral geniculate nucleus, a similar range of responses has been observed (Cleland, Dubin, and Levick, 1971). At the cortical level, the detection of the direction of the movement of a visual stimulus is first accomplished in the striate cortex (Hubel and Wiesel, 1962, 1968). Directional cells, showing a modulation of firing rate with target velocity, have been identified in area 17 and 18 (Orban, Kennedy and Maes, 1981). However, the analogues of the visual primate areas MT and MST (Komatsu and Wurtz, 1988) involved in movement perception and pursuit eye movements have not been identified so far in the cat.

Most cells in the cat's Superior Colliculus (SC) have a larger response for moving visual spots than for stationary ones and are directionally selective. The number of cells selective for a narrow range of directions increases as one moves toward the stratum griseum profundum in the cat (Straschill and Hoffmann, 1969). The velocity of the moving stimuli is important in determining the neural response of cells located in the superficial layers (MacIlwain and Buser, 1968). Syka and co-workers (Syka, Popelar and Bozkov, 1979) reported that in the intermediate and deep layers, most cells (71%) are directionally selective. The optimal velocity evoking a maximum response was between 20 and 40 deg/s for 76% of tested cells (n=42). Only a few neurons responded to velocities higher than 40 deg/s and five units responded over the whole range of velocities tested.

In the monkey, it has been shown that in the deeper layers of the SC, eye movement cells near the area centralis representation are discharging during pursuit eye movements and that this activity is probably correlated with the action of keeping the image of a moving spot on the fovea (Schiller and Koerner, 1971). Visual area MT, an essential cortical structure for the identification of target velocity, projects to the lower half of the superficial layers of the SC (Lund, 1972). Many cells of the superficial layers are selective for relative motion between a target and a background. Their response is suppressed if the background moves in the same direction and with the same velocity as the target. Damage to the corticotectal tract severely impairs relative motion selectivity in the SC. It has been suggested that, in the monkey, the superficial layers of the SC play a role in the analysis of motion, dependent on descending input from the cortex. (Davidson, Joly and Bender, 1992).

The dorsolateral pontine nucleus (DLPN) has been identified as another important relay in the smooth pursuit pathway of the monkey. This nucleus contains neurons that encode a variety of visual and ocular signals (Mustari, Fuchs and Wallman, 1988) and projects to the cerebellum. Interestingly, in the cat, it has been show that DLPN units respond selectively to single moving spots. Moreover, their main input comes from the superficial and deeper layers of the ipsilateral SC (Mower, Gibson and Glickstein, 1979; Hashikawa and Kawamura, 1977).

On grounds of these arguments, we hypothesize that the cat's SC might be one of the structures controlling smooth pursuit eye movements.

References

1. Berthoz, A., Grantyn, A. and Droulez, J. (1986) Some collicular efferent neurons code saccadic eye velocity. *Neurosci. Lett.*, **72**, 289-294, .
2. Carpenter, R. H. S. (1991) Vision and visual dysfunction. Eye movements (Vol. 8) pp. 1-10, London: Macmillan.
3. Cleland, B.G., Dubin, M.W., and Levick, W.R. (1971). Sustained and transient neurons in the cat's retina and lateral geniculate nucleus. *J. Physiol.* **217**, 473-496.
4. Collewijn, H., Erkelens, C.J. and Steinman R.M. (1988a) Binocular coordination of human horizontal saccadic eye movements. *J. Physiol.* **404**, 157-182.
5. Collewijn, H., Erkelens, C.J. and Steinman R.M. (1988b) Binocular coordination of human vertical saccadic eye movements. *J. Physiol.* **404**, 183-197.
6. Davidson, R.M., Joly, T.J. and Bender, D.B. (1992) Effect of corticotectal tract lesion on relative motion selectivity in the monkey superior colliculus. *Exp. Brain Res.* **92**, 246-258.
7. Grantyn, A. and Berthoz, A. (1985) Burst activity of identified tecto-reticulo-spinal neurons in the alert cat. *Exp. Brain Res.*, **57**, 417-421, .
8. Grantyn, A. and Grantyn, R. (1982) Axonal patterns and sites of termination of cat superior colliculus neurons projecting in the tecto-bulbo-spinal tract. *Exp. Brain Res.*, **46**, 243-256.
9. Guitton, D., Crommelinck M. and Roucoux A. (1980) Stimulation of the Superior Colliculus in the alert cat. I. Eye movements and neck EMG evoked when the head is restrained. *Exp. Brain Res.*, **39**, 63-73.
10. Guitton, D., and Munoz, D. P. (1991) Control of orienting gaze shifts by the tecto-reticulo-spinal system in the heat-free cat. 1. Identification, localization and effects of behavior on sensory responses. *J. Neurophysiol.*, **66**, 1605-1623.
11. Hashikawa, T. and Kawamura, K. (1977) Identification of cells of origin of tectopontine fibers in the cat superior colliculus: an experimental study with the horseradish peroxidase method. *Brain Res.* **130**, 65-79.
12. Hubel, D.H. and Wiesel, T.N. (1962) Receptive fields, binocular interaction and functional architecture in the cats visual cortex. *J. of Physiol.* **160**, 106-154.
13. Hubel, D.H. and Wiesel, T.N. (1968) Receptive fields and functional architecture of monkey striate cortex. *J. Physiol.* **195**, 215-243.
14. Jeannerod M (1988) The neural and behavioral organization of goal-directed movements. (Oxford psychology series, 15) Clarendon, Oxford.
15. Kapoula, Z.A., Robinson, D.A. and Hain T.C. (1986) Motion of the eye immediately after a saccade. *Exp. Brain Res.*, **61**, 386-394.
16. Komatsu, H. and Wurtz, R.H. (1988) Relation of cortical areas MT and MST to pursuit eye movements. I. Localization and visual properties of neurons. *J. Neurophysiol.* **60,2**, 580-603.
17. Lefèvre, P. and Galiana H.L. (1992) Dynamic feedback to the Superior Colliculus in a neural network model of the gaze control system. *Neural Networks*, **5**, 871-890.
18. Lefèvre, P., Galiana, H.L. and Roucoux A. (1994) Neural networks and dynamic feedback in the Superior Colliculus. In: Visual and oculomotor functions: advances in eye movement research (eds G. d'Ydewalle and J. van Rensbergen), Elsevier North Holland, pp 257-269.
19. Lefèvre, P., Missal, M. and Galiana, H.L. (1994) Modelling slow correcting gaze

movements, *J. Vestib. Res.*, **4**, 371-381.

20. Lund, R.D. (1972) Synaptic patterns in the superficial layers of the superior colliculus of the monkey. *Exp. Brain Res.* **15**, 194-211.
21. McIlwain, J.T. and Buser, P. (1968) Receptive Fields of Single Cells in the Cat's Superior Colliculus. *Exp. Brain Res.* **5**, 314-325.
22. McIlwain, J. T. (1986) Effects of eye position on saccades evoked electrically from superior colliculus of alert cats. *J. Neurophysiol.*, **55**, 97-112.
23. McIlwain, J.T. (1990) Topography of eye-position sensitivity of saccades evoked electrically from the cat's Superior Colliculus. *Visual Neuroscience* **4**, 289-298.
24. Missal, M., Crommelinck, M., Roucoux, A. and Decostre, M.F. (1993) Slow correcting eye movements of head-fixed, trained cats toward stationary targets. *Exp. Brain Res.* **96**, 65-76.
25. Mower, G., Gibson, A., and Glickstein, M. (1979) Tectopontine pathway in the cat: laminar distribution of cells of origin and visual properties of target cells in dorsolateral pontine nucleus. *J Neurophysiol.* **42,1**, 1-15.
26. Munoz, D.P. and Guitton, D. (1985) Tectospinal neurons in the cat have discharges coding gaze position error. *Brain Res.*, **341**, 184-188.
27. Munoz, D.P., Pélisso, D. and Guitton D. (1991) Movement of neural activity on the superior colliculus motor map during gaze shifts. *Science*, **251**, 1358-1360.
28. Munoz, D.P. and Wurtz, R.H. (1993) Fixation cells in monkey superior colliculus. 1. Characteristics of Cells discharge. *J. Neurophysiol.*, **70**, 559-575.
29. Munoz, D.P. and Wurtz, R.H. (1995) Saccade-related activity in monkey superior colliculus. 1. Characteristics of burst and buildup cells, *J. Neurophysiol.*, in press.
30. Mustari, M.J., Fuchs, A.F. and Wallman, J. (1988) Response properties of dorsolateral pontine units during smooth pursuit in the rhesus macaque. *J. Neurophysiol.* **60**, 664-686.
31. Olivier, E., Grantyn, A., Chat, M. and Berthoz, A. (1993) The control of slow orienting movements by tectoreticulospinal neurons in the cat: behavior, discharge patterns and underlying connections. *Exp. Brain Res.* **93**, 435-449.
32. Orban, G.A., Kennedy, H. and Maes, H. (1981) Response to movement of neurons in areas 17 and 18 of the cat: direction selectivity. *J. Neurophysiol.* **45,6**, 1059-1073.
33. Paré,M., Crommelinck, M. and Guitton D. (1994) Gaze shifts evoked by stimulation of the superior colliculus in the head-free cat conform to the motor map but also depend on stimulus strength and fixation activity. *Exp. Brain Res.* **101**, 123-139.
34. Rohrer, W., White, J. and Sparks D. (1987) Saccade-related burst cells in the superior colliculus: Relationship of activity with saccadic velocity. *Soc. Neurosci. Abstr.* 13, 1092.
35. Roucoux, A., Guitton, D. and Crommelinck M. (1980) Stimulation of the superior colliculus in the alert cat. II. Eye and head movements evoked when the head is unrestrained. *Exp. Brain Res.* **39**, 75-85.
36. Schiller, P.H. and Koerner F. (1971). Discharge characteristics of single units in superior colliculus of the alert rhesus monkey. *J. Neurophysiol.* **34**, 920-936.
37. Steinman, R.H., Kowler, E. and Collewijn, H. (1990) New directions for oculomotor research. *Vision Res.* **30**, 1845-1864.
38. Straschill, M. and Hoffmann, K.P. (1969) Functional aspects of localization in the cat's optic tectum. *Brain Res.* **13**, 274-283.
39. Syka, J., Popelar, J. and Bozkov, V. (1979) Responses of neurons in the superior colliculus of the cat to stationary and moving visual stimuli. *Vision Res.* **19**, 213-219.
40. Weber, R.B., Daroff, R.B. (1972). Corrective movements following refixation saccades: type and control system analysis. *Vision Res.*, **12**, 467-475.
41. Zee, D.S., Optican, L.M., Cook, J.D., Robinson, D.A. and Engel W.K. (1976). Slow saccades in spino-cerebellar degeneration. *Arch. Neurol.*, **33**, 243-251.

CORTICO-COLLICULAR CONTROL OF ARM MOVEMENTS

V. STUPHORN, E. BAUSWEIN AND K.-P. HOFFMANN

Allgemeine Zoologie und Neurobiologie

Ruhr-Universität Bochum,

D-44780 Bochum, Germany

1. Introduction

Orienting movements in space are governed by information from different sensory modalities. Although the sensory signals from the eyes, the ears, or from the body surface are mapped in different reference frames, the neural motor system generates a motor program which is organized within a body-centered coordinate system. This implies that coordinate transformations are a necessary step in the sensorimotor signal processing [11, 19].

The superior colliculus is a structure which receives information from different sensory pathways via both direct and indirect connections. The latter route includes brain areas like the visual and parietal cortex, the frontal eye field (FEF), the premotor (PMC), and the primary motor (M-I) cortex [5, 8, 13]. It can be assumed that activity transferred from these areas to the SC has already been subject to some kind of information processing so that signals with different levels of integration may arrive at collicular cells. Furthermore, it is known that signals from different sensory modalities can converge onto a single collicular neuron. Such cells are not only activated by the input of different receptor organs, but the response can be enhanced when spatially corresponding stimuli occur simultaneously [15], indicating

F. Lacquaniti and P. Viviani (eds.), Neural Bases of Motor Behaviour, 185–204.

mechanisms of multisensory spatial integration. The importance of spatial principles for the interaction of different sensory modalities was also demonstrated by Jay and Sparks [9, 10]. The authors showed that the activity of deep collicular neurons in response to auditory or visual stimuli is translated into a 'motor error' signal appropriate for directing the eyes to the target in space taking into account the current eye and head position.

Recently, Werner [21] demonstrated the involvement of collicular neurons in the control of arm movements of monkeys which were trained in a delayed reaching task. In view of the described connections of the SC, and the well known oculomotor mechanisms of sensorimotor integration it is probable that also arm-movement-related activity of the SC and closely linked motor areas may exhibit different stages of sensori-motor signal processing. In order to detect possible relations of neuronal activity to different stages of the coordinate transformations which are necessary for the translation of the spatially localized visual target to the body-related motor command we designed two reach tasks in which the spatial variables of the arm movement could be changed separately from the retinal and eye position coordinates of the visually perceived reach target.

2. The tasks

A monkey was trained to fixate a central light emitting diode (LED) on a planar screen in front of him as long as it was illuminated. A second LED was presented after a delay period at one of 8 peripheral locations (15° eccentricity) in a circle around the fixation spot. In 'saccade-reach' trials, the fixation light was turned off after a further delay. This was the signal to make a saccade towards the target. After a third delay period during which the monkey had to hold fixation of the target an acoustic cue triggered the arm reaching movement from the touchbar positioned at waist level to the illuminated target on the screen.

In the *fixation-reach* task the monkey had to maintain fixation of the central LED throughout the trial. After the appearence of the target light the accoustic 'go'-signal was given and the monkey had to reach to the target while fixating the central light. Therefore the retinal coordinates of the target was directly related to its position on the screen.

Horizontal and vertical eye position was measured with the search coil technique. Single units were recorded from PMC and SC, and the neuronal activity was analysed in relation to target presentation as well as to eye and arm movements towards that target. The SC electrode was lowered within the parasagittal plane with an angle of 45° relative to vertical penetrating the SC surface roughly perpendicularly. Strong multi-unit responses to small light spots at a particular site of the contralateral visual field indicated the dorsal border of the SC.

3. The cortico-collicular projection neurons of PMC

The PMC, like the primary motor cortex (M-I) is known to be involved in the control of limb movements [12, 20]. In an anatomical study, Fries [5, 6] identified a large number of cortical areas containing projection neurons to deep layers of the SC. In the PMC / M-I, the population of labelled projection cells formed a continuous band located in the hand-arm-trunk representation area of these motor fields. By antidromic stimulation in the SC, we identified electrophysiologically cortico-collicular projection neurons, and recorded their activity during performance of the tasks.

Fig. 1 A shows the relation of the current threshold for evoking an antidromic activation to the depth of the stimulating electrode in the SC for four typical neurons. They showed a minimum at a particular depth with a steep increase of current strength for higher or lower stimulating sites. The minima for all projection neurons in which the depth dependency of the stimulating current was measured (n=9, Fig. 1B) were located between 2 mm and 5 mm below the collicular surface corresponding to the deep layers of the SC and the ventrally adjoining tegmental reticular formation. The average depth for the lowest stimulation threshold was 3.5 mm suggesting a high density of cortico-collicular axons in the deepest collicular layers. This area coincides fairly well with the most effective stimulating

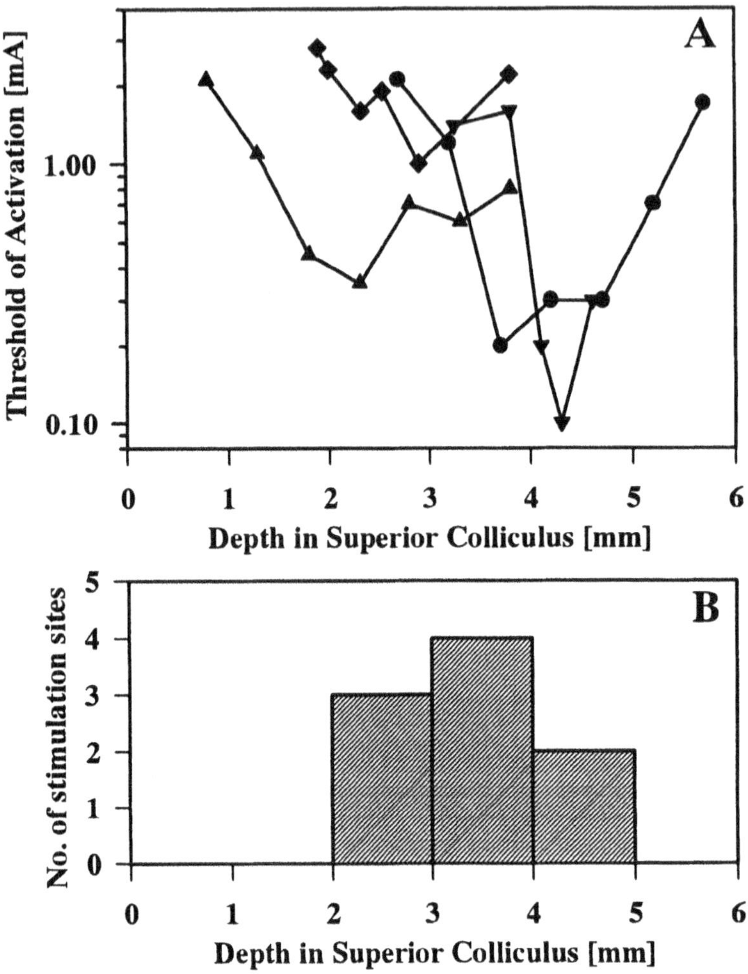

Figure 1. (A) Relation of threshold current intensity (in mA) for evoking antidromic activation to the depth of the stimulating electrode (in mm) below the SC surface. Curves of four antidromically identified neurons are shown. (B) Number of stimulation sites with minimum current (ordinate) plotted against 1 mm-steps of electrode depth below the SC surface (abszissa).

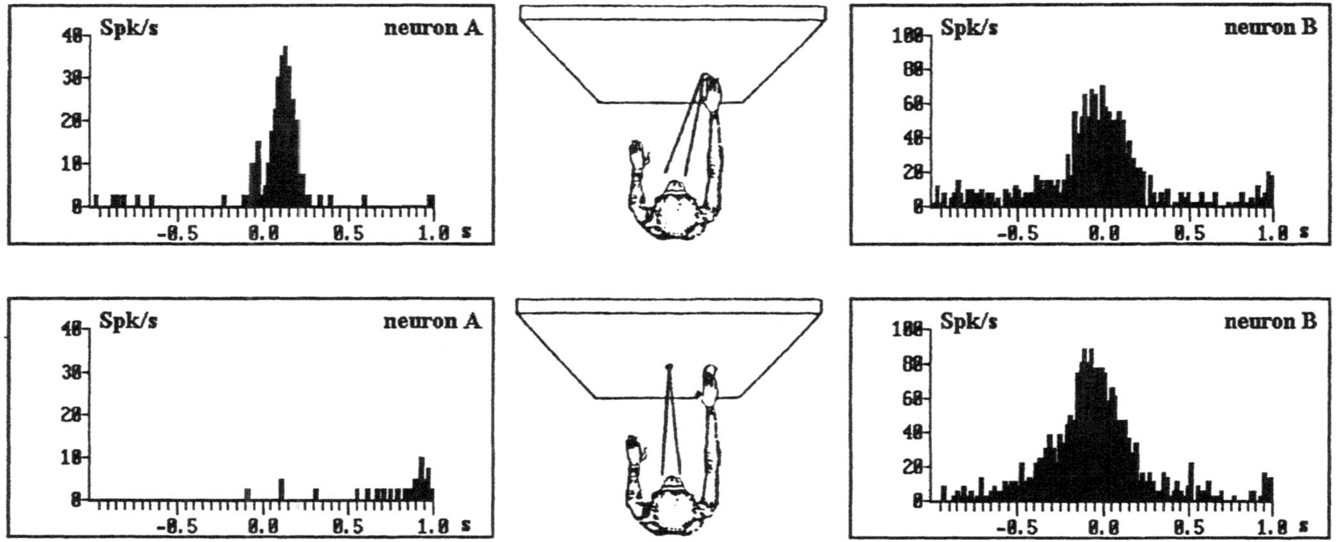

Figure 2. Activity of two cortico-collicular PMC neurons during the 'saccade-reach' task (upper row of diagrams) and the 'fixation-reach' task (lower row of diagrams). The two sketches illustrate the spatial relation between gaze and reach target in the respective task. The reach-related activity of the cells is shown in form of histograms (binwidth 20 ms) with 2 s time periods alligned to movement-onset (time=0). Neuron A is activated during the 'saccade-reach' task only, whereas neuron B shows a stronger reach activation during the 'fixation-reach' task.

site for evoking postsynaptic responses in cat motoneurons of neck, forelimb, and hindlimb in the experiments of Anderson et al. [1].

We have been able to record from ten neurons in the dorsal part of premotor cortex (PMCd), projecting to the SC. Their location matched the field in PMC, where Fries [5, 6] found the population of anatomically identified projection cells. All of our 10 projection neurons were activated during arm reaching, although a few showed activity predominantly during the return movement of the hand from the target to the touchbar. The activity increase was not a response to the auditory GO-signal, because the onset of activation closely covaried with the arm movement, which started at different delays due to variation in the reaction time.

The activity of two cortico-collicular PMC neurons (A and B) during the 'saccade-reach' task and the 'fixation-reach' task is illustrated in Fig. 2. Sketches of the two tasks illustrate the spatial relation between gaze and arm direction during reaching movements. The upper one shows the 'saccade-reach' task (matching of gaze and hand target), whereas the lower one shows the 'fixation-reach' task (gaze and hand at different targets). The left and right panels of the figure present the histograms of the cells' activity in the respective task. The trials were aligned to movement-onset (time=0). The target location was in all cases 15° to the right. The left neuron was recorded in the depth of the caudal bank of the superior arcuate sulcus. The right one was recorded more mediocaudally relative to the other one. In both cells the movement related burst was different in the two tasks, although the trajectory of the arm movement was virtually identical. Neuron A showed higher reach activity when the target was foveated, and this was consistently found during reaching to different target locations (not shown in the figure). Simultaneously with neuron A the activity of another cortical cell could be recorded which did not project to the stimulation site in the SC. This cell showed the same reach activation during both tasks, and thus revealed different properties, although both cells were located in immediate vicinity. Neuron B, also a projection cell to the SC was more active when the eyes fixated another point in space than the reach target, and showed weaker activation when both targets matched spatially. Hence, the reach-related activity of both projection neurons depended - in a different manner - on the condition whether or not the targets for the eyes and the arm were spatially different. Other cortico-collicular neurons recorded in the PMC showed a third characteristic, i.e. they exhibited no significant difference in their reach activity between the two tasks. These results are comparable to the findings of Boussaoud [2] in set-related PMC cells whose activation during the planning phase of arm movements was also influenced by eye position. Therefore, it may be concluded that at this neuronal level the control of reaching is organized partly in a gaze-dependent reference frame.

4. The reach cells in the superior colliculus

Naturally the question arises, in which reference frame the reach related cells in the SC operate, which receive the input from premotor cortex. On the one hand it could be organized in a body-centered system, in which the same arm movement would be associated with the same neural activation, irrespective of the specific relation between eye and target. Alternatively, the cell's activity might be modulated by shifts of retinal target location or eye position in the same way as the PMCd neurons.

With regard to these questions, we found at least two different functional groups of SC neurons. One group of cells showed a clear dependency of the reach-related activation on the retinal coordinates of the perceived target. An example is shown in Fig. 3. The four histograms in the upper panel are constructed from time intervals out of saccade-reach trials aligned to movement-onset at t=0. During reaching to the four targets at 15° left, right, up and down from screen center the monkey foveated the respective target. The location of the histograms indicates the location of the target on the screen. A weak activation after movement onset was observed in each histogram. The lower four histograms show the activation during reaching to the same targets, but in this case, in the 'fixation-reach' task, i.e. when the monkey held constant fixation of the central LED, and made arm movements to the targets which were projected on different retinal coordinates (see the respective sketch in the middle of the histograms). Reaching to the right target was associated with a much stronger activation than during the same arm movement in the saccade reach condition, when the gaze was directed to the reach target. Since the trajectories

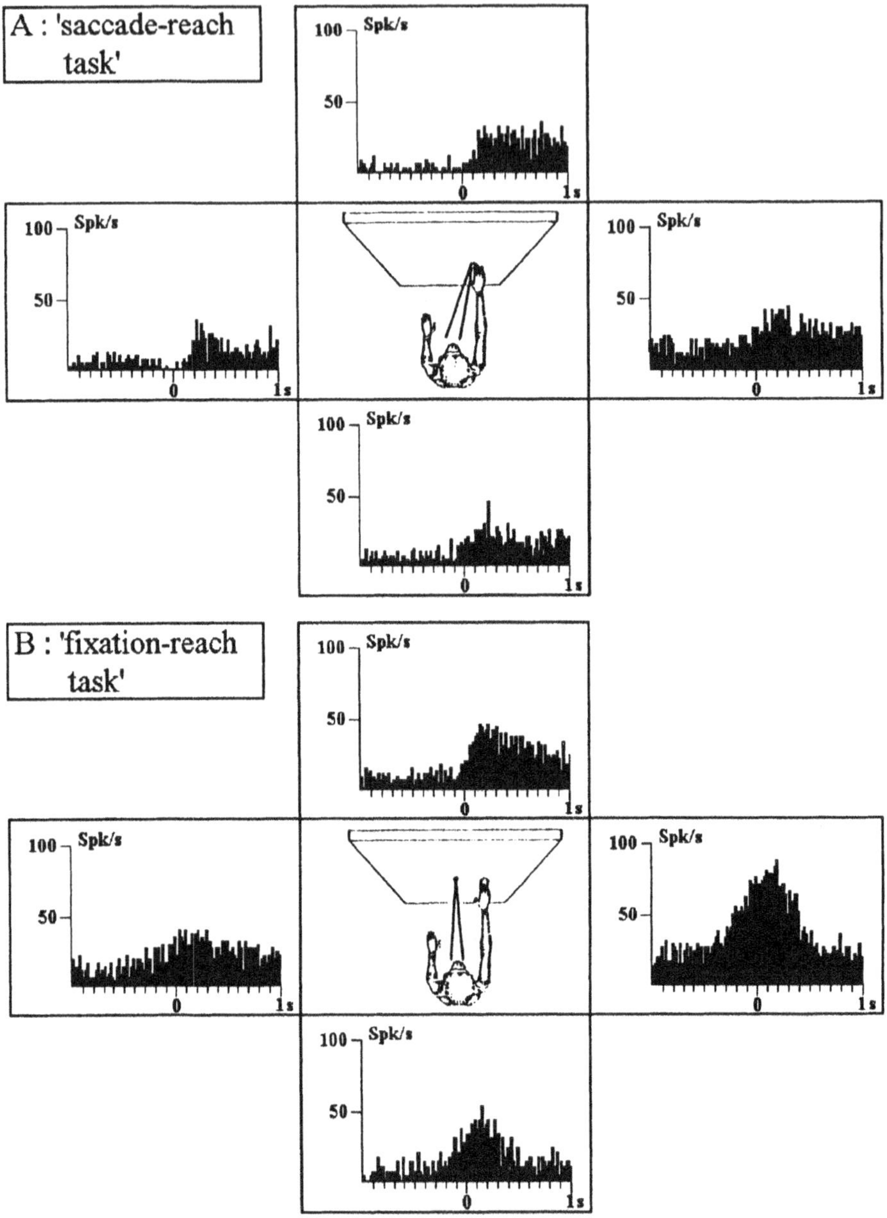

Figure 3. Activity of an SC neuron during reaching to four different targets (15° to the right, down, left and up; note the corresponding position of the histograms). The trials are aligned to movement onset (time=0) and show a period of 1 s before and 1 s after in all histograms. The activity of the cell is given in spikes/s [Spk/s] in a binwidth of 20 ms. The upper four panels show the activity of the cell in the 'saccade-reach' task (A), whereas the lower four panels show the activity in the 'fixation-reach' task (B). The sketches in A and B illustrate the spatial relation of gaze and arm direction for the respective task. Note the stronger activation during reaching to the right when the target is not foveated (B).

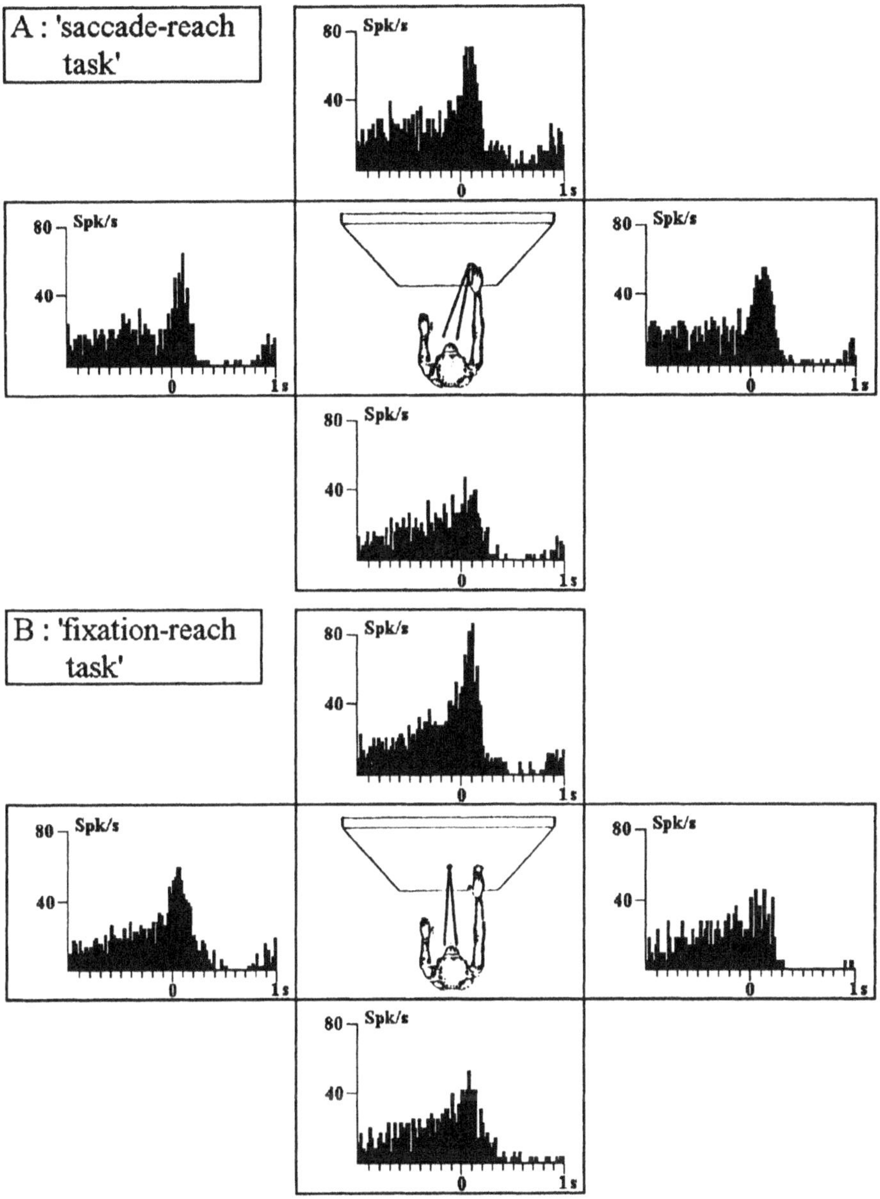

Figure 4. Activity of a reach-related cell in the SC during arm movements to four targets. The arrangement of the figure is the same as in Fig.3. This neuron shows a systematic variation of the reach activity amplitude with respect to the four targets. This relation is unaltered in the two tasks, despite the differences in eye position and retinal target coordinates.

of the arm movements were identical, the difference in reach activation must be due to a retinal, or gaze-related variable. Different eye positions per se do not alter the reach activity very much (cf. the difference between spontaneous activity and movement-related activity in the upper four histograms). Therefore, the difference of reach activity during arm movements to the same (right) target, found between 'saccade reach' and 'fixation rach' trials must be due to the specific location of the target in retinal coordinates. This means that the activity of the cell depends on two factors. Firstly, there has to be an arm movement, because the cell never showed an increase above the level of its spontaneous firing rate, if sensory or motor events other than arm reaching occurred. Secondly, the intensity of this activation depends on the location of the target on the retina. In this case, a presentation in the right hemifield causes a profound increase in neuronal activity. New experiments must be designed for testing further interdependencies between parameters of different coordinate systems. In any case, the reach activity of this SC neuron would not provide an appropriate motor command for directing the arm in space, although it was closely time-locked to the motor event. The distinct relation of the activity to retinal coordinates does not mean that it is a pure sensory response, because the cell's firing was contigent upon the arm reaching. This cell type seems to operate in an intermediate level of the transformation process.

A second, larger group of reach related SC cells showed an independency of the reach activation from gaze or retinal location of the visual reach target. Fig. 4 shows an example of this type. The figure is compiled in the same way as Fig. 3, with histograms representing 'saccade-reach' trials in the upper part, and those for 'fixation-reach' trials in the lower half of the figure. Comparing the reach activity of movements to the same target between 'saccade-reach' and 'fixation-reach' trials, the same discharge rates are found, although gaze angle and retinotopic target location differed. Thus the cell certainly does not work in a retinocentric frame of reference, like the former group of cells. The unchanged directional tuning of the reach activity during both tasks rather suggests a relation to a body-centered reference frame.

As has been outlined in the introduction, the process of planning a reach movement could be modeled as a series of transformation steps from one coordinate

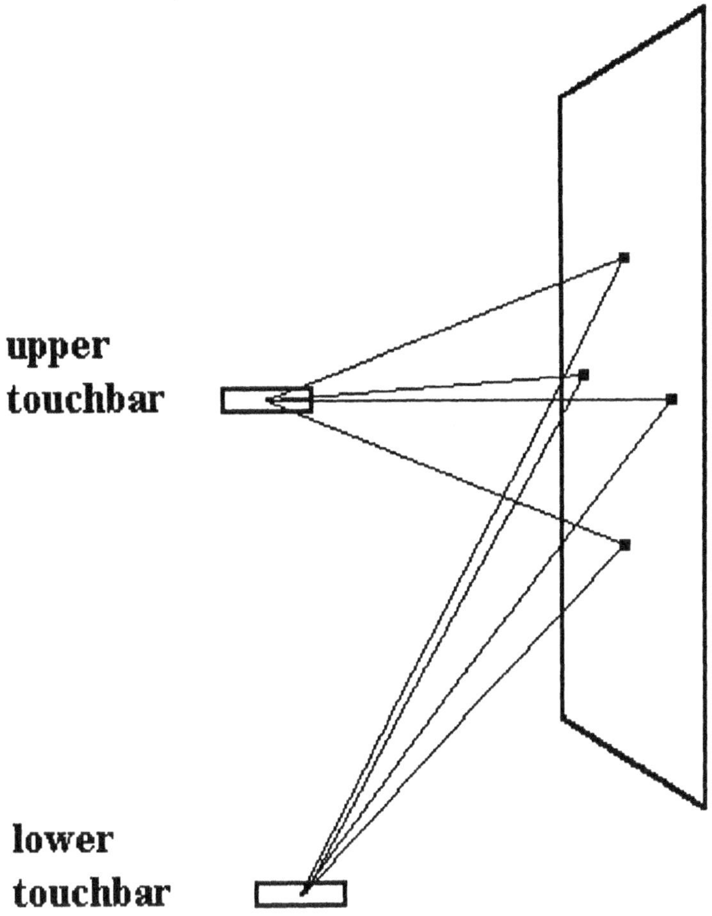

Figure 5. Experimental setup for the test of 'target position' versus 'hand trajectory' coding of reach-related cells. The two starting positions (touchbars) and four of the eight targets on the screen are shown. The lines between the touchbars and the targets show the different directions of the reach movements with respect to the starting position of the hand. The elevation angle for movements from the lower touchbar varies between 52° and 65° upward, the angle for movements from the upper touchbar is 17° upward and downward from horizontal.

system to the next. According to this hypothesis the group of neurons, to which this cell belongs, represents a more progressive level of this transformation process as compared to the first retinocentric group of reach neurons (Fig. 3). We could not distinguish between a head or body centered reference frame, because we did not change the position of head and shoulder. In any case, in a serial model of sensori-motor transformation, this cell would belong to a neural module contributing to the generation of the final motor command.

Even after the sensory signals have been translated into motor coordinates, there is a variety of possible movement parameters, to which the neuronal activity might be related, e.g. direction, distance, or endpoint of movement, as well as joint angle or muscle force. In a further series of experiments, we investigated part of these questions, and changed the parameter 'movement direction' while keeping the location of the targets constant, i.e. we changed the starting position of the hand. Fig. 5 shows a side view of the experimental arrangement with one touchbar in the lower position (approximately at waist level), as used in the experiments described above. This was exchanged for an upper metal bar for hand starting positions at eye level. The connecting lines to the targets on the screen indicate the variation of movement directions with the two touchbar positions.

A few neurons were tested in relation to movements to identical targets from the lower versus the upper touchbar. The upper panel of Fig.6 shows the reach-related activity during movements to 8 different targets starting from the upper touchbar, and the lower panel represent the reach activity when movements start from the lower position. The amplitude variation of the neurons discharge with respect to target location is the same when comparing movements from the two different starting positions. This suggests that this cell does not encode direction of movement, but rather target location in space. The result of a quantitative comparison of this cell's activity is shown in Fig. 7. The three-dimensional diagram shows the mean discharge rate of the cell during the reach-related burst plotted against the location of the target on the screen. The activity during movements from the upper starting position are shown as white dots. The activity during movements from the lower starting position are indicated by black dots. Two-dimensional regression analyses were performed

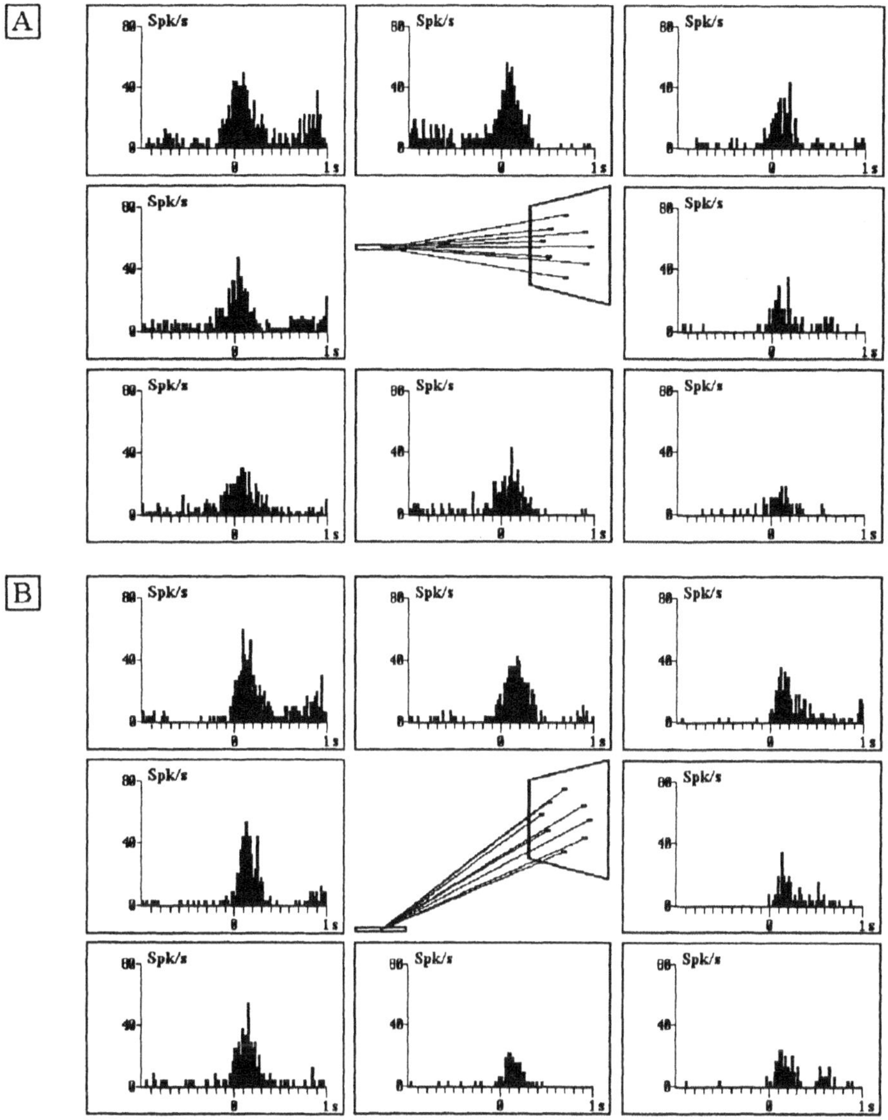

Figure 6. Activity of a reach-related cell in the SC in the 'target position' versus 'hand trajectory' test for reaching to eight different targets. The respective histograms are positioned correspondingly. The two sketches in the middle of A and B indicate the different movement directions for reaching from the upper (A) and lower touchbar (B). The histograms are constructed in the same way as in Figures 3 and 4.

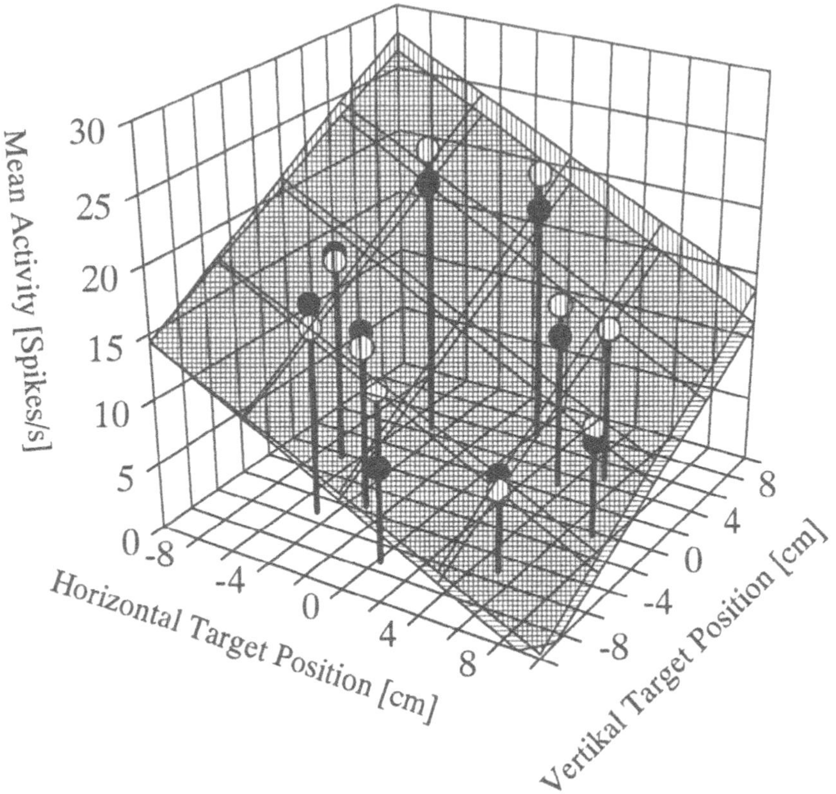

$$\text{Activity}\,(x,y)_{\text{Waist}} = 12.81 - 0.75x + 0.60y$$

$$\text{Activity}\,(x,y)_{\text{Eye}} = 14.09 - 0.70x + 0.66y$$

Figure 7. Discharge rate (Spikes/s, vertical axis) of the reach activation of the cell in Figure 6 (measured between 200 ms before and 300 ms after movement onset) in relation to the position of ten targets (horizontal axes). The retinal location of the target on the screen is given in cm (with 7 cm distance from screen center corresponding to 15° gaze angle). The activity during the movements from the upper touchbar are shown as white dots, whereas the activity during movements from the lower touchbar are indicated by black dots. In addition to the eight targets shown in Fig. 6 two targets with 10° ecentricity were used for a two-dimensional regression analysis of each group of data. The resulting regression planes are displayed in the 3-D-plot. The equations with the calculated parameters of the two planes are given below the diagramm. No significant difference for the two groups of data was found in the Kolmogorov-Smirnov-test of goodness of fit ($p > 0.65$).

separately for the two groups of values yielding two regression planes, with the equations given below the diagram. There was no statistically significant difference between the two groups (Kolmogorov-Smirnov-Test, $p > 0.65$).

Dannenberg et al. [4] compared reach-activity of SC neurons between movements with a 150g load attached to the wrist and no-load condition, and found differences in most of the tested neurons. This result can hardly be reconciled with a pure encoding of target location. So it may be assumed that several movement variables are represented in the reach-activity of SC neurons.

5. The relation between PMC and SC

A general outline of the indirect sensory inputs to the SC, and of functional connections to the motor apparatus is shown in Fig. 8. The scheme is by no means intended to be complete, but instead should highlight some anatomical and physiological results, which are of special interest with regard to our study.

A stimulus appearing in the surrounding space can be selected as a target for an orienting movement of the eyes and/or the arm. The information about the spatial location of this target could be given in one or all of three different sensory channels: the visual, the auditory and the somatosensory. These sensory signals are organized in different frames of reference and, in addition to that, do not provide immediately an adequate reference system for the planning of a movement. Therefore the incoming sensory information streams are subject to coordinate transformations.

The processed signals can be transmitted to the FEF, the PMC and the deep layers of the SC. The scheme highlights the similarity between the connections of the SC with the FEF and the oculomotor system, and correspondingly those with the PMC and the skeletomotor apparatus, as derived from anatomical studies [5, 8]. Both cortical areas have a direct connection to motor centers exerting a relatively direct influence on the respective effector system. On the other hand both have also a second connection to a group of premotor neurons in the colliculus, which provide another, indirect access to the motor centers [14]. For the oculomotor route, this parallel organization was confirmed by lesion studies in which permanent impairment of eye

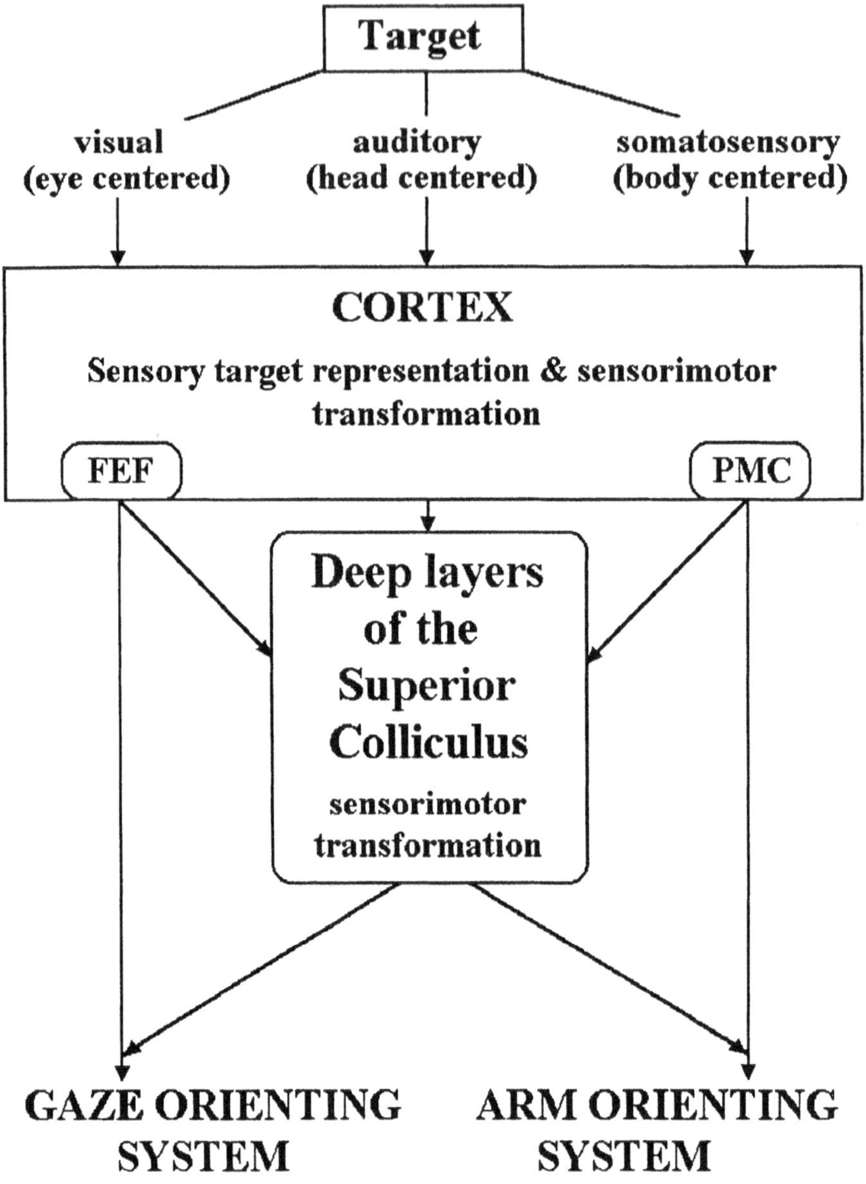

Figure 8. A schematic drawing of the anatomical and functional relations between a subset of cortical areas, which are involved in saccades and reach movements, and the SC. The diagram is restricted to the structures, which are of immediate interest with regard to the results of our study. It highlights the analogous organization of the corticocollicular projections from the FEF and the PMC. Abbrevations: FEF: frontal eye field; PMC: premotor cortex; SC: superior colliculus

movements was not observed after elimination of only one structure [17]. A similar connectional scheme might be assumed for the PMC-SC subsystem [5, 7]. Our physiological studies have shown that similar steps of the sensorimotor transformation are found in the PMC and in the SC reach neurons. This suggests for this function, that the cortico-collicular system does not operate in a hierarchical fashion, instead 'higher' and 'lower' levels of the transformation process are represented in both structures.

The cortex is known to be also involved in target selection and stimulus-response combatibility [3, 16]. A large population of neurons show both a response to target appearance and to the impending movement However, in corticotectal neurons of the FEF projecting to the deep SC layers, Segraves and Goldberg [18] found no visual responses. Also in our small sample of PMC-SC projection neurons, the reach related activity was clearly predominant. Therefore, the information transmitted from the cortex to the deep SC layers is primarily concerned with motor aspects of visuospatial processing.

6. Summary and conclusions

We developed an experimental paradigm to seperate the spatial parameters of an arm movement from the location of the visual target on the retina and from eye position. Electrophysiological recording of single-unit activity in the premotor cortex and the superior colliculus of a rhesus monkey, coupled with antidromic identification of corticocollicular projection neurons revealed the following points.

(1) Although the arm-related activity of the cortico-collicular neurons is closely time-locked to the reach movement, in some cells the activity is modulated by gaze direction.

(2) In the SC there is a population of cells whose reach activity depends on a retinocentric reference frame and a larger group which operates in an head- or body-centered reference frame.

(3) For some of the reach-related cells in the SC we could show, that their activity is better related to the target position in space than to the trajectory of the hand.

(4) The occurence of different frames of reference in the SC indicates, that this subcortical area might be involved in visuomotor transformations which operate parallel to cortical processing.

This work was supported by Deutsche Forschungsgemeinschaft (NEUROVISION) and ESPRIT (MUCOM II).

References

[1] Anderson, M.E.; Yoshida, M. and Wilson, V.J. (1972) Tectal and Tegmental Influences on Cat Forelimb and Hindlimb Motoneurons, *J.Neurophysiol.* **35**, 462-469

[2] Boussaoud, D. (1995) Primate Premotor Cortex: Modulation of Preparatory Neuronal Activity by Gaze Angle, *J. Neurophysiol.* **73**, 886-890

[3] Crammond, D.J. and Kalaska, J.F. (1994) Modulation of Preparatory Neuronal Activity in Dorsal Premotor Cortex due to Stimulus-Response Compatibility, *J. Neurophysiol.* **71**, 1281-1284

[4] Dannenberg, S.; Werner, W.; Kutz, D. and Hoffmann, K.-P. (1994) Neuronal activity in the primate superior colliculus during arm movements under varying loading conditions. *Proceedings of the 22nd Göttingen Neurobiology Conference* **22**, 290

[5] Fries, W. (1984) Cortical projections to the superior colliculus in the macaque monkey: A retrograde study using horseradish peroxidase, *J. Comp. Neurol.* **230**, 55-76

[6] Fries, W. (1985) Inputs from motor and premotor cortex to the superior colliculus of the macaque monkey, *Behav. Brain Res.* **18**, 95-105

[7] He, S.-Q.; Dum R.P. and Strick P.L. (1993) Topographic Organization of Cortico- spinal Projections from the Frontal Lobe: Motor Areas on the Lateral Surface of the Hemisphere, *J.Neurosci.* **13(3)**, 952-980

[8] Huerta, M.F.; Krubitzer, L.A. and Kaas, J.H. (1986) The frontal eye field as defined by intracortical microstimulation in squirrel monkeys, owl monkeys, and macaque monkeys. I.Subcortical connections, *J.Comp.Neurol.* **253**, 415-439

[9] Jay, M.F. and Sparks, D.L. (1987a) Sensorimotor integration in the primate superior colliculus. I.Motor convergence, *J. Neurophysiol.* **57**, 22-34

[10] Jay, M.F. and Sparks, D.L. (1987a) Sensorimotor integration in the primate superior colliculus. II. Coordinates of auditory signals, *J. Neurophysiol.* **57**, 35-55

[11] Kalaska, J.F. (1991) Reaching movements to visual targets: neuronal representations of sensori-motor transformations, *Seminars Neurosci.* **3**, 67-80.

[12] Kurata,K. (1989) Distribution of neurons with set- and movement-related activity before hand and foot movements in the premotor cortex of rhesus monkeys, *Exp. Brain Res.* **77**, 245-256

[13] Lynch, J.C.; Graybriel, A.M. and Lobeck, L.J. (1985) The Differential Projection of Two Cytoarchitectonic Subregions of the Inferior Parietal Lobule of Macaque Upon the Deep Layers of the Superior Colliculus, *J. Comp. Neurol.* **235**, 241-254

[14] May, P.J. and Porter, J.D. (1992) The laminar distribution of macaque tectobulbar and tectospinal neurons, *Visual Neuroscience* **8**, 257-276

[15] Meredith, M.A. and Stein, B.E. (1986) Visual, Auditory, and Somatosensory Convergence on Cells in Superior Colliculus Results in Multisensory Integration, *J. Neurophysiol.* **56**, 640-662

[16] Schall, J.D. and Hanes D.P. (1993) Neural basis of saccade target selection in frontal eye field during visual search, *Nature* **366**, 467-469

[17] Schiller, P.H.; True, S.D. and Conway, J.L. (1979) Effects of Frontal Eye Field and Superior Colliculus Ablations on Eye Movements, *Science* **206**, 590-592

[18] Segraves, M.A. and Goldberg, M.E. (1987) Functional Properties of Corticotectal Neurons in the Monkey's Frontal Eye Field, *J.Neurophysiol.* **58,** 1387-1419

[19] Soechting, J.F. and Flanders, M. (1992) Moving in three-dimensional space: Frames of reference, vectors and coordinate systems, *Annu. Rev. Neurosci.* **15**, 167-191

[20] Weinrich, M. and Wise, S.P. (1982) The premotor cortex of the monkey, *J. Neurosci.* **2**, 1329-1345

[21] Werner, W. (1993) Neurons in the primate superior colliculus are active before and during arm movements to visual targets. *Eur. J. Neurosci.* **5**, 335-340

SENSORY AND MEMORY INFORMATION IN THE CONTROL OF DEXTEROUS MANIPULATION

R.S. JOHANSSON
Department of Physiology,
Umeå University, S-901 87 Umeå, Sweden.

1. Introduction

Successful manipulation requires that the subject selects the appropriate pattern of motor commands based on the manipulative intent, on various constraints imposed by the task and on the relevant physical properties of the manipulated object(s). For instance, most tasks require that we stabilize the object within our grasp as we move the object or use it as a tool. To prevent slips and accidental loss of the object we apply adequately large forces normal to the grip surfaces ('grip' forces) in relation to destabilizing forces tangential to the grip surfaces ('load' forces). At the same time, excessive grip forces must be avoided because they cause unnecessary fatigue and may crush fragile objects or injure the hand. Accordingly, the term *grasp stability* entails a prevention of accidental slips as well as of excessive fingertip forces. Various types of constraints are imposed by the object. For instance, its location in space and its size and shape may influence the selected grasp configuration (for review of grasp classification schemes see [1]), whereas its friction in relation to the skin, weight and mass distribution and possible external forces imposed on the object determines primarily the magnitudes of the forces generated by the muscles.

Based on the properties of the object, manipulative tasks may be divided into two principal classes: One concerns the handling of mechanically predictable, 'passive', objects whose relevant physical properties are stable over time (i.e., objects subjected only to inertial and gravitational forces). Because the motor output is self-paced and the objects are predictable, the spatio-temporal control of the muscle commands can rely largely on anticipatory mechanisms based on memory information from previous manipulation. The second class concerns the handling of objects that are

F. Lacquaniti and P. Viviani (eds.), Neural Bases of Motor Behaviour, 205–260.

subjected to unpredictable load forces, such as when one holds a dog's leash, restrains a kid by holding her arm or when operating power tools. The role of anticipatory control strategies are obviously of limited use in such tasks; somatosensory information related to the load changes must be used to control the forces of the grasp not to loose the object.

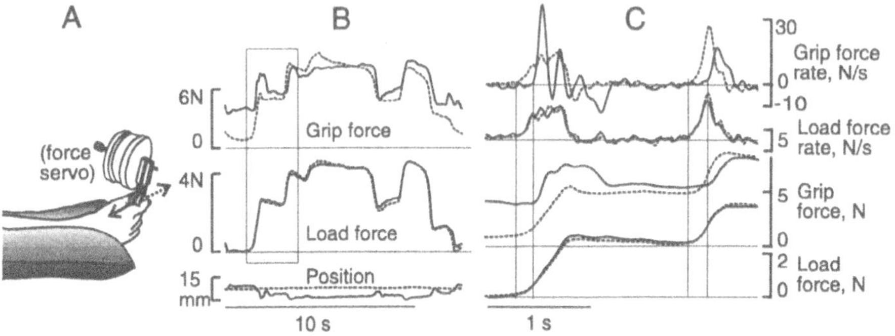

Figure 1. Comparison of the grip forces employed during load changes applied by the subject to a 'passive' manipulandum with identical loads profiles imposed on the grasp through the same manipulandum that was loaded by a servo controlled force motor ('active' object). **A.** Schematic drawing of the apparatus. The manipulandum consists of two parallel grip surfaces (30 mm diameter; spaced 25 mm apart). The servo controlled motor could deliver load forces in two opposite directions and the arrows indicate the directions of loading (D—distal direction, P—proximal direction). The grip and load forces, perpendicular and tangential to the grip surfaces, respectively, were recorded, and the position of the object in the loading directions. **B** — **C** Load and grip forces during manipulation of a 'passive' object (dashed curves); the subject only pulls manipulandum that is immovable to produce various periods of force changes mixed with hold periods. The resulting load pattern is recorded and later played back through the force servo while the subject is asked to restrain the 'active' manipulandum from moving (solid curves). In both tasks the blindfolded subject grasped the object by the tips of the thumb and index finger. The rectangle in **B** marks the period represented at a faster time base in **C** together with the grip and load force rate profiles. Pairs of vertical lines in **C** indicate onset of grip force responses during the passive and active conditions. Single subject. Adapted from [2].

Figure 1B—C compares the grip forces employed by a subject while generating load forces on a passive manipulandum (dashed curves) with those employed while identical load force profiles were imposed on the grasp through the same manipulandum (active object; solid curves) [2]. In the latter case the subject's task was to restrain the manipulandum and not to lose it. Note that in both tasks the load and grip forces increase and decrease in parallel, ensuring that adequately strong grip forces are used to prevent frictional slippage at any load force. This parallel coordination, supporting grasp stability, emerges automatically and never requires any instructions. However, there are differences in the behavior: (i) The change in the grip and load forces are nearly synchronous when the manipulandum is 'passive' and the grip force changes are programmed in parallel with rather than as a reaction to the self-induced load changes. In contrast, with the 'active' manipulandum the grip force changes lag those of the load force because they are reactively generated (Fig. 1 C). Subjects automatically also produce reaction forces that prevented excessive displacement of the manipulandum in the direction of the loading. (ii) Higher grip forces are generally used with the "active" manipulandum than with the "passive", in particular at low load forces. This strategy prevents frictional slips during the delay between the onset of the load force increase and the onset of the grip force response. (iii) Finally, with the 'passive' manipulandum the force rate profiles are generally unimodal indicating that the target forces are determined in advance (see 'anticipatory parameter control' below). In contrast, with the active object the grip force responses to extended periods of load force increases consist of an initial strong grip force rate pulse followed by a period of slower, often stepwise, grip force increases (Fig. 1 C).

The first part of this presentation concerns the sensorimotor control of manipulation while handling passive objects, whereas its second part deals with handling active objects subjected to unpredictable load forces. These sections mainly review control principles as revealed by analyses of human behavior and signals in peripheral afferents. In particular, I will consider mechanisms for the control of grasp stability during actual object manipulation, i.e., while the digits mechanically contact the object(s). Finally, I will comment on central neural mechanisms that may be involved in the sensorimotor control of manipulation.

2. Handling of Passive Objects

2.1. BASIC SEQUENTIAL AND PARALLEL COORDINATION

Sequential coordination of series of phases. Most manipulative tasks evolve in a series of phases, each of which is characterized by a particular goal (e.g., to establish contact or to lift the object off the supporting table), unique patterns of muscle activity, and typically a transient mechanical event that marks the goal completion and the transition to the next phase. This certainly applies to lifting tasks in which subjects grasp an object between a finger and thumb and impose various action forces on it, e.g., to lift it from a support table, hold and move it in the air and then replace it on the table (Fig. 2) [3-5].

When we reach out to grasp and lift a small object, the first goal is to adequately position the tips of the digits onto its surfaces so that a stable grasp can be established. The transport and shaping of the hand reflect a precise coordination to achieve the necessary spatio-temporal features for arresting the reach, and for an appropriate closure of the fingers around the object [6-8]. Visuomotor mechanisms are used to preshape the hand to object shape and position the digits on its surfaces so as to promote grasp stability and the achievement of further action goals [6, 7, 9-13]. Importantly, the kinematics of these movements are largely determined by the initial view of the object before the movement onset, i.e., visual feedback signals seem of little importance during the movement itself [10, 14] (also see [6, 15-17]). This points to the importance of implicit memory control of relevant motor program parameters in manipulation (see below).

Once the digits contact the object the goal is to generate the necessary finger tip forces to lift it (Fig. 2B). The initial contact marks the beginning of the first phase, the *preload phase*. After contact has been established by a small increase in the grip force (normal to the grip surface), the subsequent *load phase* is characterized by the grip force increasing in parallel with the load force (i.e., the lift force tangential to the contact area). When the load force has overcome the weight of the object it starts to move (*transitional phase*) into the desired vertical position (*static(-hold-)phase*). Similarly, at the end of a lifting task, a parallel decrease in grip and load force begins shortly after the object makes contact with the table (*unload phase*).

Coordinative constraints supporting grasp stability—parallel coordination. Grasp stability is primarily obtained by a parallel change in the grip and load forces applied to each contact surface (Figs. 1, 2B—C) [3, 19]. A parallel change in the grip and load forces represents a general control strategy during prehensile tasks that require grasp stability and is not specific to any particular task (e.g., mode of object transport) or grasp configuration. This coordinative constraint ensures adequate grip forces during the considerable changes in load forces that may occur in many manipulative tasks. For instance, this way of linking the forces offers considerable flexibility when lifting objects of different weights (e.g., Fig. 9A) [5]. With a heavy object the load force reaches high values before the weight is counterbalanced and the object is lifted, while the proportional increase in grip force ensures appropriate grip forces. Accordingly, with a lightweight object the grip force will be low. Likewise, the grip forces are modulated with the fluctuations in inertial loads that arise from moving a grasped object in space as the object is accelerated and decelerated by the arm [20-22], from the variations in object acceleration when objects are lifted at various speeds [23], while operating against spring loads [3] and while applying pushing or pulling forces on immovable objects (cf. Fig. 1; passive object). The grip forces also modulate approximately in phase with changes in load force that are induced by whole-body jumping even though the arm's joint angles are fixed [24] and during walking and running [22]. This basic coordination of grip and load forces shows effector invariance in the sense that it applies for a variety of grips, including one- and two-handed grips, 'inverted' grips [24, 25] and multidigit grips [26]. Moreover, this coordinative constraint is expressed in anticipatory adjustments of the grip force in more complex bimanual actions such as when weights are transferred between the hands [27, 28].

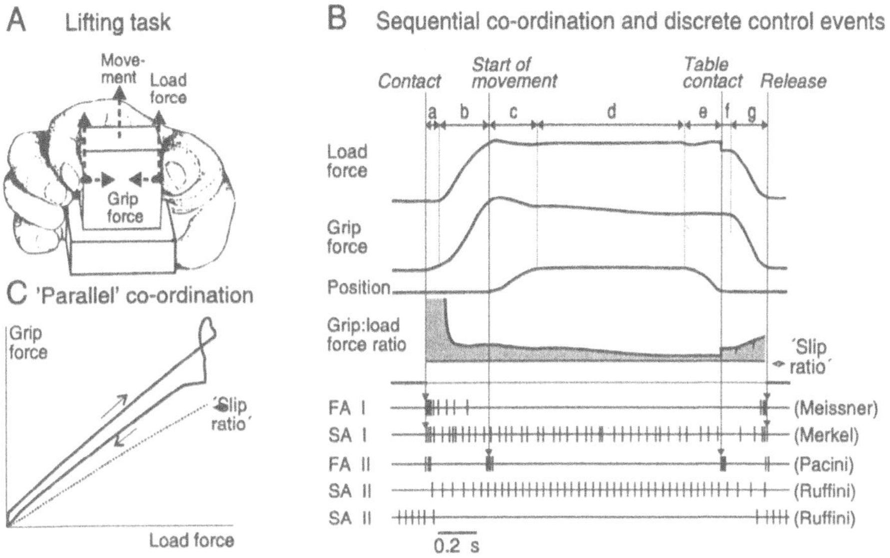

Figure 2. Schematic illustration of the lifting task. A Subject lifts an instrumented test object from a table, holds it in the air and then replaces it. The tangential (load) force applied to the object to overcome gravity and inertia and the normal (grip) force are measured as well as the vertical movement. B Shown schematically as a function of time are the grip and load forces, vertical movement, grip:load force ratio and obligatory afferent responses present in the four types of tactile afferents from the human glabrous skin. Minimum grip:load force ratio to prevent slips is indicated by slip ratio. The safety margin to prevent slips is indicated by hatching, i.e., the difference between the employed ratio and the slip ratio. After contact with the object the grip force increases by a short period (a—preload phase) before the command is released for a parallel increase in grip and load force during isometric conditions (b—load phase). This parallel increase continues until the start of movement when the load force overcomes the force of gravity. The object is lifted to the intended position (c) by wrist and/or elbow flexion, and a static (-hold-) phase is reached (d). After the replacement of the object (e) and table contact occurs there is a short delay (f) before the two forces decline in parallel (g—unload phase) until the object is released. Apart from the transient event-related responses in tactile afferents (se text), there also are responses characterized by an ongoing impulse activity in the SA II afferents and many of the SA I afferents. Some spontaneously active SA II's are unloaded during the lift and cease firing. C Grip force as a function of the load for a similar lift as in B. Dashed line indicate the minimum grip:load force ratio to prevent slips. Adapted from [18].

Apparently, this coupling of grip and load forces in manipulation forms the basis for an efficient sensory control of the motor output to accommodate the physical properties of the object, such as its weight and the friction in the hand-object interface (see further below). Indeed, it is commonly believed that various task related coordinative constraints (or synergies) represent strategies that simplifie the demands on the control mechanisms by reducing the number of degrees of freedom of the musculoskeletal apparatus that have to be explicitly controlled [29-31].

2.2. PARAMETRIC ADJUSTMENTS OF FINGER-TIP FORCES TO OBJECTS' INTRINSIC PROPERTIES

2.2.1. *Grip:load force ratio and friction in the digit-object interface*

The approximately proportional relationship between the changes in grip and load force is functional only if they are appropriately scaled to the friction between the skin and the object: at a given load force (tangential force) there is a minimum grip force (normal force) required to prevent slip. Thus, to prevent slips, the employed grip:load force ratio must exceed a certain minimum determined by the inverse of the coefficient of friction, i.e., the *slip ratio* (Figs. 2B and C) [3]. In everyday situations there may be substantial changes in friction related to the various surface materials of objects (Fig. 3A), objects may be soaked and the sweating rate may vary, etc. Indeed, people automatically adjust the employed grip forces and thereby the grip:load force ratio to these changes in friction, i.e., the more slippery the object the higher the employed force ratio (Figs. 3B—C) [3, 4, 19, 27, 32, 33] (also see [34-36]). Figure 4A show frictional adjustments in an everyday task [28]. An empty 'transducer-equipped' glass was held by one hand and the subject was asked to take a jug of water with the free hand and fill up the glass. The grip surfaces were covered by each of the three surface materials showing different frictional properties in relation to the skin: silk, suede and sandpaper. As can be seen, the grip force increases in parallel with the increasing load force while the glass is filled ('parallel' coordination), and the force ratio is maintained at a level adequate for the actual frictional demands. One expression of the safety margin against slip is the difference between the force ratio employed by the subject and the slip ratio (Figs. 2B, 3B and 4A). The safety margin employed while objects are held stationary usually constitutes some 10—40 % of the grip force depending on subject and context.

It seems clear that this adaptation is made to the friction per se, rather than on the basis of different texture properties of the touched materials [33, 37] . Thus, for example, Johansson and Westling [37] showed that subjects compensate for the drop in friction caused by washing the hands even though there is no change in surface texture of the grasped object (Fig. 4B).

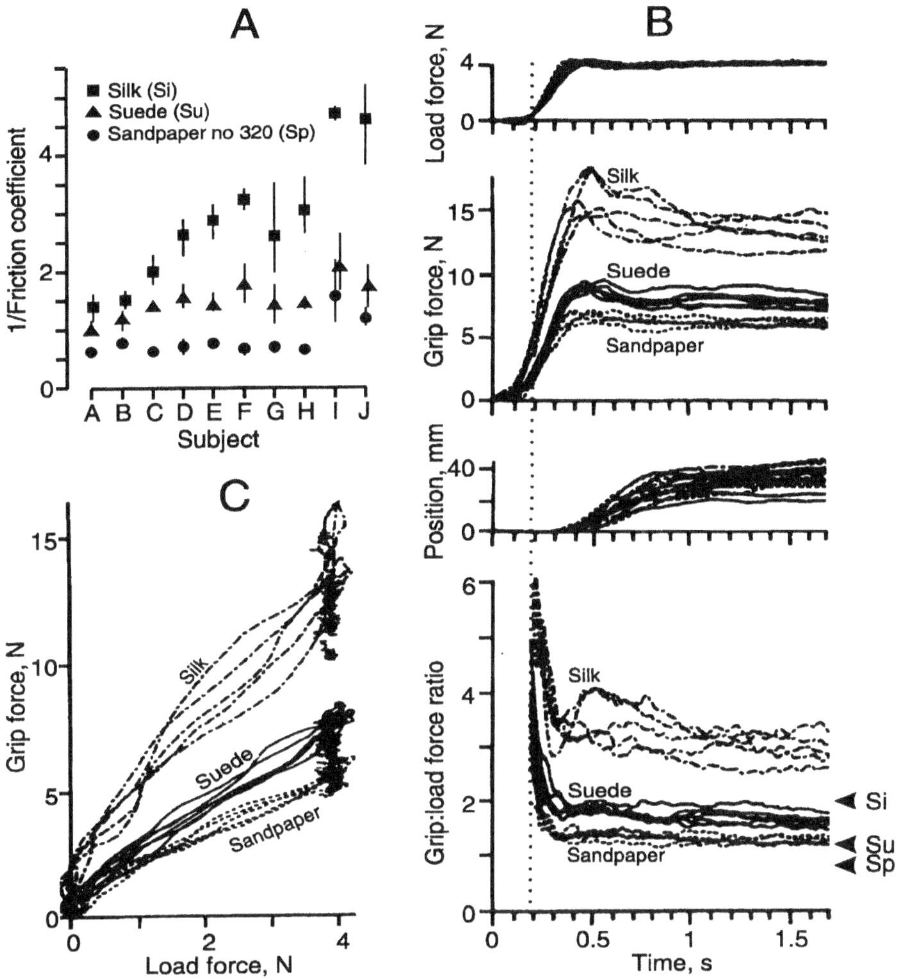

Figure 3. Influence on force coordination during the initial part of lifting trials of different surface structures: finely textured silk—most slippery, suede—less slippery, and fine grain sandpaper—least slippery. **A** Friction between the fingertip skin for the three surfaces in ten subjects (A—J) plotted as the inverse coefficient of friction. Symbols represent median values of 5 trials and bars the corresponding range. Note, these materials exhibited different friction but there were large interindividual variations. Some of this variation may be explained by differences in the sweating rate between individuals. Indeed, the two extreme subjects showing the lowest friction were postmenopausal women (subject I and J). **B** Load force, grip force, vertical position and ratio between grip and load force as a function of time; fifteen trials by single subject superimposed. Arrowheads indicate mean slip ratios for the three structures. A comparison between the force ratios employed for the three materials and the corresponding slip ratios clearly indicate that the force coordination is adapted to the frictional condition. **C** Coordination between grip force and load force during the same trials as in A illustrated by displaying the grip force against the load force. **B—C** Illustrations represent standardized conditions in which the previous trials were carried out with the same friction. Weight constant at 400 g. **A** modified from [37], and **B—C** from [3].

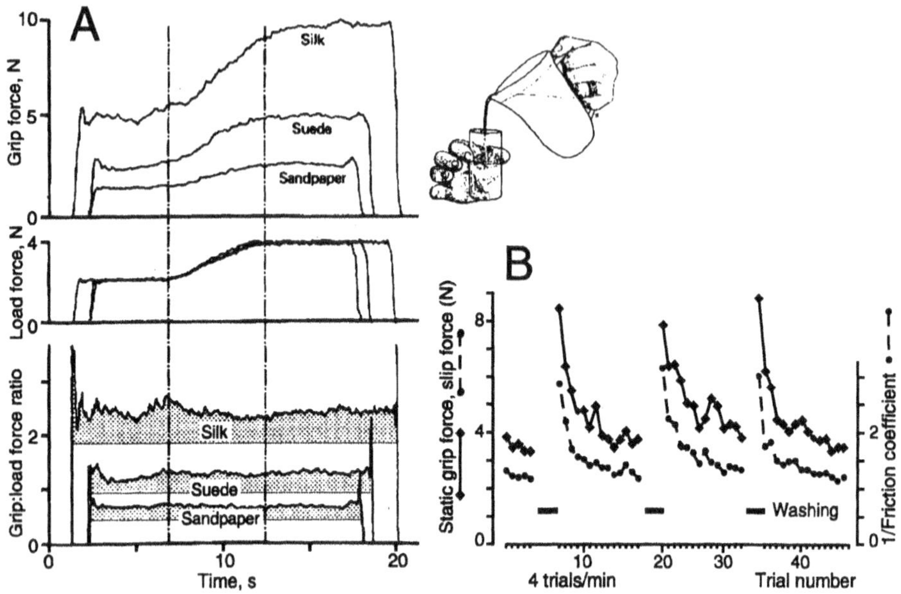

Figure 4. Adjustment of the grip force to the frictional changes in everyday situations. **A** Coordination between grip and load force during a bimanual "every day" tasks. An 'instrumented' empty glass was lifted and held stationary in space (weight: 200 g). The changes of the load force taking place during the period between the two vertical lines was accomplished by the subject pouring the glass full with water (ca 0.2 l) using a jug held by the contralateral hand. Three superimposed trials with three different surface materials in contact with the skin. Load force, grip force and ratio between grip and load force shown as a function of time. Time scale with an arbitrary origin. Dashed areas illustrate safety margins. **B** Static grip force (diamonds) and the corresponding minimal grip force to prevent slips (slip force - dots) before and after the subject washed and dried her hands using soap and water and a towel. Forty-eight consecutive lifts with suede as surface structure (single subject, weight constant at 400 g). The right ordinate only refers to the data represented by dots and provide the inverse coefficient of friction estimated from the slip forces. **A** adapted from [28] and **B** from [37].

2.2.2. *Parametric adjustment to objects' weight*

People parametrically adapt their force output to the weight of a lifted object. This is most clearly observed during the isometric load phase prior to lift-off [5, 38]. To get a predictable, smooth and critically damped vertical lifting movement the lifting drive must be decreased and appropriately adjusted to match the weight of the object before the moment of lift-off. In fact, with an adequately programmed lift, the first time derivative of the grip and load forces have their maximum when the load force matches about half the weight of the object; the force rates are reduced prior to lift-off to harmonize with the expected weight (see Figs. 5A and trials with adequate weight anticipation in

Fig. 9). In addition, the desired speed (acceleration) and height of the lifting movement play a significant role for parameterizing the force output during the isometric load phase [23].

2.3. ANTICIPATORY PARAMETER CONTROL

There are at least two important determinants of the parametric changes described in the previous section. One, considered in this section, is anticipatory control based on memory information acquired during previous manipulation. The other, considered in the following section, is the role of afferent information about various events in the hand-object interface during execution of the task.

It has been repeatedly shown that the initial forces applied to an object reflect the requirements of the previous lift. The frictional conditions in the preceding lift with an object are reflected in the development of the grip and load forces immediately after the object is touched (see grip forces in Figs. 3B and C) [3, 19, 35]. Likewise, memory representations of the object's weight acquired during previous lifts are used to adapt of force development to an object's weight during the load phase because explicit information about object weight is not available until lift-off [5, 38, 39]. Thus, one must hypothesize that there exist sensorimotor memories that represent both important physical properties of the objects to be manipulated and the appropriate magnitude parameters of the motor commands. Furthermore, the fact that information about both frictional conditions and weight is transferred from one hand to the other hand in subsequent manipulation of the same object further supports the idea that relevant physical properties of objects as such are retained [3, 27, 40] (also see [36]).

2.3.1. *Use of vision in recall of relevant sensorimotor memories*

A variety of sensory-based identification processes, including the retrieval of relevant object properties based on visual object identification are involved in the mechanisms for anticipatory parameter control. During everyday activities we commonly handle objects of different shapes, weights and densities, and often the object is lifted only once. In the very first lift of such objects, and before sensory information related to the object's weight is available, the rate profiles of the grip and load forces during the load phase are appropriately targeted for the weight of the current object (Fig. 5) [38].

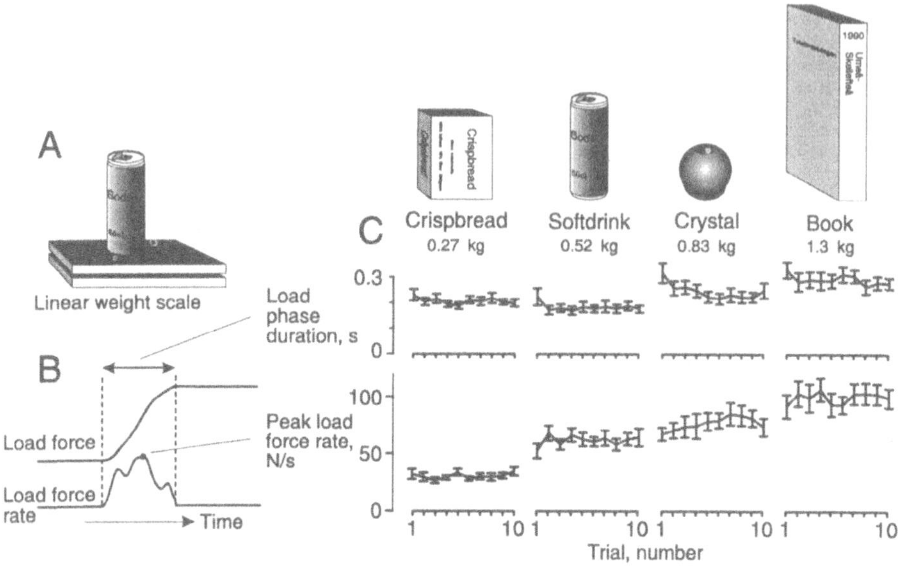

Figure 5. Anticipatory parameter control for the weight of commonly handled objects of different shapes, weights and densities. **A** The isometric lifting force (load force) was recorded while subjects lifted from an analog weight scale a variety of objects, i.e., crispbread, soft drink, crystal, and telephone book. **B**. Duration of load phase and load force rate were measured from the unloading of the weight scale while subjects lifted the test object. **C** Load phase duration and peak load force rate for each of then consecutive lifts of each common object (means ± 1 SEM; data averaged across subjects). During the very first lift, and before sensory information related to the object's weight was available, the load force output was scaled differently for the various objects with force rate profiles targeted for the weight of the current object. Also note the similar performance in the 1st and last lift. This indicates that the force output is successfully specified from information in memory related to the weight of the common objects, retrieved by visual identification of the target object prior to the first trail; the subjects had no practice prior to the lift series. Adapted from [38].

This indicates that physical properties of common objects are indeed represented in memories that are used for anticipatory parameter control of the force output. The ability of humans to learn to identify novel unfamiliar objects for anticipatory parameter control of the force output for objects' weight has also been investigated [38]. With unfamiliar objects, it appears that subjects initially obtain weight estimates by size-weight associations using a default density estimate that is in the range of common densities. Indeed, size-weight associations is efficiently used for classes of related objects [39, 41, 42]. This obviously makes anticipatory control of similar objects of different size possible without vast amounts of previously stored data. With novel objects of a density that is unusually high (4.0 kg/l) two to four trials are required to form

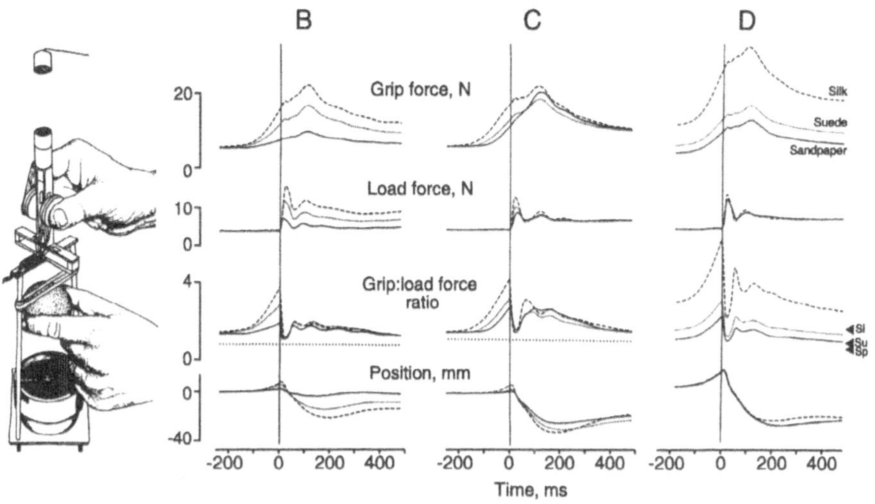

Figure 6. Predictive feed-forward control during manipulation in a bimanual task. **A** A ball is dropped by the subject from one hand into a receptacle (400 g weight; instrumented for grip force, load force and position measurements) held in a precision grip by the other hand. **B** The weight of the ball was randomly varied between trials: 100 g (——), 300 g (·········) and 500 g (-----). **C** The distance the ball fell was randomly varied between trials: 2 cm (——), 4 cm (·········) and 8 cm (-----) drop. Weight of ball 300 g. **D** The surface material was randomly varied between trials: sandpaper (——), suede (·····) and silk (-----). Arrowheads indicate slip ratios for these grip surfaces, respectively. Weight of ball 300 g. **B—D** Data averaged from totally 22 trials, respectively. Vertical lines indicate the moment the ball hits the receptacle of the grip apparatus. **B—C** Horizontal dotted lines indicate estimated slip ratio. Note that the preparatory force increase is precisely scaled such that the minimum force ratio at the impact is always the same in **B—C**, and adequately adapted to the frictional condition (**B—D**). Adapted from [27].

an adequate memory representation of the weight. When lifting the same object 24 hours later, the force output is adapted to the object's weight in the first lift.

2.3.2. *Anticipatory control in bimanual manipulative tasks*

Anticipatory control during manipulation is also important in coordinating the hands during bimanual tasks. Consider a task in which a ball is dropped by the subject from one hand into a receptacle held in a precision grip by the other hand (Fig. 6A) [27]. Whether visual control is allowed or not, there is a task dependent graded preparatory grip force increase that automatically appears some 150 ms prior to the impact (e.g., Fig. 6B). In addition, the target object is simultaneously lifted to meet the impact of the ball. These preparatory actions are precisely scaled to the peak in load force at impact: the former to prevent dropping the target apparatus due to slips, the latter to avoid excessive position deviations. As such, these preparatory actions are scaled to the magnitude of the perturbation which is influenced by the weight of the dropped object (Fig. 6B), the weight of the target apparatus (not illustrated) and the length of the drop (Fig. 6C). In addition, the preparatory grip force responses are also properly adapted to the frictional condition between the skin and the grip surfaces of the receptacle in ways that conform to the principles described above (Fig. 6D). The somatosensory information used to adjust the preparatory action is obtained while the ball and target object are handled before the drop by the separate hands, and combined in a predictive feed-forward manner with other task related information such as height of drop.

Electromyographic (EMG) recordings during the ball-drop task reveal that the entire arm-hand system is appropriately stiffened by cocontractions in agnonistic and antagonistic hand and arm muscles during the preparatory grip force increases [27]. Likewise, the impact elicits some reflex coactivation appearing 35—40 ms and 55—65 ms after impact in the proximal arm and distal hand muscles, respectively [28]. However, if the ball is dropped unexpectedly (e.g. by the experimenter), in the absence of the preparatory actions there are much stronger triggered responses. This is true even though they appear too late to be useful to prevent slips. Thus, the motor program executing the preparatory actions mediates a pronounced depression of this reflex activation. In the prepared case, though, there is no obvious need for these responses. These findings are reminiscent of patterns of preparatory and triggered actions recently observed when humans catch a freely falling ball (e.g., [43, 44]). Evidently, the central nervous system (CNS) dynamically regulates both the preparatory muscle activity and spinal and supraspinal reflex pathways in a predictive manner. Hence, this control represents an anticipatory adaptation of limb mechanics according to task demands and implies that the CNS entertains global models of relevant object and task properties during manipulation (e.g., [45-47]), including the associated postural control [48-50].

2.4. TACTILE AFFERENT CONTROL DURING TASK EXECUTION AND UPDATING OF RELEVANT MEMORIES

To grasp and further manipulate small objects, visual information is essential for accurate reaching and preshaping of the grasp and for anticipatory control of initial force output as described above. Yet once the object is touched, the most relevant afferent information for the final adaptation of the motor commands to the intrinsic physical properties of the object is provided by the mechanoreceptors located close to the interface between the hand and the object, i.e., tactile receptors in the glabrous (non-hairy) skin of the digits. The general importance of cutaneous afferent input in manipulation is well illustrated in people with impaired finger sensibility. They show clumsiness during precise manipulation and difficulties with gripping and holding. Objects are frequently dropped or fragile objects may be crushed. Even with visual compensation, their manipulative functions are hampered severely. This also applies when the efferent and afferent muscle innervation is intact. The fundamental dependence on cutaneous input was identified early by Mott & Sherrington, 1895 [51], who also remarked that, with spared distal cutaneous innervation, there was little functional impairment in the forelimb even if the rest of the arm was deafferented. Indeed, the general importance of sensory input from the digits in the control of finger movements has been extensively documented over decades (e.g., [52-62]).

One reason why cutaneous sensibility plays such a decisive role during precise manipulation is probably because cutaneous surface deformations directly and discretely reflect the functional end products of many manipulative actions (e.g., making and breaking contact with objects, successful grasp, strength and directions of finger tip forces, the distribution of contacts between the hands and objects). Information about these deformations may be preconditions for the release of motor commands accounting for further manipulative actions. Likewise, tactile afferent signals efficiently can provide information about a mismatch between the expected and the true mechanical properties of an object. Information from such mismatches is used to trigger pre-programmed patterns of corrective responses and to update the relevant sensorimotor memories used in anticipatory parameter control. Moreover, these skin deformations occur only with equivocal correlates in muscle length, joint angle, or muscle force (cf. [63, 64]). Hence, information about successful completion of manipulative acts is most likely provided by the activation patterns in cutaneous mechanoreceptors in the glabrous skin rather than by proprioceptive signals. Our findings regarding the actual use of tactile afferent information (see below) suggest that the motor program generates a set of predicted afferent signals which are compared to the actual afferent signals (cf., [65-67]). When a particular situation is sensed an appropriate motor response appears to be generated in a conditional manner according to a rule based scheme (see further the discussion of Finite State Control in [68]).

2.4.1. *Tactile afferent innervation of the glabrous skin of the human hand*

The tactile afferent units in the glabrous skin (non-hairy skin) on the inside of the hand signal information related to the mechanical status of the skin to the central nervous system through fast myelinated fibers conducting at 35 - 80 m/s [70]. As illustrated in Fig. 7, four different types of tactile afferents can be distinguished on the basis of the adaptation to sustained skin deformation and the structure of the cutaneous receptive field (for references see [70, 71]). All four types have strong similarities with well defined classes described in other species (e.g., [72-75]).

Two types show fast adaptation (FA) to maintained skin deformation, the FA I and FA II afferents, i.e. they only respond to skin deformation changes. The other two types adapt slowly (SA), the SA I and SA II afferents. In addition to being dynamically sensitive (in particular the SA I's) the SA afferents exhibit a response related to the strength of maintained skin deformation.

The FA I and SA I afferents have small and well-defined cutaneous receptive fields (typically 10 mm^2). Their field properties and their high

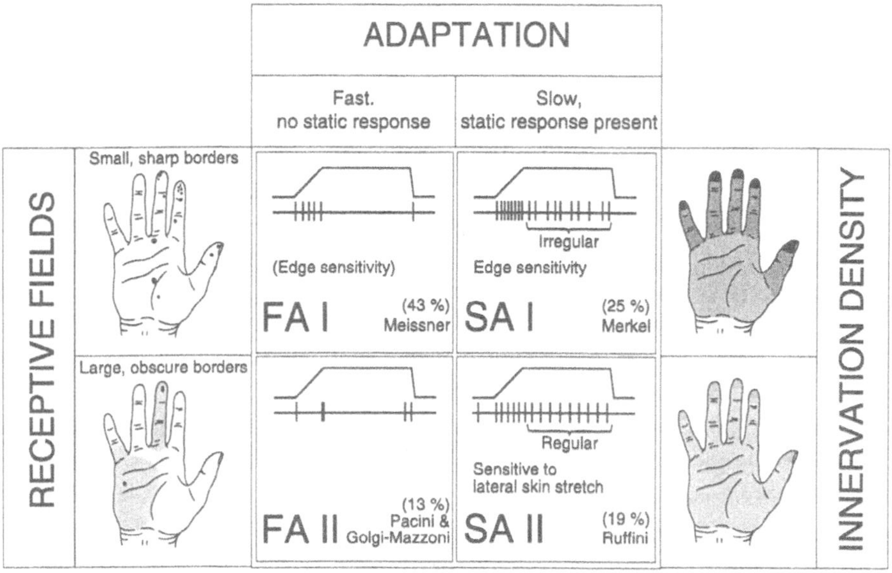

Figure 7. Types of mechanoreceptive afferents from the glabrous skin of the human hand, and some of their distinguishing properties. The graphs in the middle schematically show the impulse discharge (lower traces) to perpendicular ramp indentations of the skin (upper traces) for each afferent type. Two types (FA) show fast adaptation to maintained skin deformation, i.e. they only respond to skin deformation changes. The other two types adapt slowly (SA), i.e. in addition to being dynamically sensitive (particularly the SA I's) they exhibit a response related to the strength of maintained skin deformation. The type I afferents (FA I and SA I) have small and well-defined cutaneous receptive fields (typically 10 mm^2; see the black patches in the left drawing of the hand). This together with their high densities in the skin of the fingers, especially in the tips, makes them suitable to encode detailed spatial information. (The degree of shading in the right drawings of the hand schematically represent the densities of the tactiel afferents.) In contrast, the FA II and SA II afferents, whose densities are lower and more uniform are responsive to more remote stimuli, i.e. they have receptive fields that are larger and less well defined (cf. shaded areas in the left drawing of the hand). The FA II's are responsive to transient mechanical stimulation, whereas the SA I's are sensitive to lateral stretching of the skin. The relative frequency of occurrence in the glabrous skin and the probable morphological correlate are indicated for each type of afferent, respectively. Adapted from [69].

densities in the skin of the fingers, especially in the tips (140 and 70 afferents/cm^2, respectively), indicate that they can encode detailed spatial information. In contrast, the FA II and SA II afferents, whose densities are lower and more uniform, have receptive fields that are larger and less well defined, i.e. they are responsive to more remote stimuli. The FA II's are responsive to transient mechanical events (ca. 50 - 500 Hz 'vibrations') often originating at remote sites, whereas the SA II afferents show a considerable sensitivity to lateral stretching of the skin often with an appreciable directionality: their response may increase for stretch in one direction and decrease for stretch applied at right angles or opposite to this direction. Thus, the population of SA II afferents constitutes a set of sense organs which provides information about direction and magnitude of lateral tension in the skin, as well as between the skin and deeper structures [76].

As schematically indicated in Fig. 2B, during the static and replacement phases of the lifing task there was a tonic input from slowly adapting tactile afferents. In agreement with the directional sensitivity of the SA II afferents to skin stretch [76-78], they could be were strongly influenced by load (shear) forces with directional preferences [79, 80]; the load force was either excitatory or inhibitory. Figures 8 A & B and C & D illustrate the complex sensitivity to grip and load force for two SA II afferents. In Figs. 8 A and C the subjects were asked to perform stepwise increases and decreases in their grip force to the following targets: 0.5, 1, 2, 3, 4 and 5 N (visual tracking paradigm) while the load force was kept near zero. Note that, although its dynamic responsiveness was reduced at the higher force levels, the afferent in A showed a clear modulation of discharge rate by grip force. Conversely, the afferent in C showed a weaker representation of grip force and did not respond at all during the period of stepwise decreases in grip force. In Figs. 8 B and D, the subject was asked to maintain the grip force constanr at ca 5 N with visual feedback, while loads were applied in the distal and the proximal direction of the finger by a servo regulated force motor attached to the handle (cf. Fig. 1A). Even though the subject had some difficulty in voluntarily suppressing the automatic grip-force responses to the applied load forces (cf. Fig. 1), it is clear that the modulation of discharge rate in B and the evoked discharges in D were driven largely by the applied load forces. Moreover, for both these afferents there was a clear directional effect: responses were greater with distally-directed loads.

Tonic input from receptors in the fingers are considered to furnish general facilitation of motor commands accounting for finger movements, particularly for flexion movements [81-86]. However, the pronounced three dimensional force sensitivity of the SA I's suggests more specific functions. For instance, during manipulation the population of SA II afferents may play a role in registering the magnitudes and directions of load forces and the balance between the grip forces and other manipulative forces.

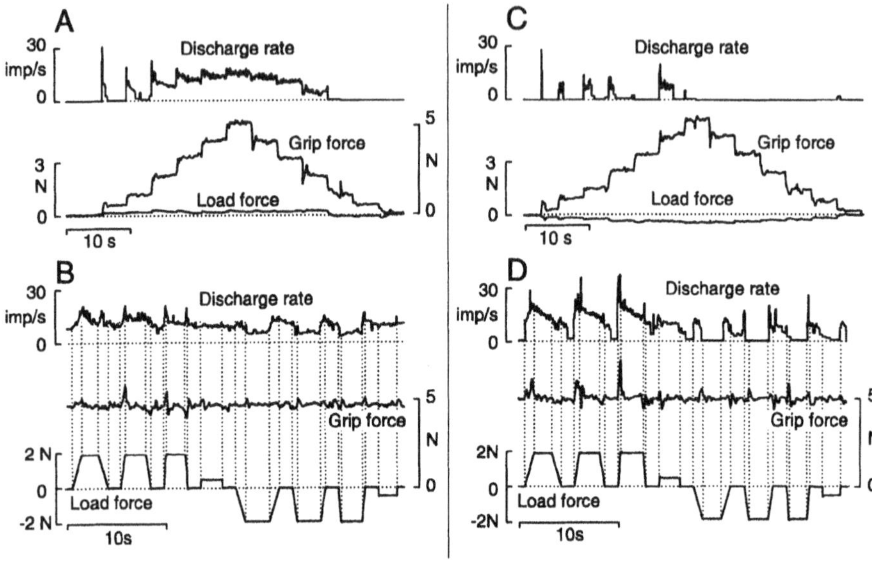

Figure 8. Sensitivity to grip and load force of two single SA II afferent represented in **A** and **B**, and in **C** and **D**, respectively. In **A** and **C** the subject increased and decreased the grip force in a stepwise fashion while the load force was kept at near zero. In **B** and **D** the subject was asked to maintain the grip force at ca 5 N while the 'rate-series' of trapezoidal load profiles (2 N amplitude) was delivered in the distal and the proximal loading direction; the subject had some difficulty suppressing automatic grip force responses to the load force changes. Note that for both afferents there was a clear directional effect; the discharge rates tended to increase during distal loading and decrease during proximal loading. However, for the afferent in **D**, low load forces in the proximal direction could also excite the afferent. **A—D**, the grip plate was centered over the receptive field. Adapted from [80].

Information about lateral tension in the glabrous skin may be significant to monitor not only shearing forces between the skin surface and hand held objects but also to sense joint position [87]. However, it is now clear that proprioceptive information obtained by cutaneous afferents is primarily furnished by receptors in the flexible non-glabrous (hairy) skin of dorsum of the hand [88-90]. These receptors appears much more sensitive to changes in the strain patterns of the cutaneous sheet than those of the glabrous skin. For instance, they signal positions of the MCP joint with a sensitivity at least on par with that of muscle spindles [89]. In contrast to the cutaneous afferents, muscle and joint afferents strongly modify their responses during active movements compared to passive movements [91-93]. Hence, cutaneous afferents from the dorsum of the hand may thus contribute to resolve ambiguous signals from other proprioceptors, such as for control of the closing movements of the hand in precision grasp to achieve a consistent contact behavior (cf. [94]).

2.4.2. *Sequential phase coordination*

One specific role of sensory signals from the hand during the lifting task is to link the various phases of the lifting task by informing the CNS that particular mechanical events have occurred, e.g. that the digits have made a stable contact with the object, or that the object has started to move (Johansson & Westling, 1991). Figure 2 B schematically shows typical responses in the four types of tactile afferents from the glabrous skin of the human digits during a lift. At the four points corresponding to phase transitions, there are distinct burst discharges in tactile afferents [79]: (i) initial responses appear during the preload phase when the object is first contacted, (ii) burst responses in the FA II afferents occur both at the start of the vertical movement when the object leaves the table and (iii) near the end when contact is again made with the table, and finally, (iv) responses appear at the end of the unload phase when the object is released.

Contact and release responses. Initial (contact) and release responses occur in slowly adapting (SA) type I and fast adapting (FA) type II afferents but most distinctly and reliably in the FA I afferents (Fig. 2B, also see Fig. 10A). The propagating front of skin deformation across the finger tip as it conforms to the flat surface of the object is probably the major determinant of the responses of type I afferents during contact. Due to the curved shape of the fingertip and its non-linear mechanical properties, an increase in grip force results in an non-linear increase in the area of contact between the fingers and the object. With a flat contact surface this increase is steepest at low grip forces, i.e., when an object is initially contacted. It has been estimated that at a grip force of < 1 N approximately 300 FA I and 150 SA I afferents are engaged at each digit [79].

The contact responses evidently provide information that a secure contact has been established. Such information must exist for the subsequent release of motor commands that drive further manipulation, i.e., a parallel increase in the grip and load forces during a lifting task (Fig. 2B). Quite predictably, when sensory information is lacking as a result of nerve damage or an anesthetic block of the digital nerves, the contact force becomes excessive before the parallel increases in load and grip forces commence [3]. Release responses, on the other hand, provide information about disengaged parts of the fingers which are free for further tasks. Or it may be used to forewarn that contact may be lost, which, in turn, may be used to elicit various compensatory responses.

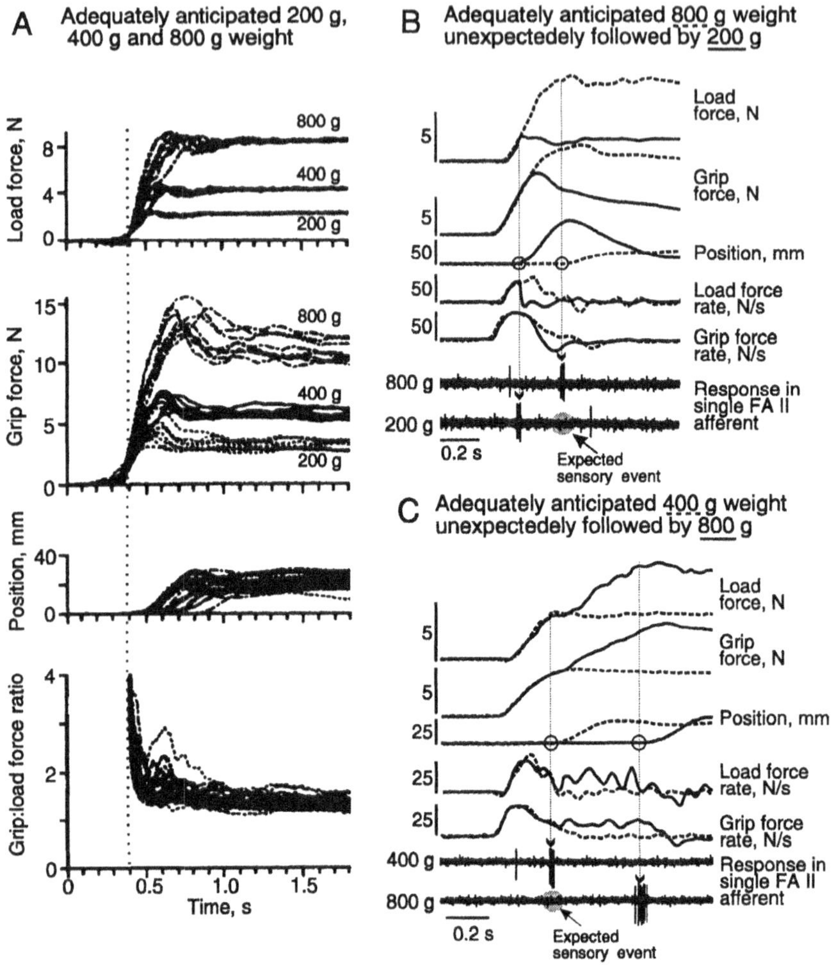

Figure 9. Adaptation of the force output to the weight of objects. **A** Force coordination during the initial part of adequately-programmed lifts with three different weights of the object (data from 24 single trials superimposed; single subject). **B** Initial parts of an adequately-programmed lift with a 800 g weight and a lift with 200 g that was erroneously programmed for the heavier 800 g weight which was lifted in the previous trial. The burst responses in the FA II (Pacinian) afferent at the start of movement were recorded using microneurography (Westling & Johansson, 1987). The sensory information about the start of movement at the unexpected point in time for the erroneously programmed 200 g trial is used to terminate the load phase. Note however overshoots in position and forces due to the 'reflex' delay. **C** Initial parts of an adequately-programmed lift with a 400 g weight and a lift with 800 g that was erroneously programmed for the lighter 400 g weight lifted in the previous trial. The absence of burst responses in the FA II afferents at the expected point in time for the erroneously programmed 800 g trial (see expected sensory event) is used to initiate a new control mode. This involves slow, more discontinuous probing force increases, until terminated by the sensory input at the actual take-off. Same FA II afferent in **B** and **C. A** adapted from [3] and **B—C** from [46].

Afferent responses to object lift-off and touch-down and somatosensory updating of weight related memories. The most striking feature of the FA II afferents (i.e., Pacinian corpuscle afferents) was distinct burst responses at the object's lift-off and at the sudden cessation of the movement at touch-down when an object is replaced on a support (Fig. 2B and Fig. 9B and C) [79]. Remotely located FA II afferents also respond readily to these transient mechanical events. Even FA II afferents with endings located at the transition between the palm and the wrist may respond. The other three types of tactile afferents in the glabrous skin are virtually indifferent to this type of mechanical events, and so are the musculotendinous receptors (cf. [93]; also see [95, 96]). The FA II responses at the start of the movement were probably caused by vibrations caused by the small lateral sliding movements between the object and the table occurring during the period while the lifting force almost counterbalance the force of gravity; we have estimated that more than 500 FA I's respond at lift-off [79].

The parallel increase in the grip and load forces during the load phase terminates shortly after the object starts to move, i.e., soon after the first of the two point at which the FA II's responses most reliably and distinctly. However, rather than being triggered by this afferent response, termination could have been driven by anticipatory mechanisms as discussed above. As such, the possible importance of these signals is not fully revealed unless there is an erroneous weight anticipation.

With an unexpected change to a lighter weight, the load and grip force rates are excessively high when the load force suddenly overcomes the force of gravity (Fig. 9B; solid curves). However, an abrupt triggered termination of the muscle commands driving the load phase takes place some 80—110 ms after lift-off (depending on the muscle) [5]. Burst responses in FA II afferents, which effectively indicate that the object has started to move, are most likely used to trigger this (Fig. 9B, afferent response during the 200 g lift). But the delays in the control loop (due to receptor and effector delays, axonal conductance and CNS processing delays) are still long enough to cause a pronounced position overshoot (a common experience when lifting an unexpectedly light object) and a high grip force peak.

In contrast, if the object is heavier than expected and the lift-off does not occur at the predicted load force (Fig. 9C; solid curves), the absence of motion is indicated through the lack of a transient sensory response at the expected moment of lift-off (Fig. 9C, afferent signal during the 800 g lift). In this case the CNS seems to use the absence of the expected sensory signal to quickly initiate a new control mode. This is characterized by slow, discontinuous increases in force that, in effect, probe for the lift-off (see force rate signals in Fig. 9C). The 'probing' continues until somatosensory information confirming movement is eventually obtained (Fig. 9C, burst response during the 800 g trial). These observations reinforce the notion that the employed motor program generates a set of predicted afferent signals which are compared to the actual afferent signals, and appropriate motor response are generated in a conditional manner.

Hence, whether the object's weight is correctly anticipated or not, somatosensory signals apparently trigger the termination of the load phase and presumably simultaneously update the memory system representation of the weight of the object. Indeed, with erroneous weight anticipation, only one lift is typically required to efficiently update the weight related memory system. Due to the remote sensitivity of the FA II afferents there are still afferents responding even if the digits are anaesthetized, and the sensory events at the transition between phases of lifting continue to take place as described. The termination of the load phase will only be disrupted during finger anesthesia if the lifting is too gentle [18] because the remote (un-anesthetized) afferents may fail to respond to the weak transients.

2.4.3. *Adjustments the grip:load force ratio for grasp stability*

Three independent sets of evidence indicate that inputs from tactile afferents with endings in the areas in contact with the manipulated object are decisive for the proper adaptation of the grip:load-force ratio to the friction between the object and the skin: (i) the adaptation fails if the afferent signals from the skin areas in contact with the object are abolished by local anesthesia [3, 4, 19]; (ii) force ratio changes can during certain experimental conditions be triggered by electric stimulation of tactile afferents [32]; and (iii) microneurography recordings have revealed that there are signals in tactile afferents to account for this adaptation, i.e. distinctive afferent responses were observed prior to adaptive changes in the force balance [32, 79].

The most important (primary) adjustments to a change in the frictional condition takes place shortly after the initial contact with the object and can be observed about 0.1 s after contact. The initial burst responses in subpopulations of excited FA I afferents are markedly influenced by the surface material [32]. The more slippery the material, the stronger the response (Fig. 10A). Hence, this afferent information most likely accounts for the early adjustments to a new frictional condition (cf. grip force traces in Fig. 10A).

If these initial adjustments are inadequate, secondary adjustments take place that always increase the grip:load force ratio. In less than 10 % of trials in series of lifting trials in which the surface friction is varied in an unpredictable manner, the initial adjustments are inadequate, or leave only a minute safety margin against slips so that the force ratio may approach the critical ratio at a later point. The resulting small slips (which typically occur at one digit only; see below) are rarely felt, but they promptly trigger an automatic increase in the

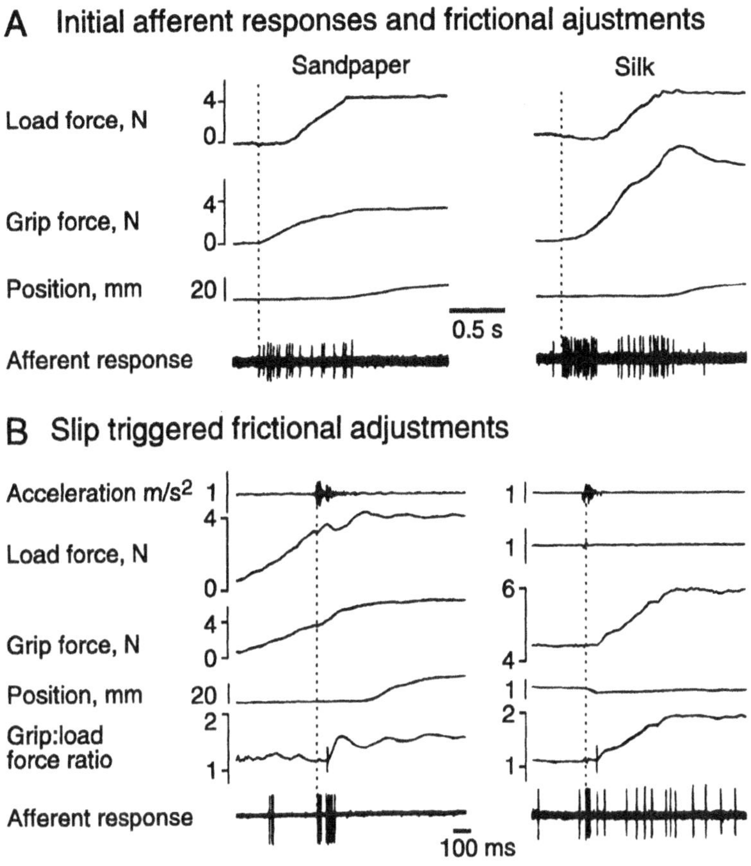

Figure 10. Single tactile afferent responses and adjustments to the frictional condition between the object and the digits. **A** Examples of initial afferent responses in a FA I afferent, and the influence of the surface structure (single trials). Note the early scaling of the rate of grip force increase to the frictional condition. Vertical lines indicate the initial touch. **B** Examples of afferent slip responses and upgradings of the grip:load force ratio elicited by small but distinct slips. Vertical dashed lines indicate the onset of the slips as revealed by vibrations in the object (acceleration signal; 10 - 600 Hz). Short vertical lines indicate the onset of the upgrading of the force ratio. Left panel: slip during the load phase, FA II afferent. The middle burst is the afferent slip response, whereas the third represents dynamic responses to the force changes following the slips and the first is the result of mechanical events not reflected in our mechanical recordings, respectively. Right panel: slip during the static phase, SA I afferent. Adapted from [32].

grip force to a higher maintained level (Fig. 10B). The slips are encoded as brief burst responses in FA I, FA II and SA I afferents which trigger an active upgrading in the grip:load force ratios appearing about 70 ms after the onset of the slip. The new higher and stable ratio restores the safety margin preventing further slips. We consider that the maintained upgrading of the ratio results from updating of a frictional memory, which, in turn, controls the force ratio [3]. Small slips localized to only a part of the skin area in contact with the object elicit similar effects, mediated by localized slip responses in FA I and SA I afferents [32].

Slip events during the load phase trigger changes in both the load (decrease) and grip (increase) force rates (Fig. 10B, left panel). During the static phase, however, just the grip force is influenced (Fig. 10B, right panel). This phase dependence is functional since gravity restrains the response alternatives preventing efficient load force adjustments during the static phase. A similar dependence on the phase of movement or postural situation has been described with other multiarticulate actions triggered by somatosensory input (for review see [97]).

With local finger anesthesia, the grip and load forces still change in parallel after a prolonged preload phase. However, the frictional adjustment is disrupted [3]. When tactile sensory information from the digits is blocked, subjects use strong grip forces. These are typically unnecessarily high during less slippery frictional conditions, yet they may be inadequate with a slippery material, and the object is dropped. Similar, but less pronounced impairments are observed in patients suffering from moderate degradation of tactile sensibility [18, 98] and also in elderly subjects [99, 100].

2.5. INDEPENDENT SENSORY CONTROL OF THE DIGITS FOR GRASP STABILITY

It has been shown that the grip:load force ratios employed at each of the digits engaged are controlled independently [19]. In these experiments the frictional condition was varied independently at each digit (thumb and index finger) engaged in lifting tasks. As illustrated in Fig. 11A—C, during the load phase (before the lift-off) the load force is already actively distributed between the digits in a way that reflects the frictional conditions at the separate digits. As a result, the safety margins against slips are similar at the two digits regardless of the surfaces they contact (shaded areas in Fig. 11A—C). The observed load distributions are established using tactile information from each digit combined with anticipatory parameter control that is specific for each digit (cf. the scheme outlined above). Hence, the initial forces at each digit reflect the frictional characteristics present at each digit for the previous lift (cf. Fig. 11A vs. B vs. C, which represent three consecutive trials; the trial preceding that in A was made with sandpaper at both contact surfaces).

Figure 11D illustrates initial adjustments to a new frictional condition (based on sensory information originating from each of the contact areas). Prior to this lift, the contact surface at the index finger was unexpectedly changed from sandpaper to silk, i.e., to a more slippery material than the sandpaper experienced in the previous trial. After a period of similar force development as in the previous trial less load force is applied to the more slippery silk surface and more to the sandpaper surface of the opposing digit. A somewhat higher grip force is also used compared to that in the previous trial since on average the object became more slippery (cf. grip forces in Fig. 11A and B). With similar frictional conditions at the two digits the force ratio increase upon changing to a more slippery material is realized only by increasing the grip force (cf. Fig. 3B).

Slips that may occur later in a lift (Fig. 11D), typically at one digit only, result in an abrupt grip:load force ratio increase on the slipping digit (as it suddenly becomes unloaded). Simultaneously, the increase in load force causes the ratio to fall suddenly at the other (non-slipping) digit and it may approach the slip ratio (Fig. 11D). The triggered upgrading in the grip force restores an adequate safety margin on the non-slipping digit to the levels that existed before the slip, and increases appropriately the safety margin on the slipping digit..

Further evidence that the digits are controlled by parallel, but independent (digit-specific) mechanisms for grasp stability, were provided in experiments where subjects lifted a test object between the fingers of the right and left hand, but also when two subjects shared the task, each subject contributing one finger [25]. The performance during precision lifting using digits belonging to one hand, two hands or two subjects was remarkably similar. Except for a slight protraction of the dynamic phases of the lift carried out by two subjects (lifting was synchronized by the experimenter by a verbal count-down), the same basic patterns of spatial and sequential coordination were observed. Importantly, the safety margins at each digit were adjusted adequately to reflect the local frictional conditions.

These findings reinforce our theory that the grip:load force ratio required for a stable grasp is controlled by a 'low-level' controller that is digit specific. Accordingly, important aspects of the distribution of forces among digits in manipulative tasks would be emergent property of the proposed digit specific controllers that communicate through sensory inputs concerning mechanical events at the separate digit-object interfaces. That is, grasp stability as such is not dependent on a mechanism that explicitly coordinate the engaged digits.

This notion is particularly attractive while considering that grasp stability applies to a variety of grips, e.g., one- and two-handed grips, inverted grips and multidigit grips [24-26]. However, this 'low-level' control is subordinated to a higher level of control related to the total load and grip forces required by the physical properties of the object (weight and average friction) [19], and to the overall manipulative intent, choice of grasp configurations etc. How these various levels of control are integrated remains to be investigated.

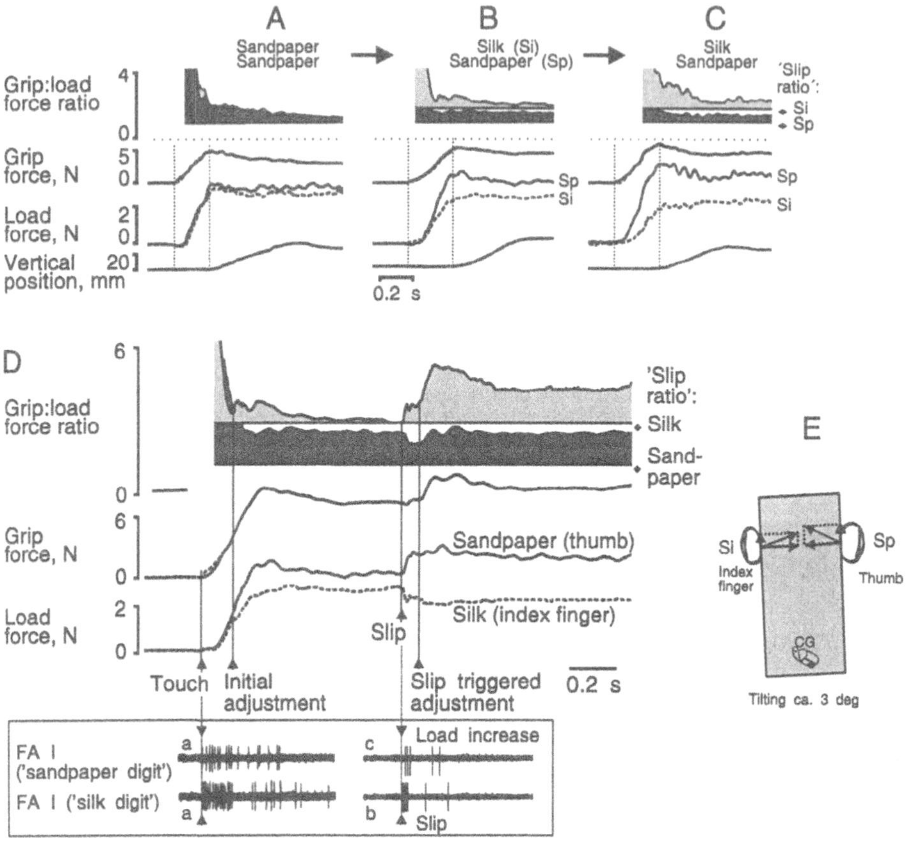

Undoubtedly, a key goal for analyzing the brain, with its many different processes active at the same time, is to determine how local interactions are integrated to yield purposeful overall results without explicit executive control.

2.6. COMMENT ON ONTOGENETIC DEVELOPMENT OF SENSORI-MOTOR CONTROL IN PRECISION LIFTING

The ontogenetic development of the control of the precision grip has been investigated, starting with 6-10 month old children (for a brief overview see [101]). Precision grip first emerges around 10 months of age but the mature pattern of grasping, lifting and holding objects is not observed before the age of about eight years. In many respects the behavior of small children are consistent with a poor use of somatosensory information. Although, the various control functions detailed in the previous sections emerge at different periods of the ontogenesis they all show a gradual development that extend over several years. This development seems to parallel that of central descending and

Figure 11. Digit specific adjustments of force coordination to the frictional condition. The subject lifted an instrumented test object (**E**) with low center of gravity (CG) from a table, and held the object still in the air before replacing it. **A—C** Initial parts of three consecutive lifts. The grip:load force ratio and corresponding slip ratios are shown for each digit; the safety margin to prevent slips is indicated by soft hatching for the index finger, heavy hatching for the thumb. In A the index finger and thumb both contacted sandpaper (Sp). The total load force was approximately uniformly distributed between the index finger and thumb. **B** The contact surface at the index finger had been unexpectedly exchanged to the more slippery silk. An adjustment to the new frictional condition occurs initially during the trial; the tangential (load) forces were distributed so that an adequate safety margin was maintained at each digit (also see **D**). **C** In the subsequent trial, the index finger which contacted the more slippery surface in **B** already picked up less load force than the thumb at the onset of the load force increase (i.e., reflecting anticipatory parameter control). These frictional adjustments resulted in an adequate safety margin at each digit despite the different friction. Due to the uneven load force distributions (in **B—D**) the test object tilts a bit while held aloft (**E**). **D** An initial adjustment to a new frictional condition followed by a secondary adjustment triggered by a slip event. The slip at the index finger, which contacted silk (Si), resulted in a rapid decrease of the tangential (load) force at that digit and a concomitant increase of the load force at the non-slipping digit. About 70 ms later the normal (grip) force increased such that the ratio at the digit that had slipped (Si) increased and the ratio at the non-slipping digit was effectively restored. Below is shown signals in examples of tactile sensors: initial responses in a FA I sensor (a) showing influence of the surface structure on the impulse rate; responses in a FA I sensor related to a distinct slip (b) associated with rapid decrease in the load force at that digit, and responses in an FA I sensor to an increase of the load force (c) corresponding to that occurring at the non-slipping digit in **A**. Modified from [19].

---

ascending and pathways in man (e.g., [102]), and takes place in conjunction with qualitative improvements of independent finger movements [103]. It is particularly noteworthy that in all the developmental studies regarding the precision grip, there were large intertrial variations reported in employed forces and in the temporal aspects of performance for tasks carried out by individual children. This large variability probably reflects an important principle for motor development [104]; it has been hypothesized that by monitoring and evaluating the effect of various spatio-temporal patterns, the CNS will eventually be able to select the best pattern [31]. For further information on the ontogenetic development of the precision grip the reader may refer to original articles [35, 40, 105-108]. Curiously, in many respects, impaired coordination of the motor commands underlying grasping in six- to eight-year old children with cerebral palsy is similar to the immature coordination seen in young children just beginning to use the precision grip [109-111].

3. Handling of Active Objects

The second prototypical prehensile task in dextrous manipulation that I wish to consider is the handling of 'active' objects, which impart more or less unpredictable forces to the hand (as when holding a dog's leash) (cf. Fig. 1B). Since the tangential forces in the grasp are not self generated and cannot be fully anticipated in such tasks the utility of anticipatory control mechanisms should play a less prominent role. That is, grasp stability must largely be maintained by a reactive control based somatosensory signals related to the time varying load forces. Indeed, it was demonstrated in the late 80-ties that abrupt load force increases imposed on objects held by the precision grip may elicit automatic grip force increases [27, 112].

3.1. GENERAL FEATURES OF THE EMPLOYED SENSORIMOTOR TRANSFORMATION

In a series of recent studies we have characterized basic sensorimotor mechanisms underlying the control of the precision grip in humans restraining an object subjected to unpredictable load forces [2, 34, 80, 93, 113-118]. We used an apparatus similar to that schematically illustrated in Fig. 1A. Trapezoidal load-force profiles with unpredictable onsets, amplitudes and rates of loading and unloading were delivered tangential to the grip surfaces.

Figure 12A shows basic features of the observed grasp response pattern. The sensory signals from the loading initially trigger a prompt grip force increase lasting for 0.2—0.3 s. This force increase, the 'catch-up' response, is initiated after a short delay ($\geq$70 ms) that varies with the load force rate (see Fig. 13A). The catch-up response (C in Fig. 12 A) compensates for the response delay in the sense that the grip force adequately matches the load demands (i.e. it ‘catches up’) and reestablish an adequate safety margin against frictional slips by the end of this response, i.e., the grip:load force ratio slightly exceeds the required minimum determined by the coefficient (‘slip ratio’ in Fig. 12A). With further changes in the load force the grip force increases in parallel with load force and maintains an approximately constant safety margin. This ‘tracking’ response (T in Fig. 12A) continues until the load cease to increase; accordingly, at the end of the load plateau phase the grip force is approximately proportional to the load amplitude (Figs. 12 A and 13 B).

Although the tracking response may rely on sensory data sampled more or less continuously, grip tracking of an insufficient rate is corrected by actions best described as additional catch-up responses to offset accumulated tracking error (Fig. 12B) [34]. These are probably driven by tactile afferent responses to localized slips in the object-digit interface since they occur when the grip:load force ratio falls too briskly [32]. Accordingly, the tracking and the catch-up responses may share a common control mechanism; there is also a critical drop in the grip:load force ratio during the grip response latent period (cf. Fig. 12A). Furthermore, that the grip tracking exploits short term prediction mechanisms similar to those of the catch-up responses is also suggested by the fact that it

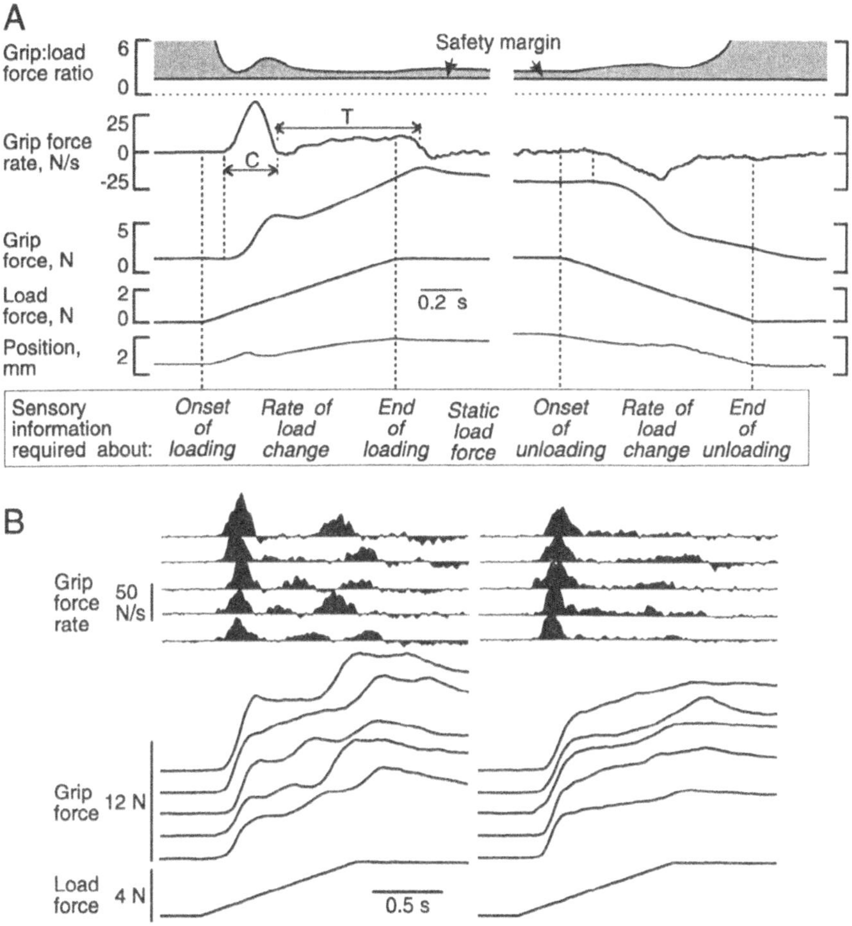

Figure 12. **A** Basic features of the employed 'load-to-grip' force sensorimotor transformation. The precision grip was loaded by distal pull to 4 N amplitude at 4 N/s. Top traces show the grip:load force ratio. Horizontal solid line indicates the minimum ratio to prevent slips ('slip ratio') and the hatched zone the safety margin. 'C' and 'T' indicate 'catch-up' and 'tracking' responses. Vertical lines indicate start and end of the loading and unloading phases, and the onsets of the reactive grip force changes. **B** Repeated trials from a single subject are shown stacked (zero values are not indicated). Left; initial and 'secondary' catch-up responses are evident. Right; tracking responses are present without secondary catch-up responses. **A** and **B** adapted from [2] and [34], respectively.

goes on for some 0.2 s after ramp termination, i.e., a period similar to the duration of the catch up response.

3.2. PARAMETRIC ADJUSTMENTS OF GRIP RESPONSES

3.2.1. *Adjustments to rate and amplitude of load force*

Grasp stability during this task requires adapting the force output to the rate and amplitude of the load force increase. These requirements are met automatically, rapidly and parametrically (Figs. 13A and 14B). Sensory information reflecting the initial load force rate during the latent period before the onset of the grip response specifies the rate of the catch-up response in a feed-forward predictive manner (Fig. 13A—B) [114]. Already from its onset, the amplitude of the catch-up response varies linearly with the load force rate, but the duration of the response stays constant (Fig. 13B). Hence, the catch-up response is a centrally programmed default response whose size is controlled by sensory information according to a 'pulse height control policy'. This type of control policy may simplify a rapid scaling of the force output and has also been described for other types of isometric motor tasks [119-121].

3.2.2. *Adjustments to friction in the objet-digit interface*

The friction between the skin and the contact surface modifies the 'gain' of the entire load-to-grip force sensorimotor transform and this provides an adequate safety margin to prevent frictional slips during different frictional conditions (Figs. 13C—D) (Cole and Johansson 1993). The background grip force and the grip responses to the load increase (catch-up and tracking responses) are all scaled approximately in proportion to the inverse friction ('slip ratio' in Fig. 13C) resulting in a small but adequate safety margin against frictional slip regardless of surface material (cf. Johansson and Westling 1984).

Hence, information about friction apparently adjusts a central scaling factor that is applied uniformly, or 'globally', to motor commands controlling the various grip force components. This 'gain' adjustment appears to be controlled in a feed-forward manner: As when lifting 'passive' objects subjects extract friction related information from signals in cutaneous sensors during the initial skin-object contact and use this information to set force coordination parameters for future actions [3, 32].

3.2.3. *Analogies with oculomotor control*

Interestingly, the overall grip response pattern is clearly reminiscent of the eye movements produced during smooth pursuit of a moving target when the head remains stationary [122]. The catch-up response would correspond to the 'catch-up' saccade, and the 'tracking response' to visual smooth pursuit. Likewise, in a manner similar to the way saccades punctuate visual smooth pursuit to offset accumulated gaze retinal errors, grip tracking of an insufficient rate is interrupted by additional catch-up responses (Fig. 12B). Short-term prediction may play a role not only in catch up responses but also in grip

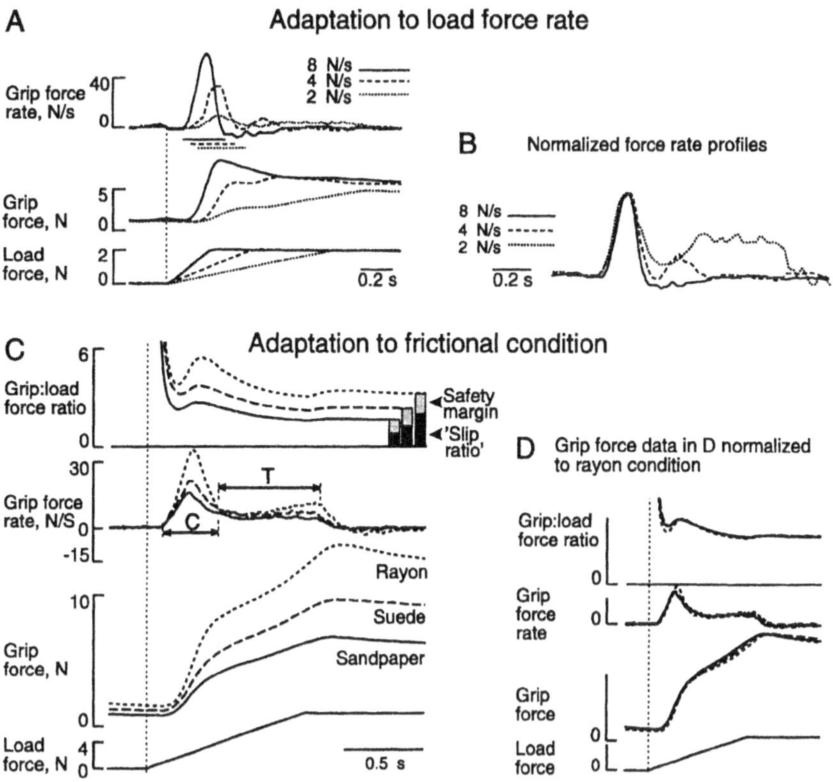

Figure 13. Sensory based adaptations of grip force responses. **A** Loading at three different force rates (2 N/s, 4 N/s and 8 N/s) to a constant force amplitude (2 N). Horizontal bars indicate the period of the catch-up responses. **B** Grip force rate profiles in **A** normalized in amplitude and horizontally aligned to peak rate. Note the similar catch-up response profiles and the brief and lengthy tracking responses at 4 N/s and 2 N/s loading, respectively. **C—D** Scaling by the frictional condition of the overall sensory-motor transform. Data from trials with sandpaper (least slippery), suede, and rayon (most slippery) surfaces are indicated by different lines. **C** The solid portions of the bars imposed upon the grip:load force ratio signals indicate the force ratio at which the finger slipped; the stippled portions represent the 'safety margin'. **D** Data shown in **C** for the sandpaper and suede conditions have been normalized in amplitude to the rayon condition using the peak grip forces. Note that the traces nearly superimpose indicating that the various response components maintained proportional relationships under the different frictional conditions. **A—B** adapted from [114] and **C—D** from [34].

tracking responses, just as in visual smooth pursuit [123]. Finally, for both the hand and visual systems, changes in the relationship between the sensory stimulus and the required motor response are met parametrically by changes in the gain of underlying sensorimotor transformations [34, 124].

3.2.4. *Body- and gravity-referenced intrinsic task variables*

Interestingly, the employed 'load-to-grip' force sensorimotor transformation is tailored to the diverse consequences of frictional slippage in different loading directions [117]. For instance, the grip response onset latencies are some 10 ms shorter for loads in the direction away from the palm and in the direction of gravity than for loads in the opposite directions (Fig. 14A). That is, the subjects react faster with loads in the direction in which there are no other potentially

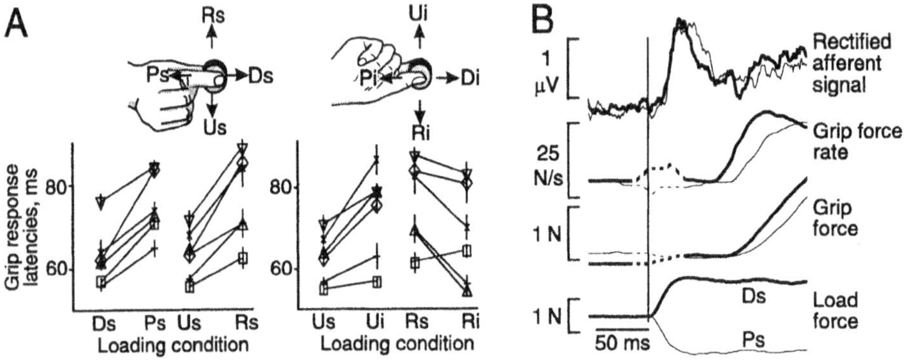

Figure 14. Grip force latencies during various loading conditions. **A** Latencies of grip force responses to 1 N 'step' load increases. Symbols joined by lines represent individual subject means; vertical bars indicate ± SEM. Left panel: distal (Ds), proximal (Ps), ulnar (Us) and radial (Rs) loads during standard hand position. Right pane: ulnar (Us, Ui) and radial (Rs, Ri) loads during standard and inverted hand position, respectively. Note that the latencies are shorter regardless of hand position while the load act in the direction of gravity (Us, Ri). **B** Averaged multi-unit activity recorded from cutaneous fascicles in the median nerve with tungsten needle electrodes during step load trials. Note that the latencies of afferent responses were similar for loads in the distal direction (Ds; thick lines) and the proximal direction (Ps; thin lines), but the grip force latencies were prolonged for loads in the proximal direction. The interrupted line of the grip force curves indicate mechanical crossover effects on the grip force from the load stimulation. The apparent time shifts between the grip force and its time derivative (grip force rate) are due to the time differentiation procedure. Adapted from [117].

available forces opposing the load than those related to friction; proximal loads could be met by the digits and eventually by the palm and loads against the direction of gravity are generally opposed by frictional forces plus gravity. These directional effects reflect different central delays for triggering grip responses because there are no directional dependent difference in the latency of the afferent signals triggering the grip responses (Fig. 14B) (see [117] for details). We interpret the shorter grip response latency for loads in certain directions to reflect a default response preparation by the CNS for rapidly issuing a response to loads in these directions (cf. [125]). The response latency during loads in other directions would be longer, since these loads would require implementation of a different grip force scaling to the frictional conditions because the finger tips show frictional anisotropies [117], and the implementation of a different muscle activation pattern to produce the adequate reactive forces in the axis of loading. Such influences by intrinsic task variables indicate that the input-output relations of sensorimotor transforms employed by manipulative tasks cannot be captured by traditional concepts of stereotyped reflexes; rather, they depend on the task and its intrinsic properties.

3.3. PERIPHERAL AFFERENT CONTROL OF THE GRIP REACTIONS TO TANGENTIAL LOADS

Using tungsten micro-electrodes inserted percutaneously into the median nerve of awake human subjects [126], we have recently recorded signals in various potentially useful mechanoreceptive sensory channels from the hand [80, 93].

Tactile afferents. Signals from tactile afferents with receptive fields in the glabrous skin areas in contact with the manipulandum provide rapid and accurate information about load force changes and therefore could be responsible for both the initiation and the scaling of the grip responses (cf. list of 'required sensory information' in Fig. 12 A) [80]. The rapidly adapting FA I afferents respond during the loading (and unloading) phase and they faithfully encode the rate of load force change (Fig. 15). In contrast, these afferents are surprisingly insensitive to the subject's grip force responses. The FA II (Pacinian) afferents are largely insensitive to the type load trials delivered: only with rapid 'step-loads' do some afferents respond.

Both classes of slowly-adapting (SA) afferents are sensitive to the load and grip forces and their chances. The discharges of the SA I afferents appeared to be relatively more influenced by the subject's grip force response than those of the SA II afferents, which are more influenced by the load-force stimulus (cf. Fig. 8). During the hold phase slowly adapting afferents exhibit tonic discharges that probably provide information about the load force during the hold phase; at the level of individual afferents the tonic discharge is differentially influenced by the grip force and by the load force. In addition, information about the maintained load may have been indirectly provided by the FA I afferents, which by virtue of their silence during the hold phase would have indicated that changes in tangential forces had not occurred. The direction in which the tangential load force is applied to the skin influences the firing of most afferents, and in particular the SA II afferents (Fig. 8) [79, 80] (also see [127-129]). However, as a population the tactile afferents would efficiently encode load changes in any direction.

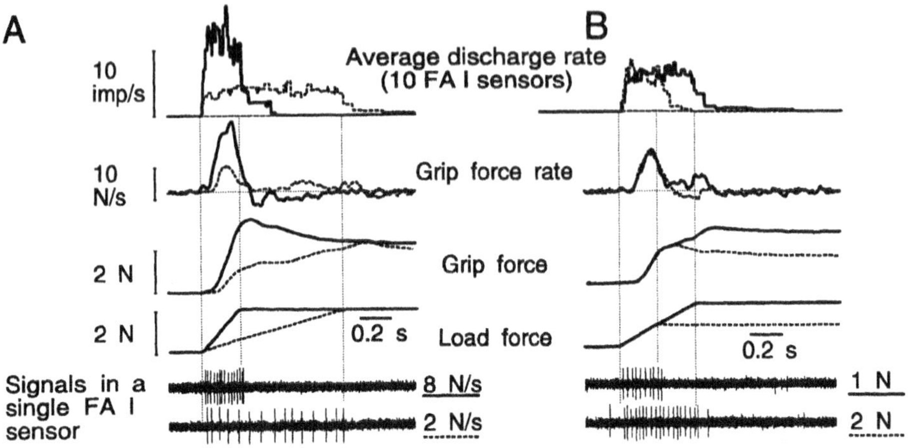

Figure 15. **A** Average instantaneous frequency of the impulse response in 10 FA I sensors to 2 N loads delivered to the receptor-bearing digit at 2 N/s (dashed lines) and 8 N/s (solid lines). **B** Average instantaneous frequency response of the same sensors to 1 N (dashed lines) and 2 N (solid lines) loads delivered to the receptor-bearing digit at 4 N/s. **A** and **B** Below the load force signals are shown example of impulse responses of a single FA I sensor.

Individual afferents within each class (except for the FA II afferents) responds to the loading ramp before the onset of the subject's grip response, and may thus be responsible for initiating the automatic increase in grip force. However, nearly half of the FA I afferents recruited by the load trials responds to the loading phase early enough to trigger the subject's grip force response (Fig. 15), whereas only ca. one fifth of the SA I's and SA II's do so. These observations, together with the high density of FA I receptors in the digits [130] and their encoding of the load force rate, might place the FA I afferents in a unique position to convey the information required to initiate and scale the reactive grip force responses to the imposed load forces.

Muscle and joint afferents. In contrast to the tactile afferents, the afferents from the intrinsic and extrinsic hand muscles do not respond to load increases early enough to allow them to contribute to the initiation and the initial scaling of the grip responses during normal digital sensibility; these respond during load trials essentially after the onset of the subject's grip response to the imposed load [93]. Likewise, the ensemble pattern from the joint afferents revealed no significant encoding of changes in finger-tip forces before the onset of the increase in grip force. Furthermore; joint receptors are much more sparse in the digits than cutaneous mechanoreceptors [130-132], and are certainly less sensitive to mechanical changes at the object-digit interface.

That mechanoreceptors in the flexors of the digits and in the interphalangeal joints cannot be awarded a significant role in triggering the automatic changes in grip force agrees with our finding that subjects with experimentally induced impaired digital sensibility often do not respond with grip force changes to object loading [113]. If present, the grip force responses are weak and markedly delayed compared to normal sensibility, and the normal pattern of catch-up and tracking responses is disrupted as well as the scaling of the responses to the load force rates and amplitudes. Higher grip forces than normal are used during the periods between load trials as a necessary strategy to prevent slips during the prolonged grip response latencies. Furthermore, during digital anesthesia much of the automatic character of the control is lost and an enhanced mental attention is required to complete the restrain task. Interestingly, grip control may improve somewhat during postural conditions which afford large wrist and arm movements caused by the imposed loads [118]. Thus, signals from different kinds of non-digital mechanoreceptors could have contributed to some grip control during impaired digital sensibility, e.g., muscle and joint receptors [93], and skin stretch sensitive cutaneous mechanoreceptors innervating the dorsum of the hand [88]. Presently we cannot determine their relative importance. Nevertheless, the increased grip response latencies during digital anesthesia would be partly explained by the higher thresholds of receptors for load changes compared to those of tactile afferents from the digits.

4. **Central Neural Mechanisms in Manipulation**

In general terms the control of grasping and manipulation appears to rely on distributed processes in the CNS, engaging most areas known to be involved in sensorimotor control. These parallel processes may to some extent be responsible for specific control functions such as selection of task-related sensorimotor transforms, initiation of motor actions, and specification of various motor command parameters based on memory systems. However, little is yet known about the exact implementation, at the neural network level, of the specific sensorimotor control functions used by humans in manipulative tasks.

4.1. PRIMARY MOTOR CORTEX

Numerous lines of evidence indicate that the primary motor cortex as well as the corticospinal pathways (including the cortico-motoneuronal pathway) are of fundamental importance for the execution of skilled hand tasks in primates, particularly those that involve precision grip [61, 133-136]. Through its descending influences upon the spinal cord, the motor cortex can modulate activity in all of the motoneuron pools involved in reach and grasp. This may

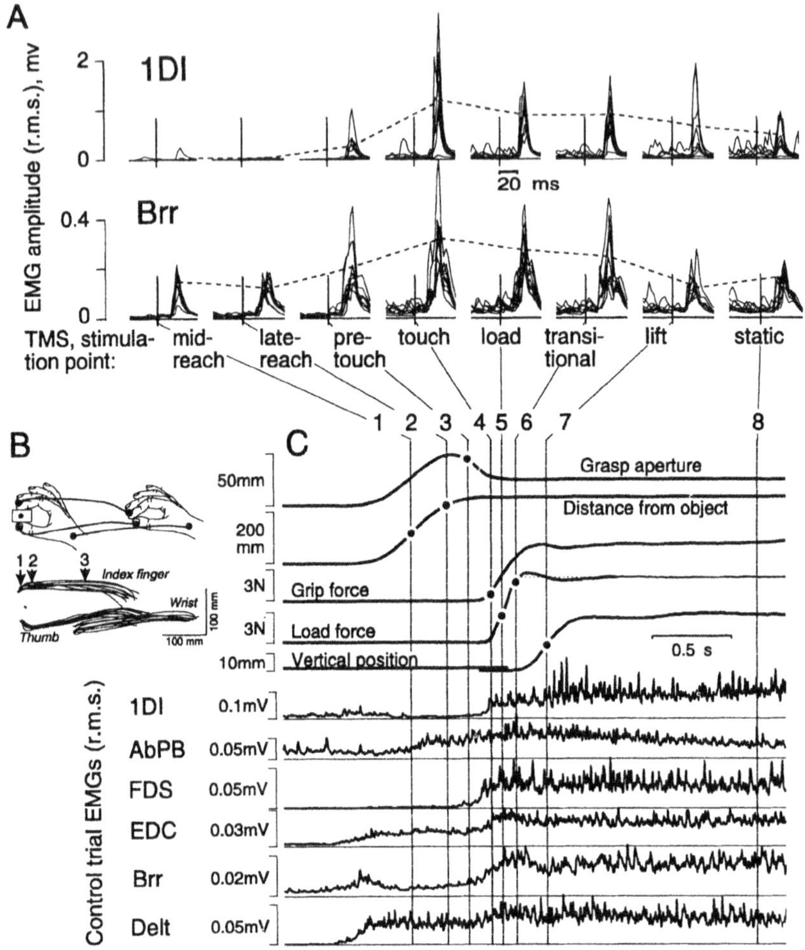

Figure 16. A Examples of TMS (0.8 T) evoked EMG responses in 1DI and Brr obtained from a single subject during the reach-lift task. Data from nine single trials superimposed for each of the eight point of stimulation during the trials (i.e. a complete data set from one series of trials is represented). The vertical lines indicate the timing of TMS delivery. The dotted line shows the mean values of the peaks of the TMS evoked responses. EMG has been r.m.s. processed. **C** Ensemble averages of kinematic, force and EMG recordings across all control trials (no TMS) in a series of trials by a single subject (n=9) R.m.s. processed EMGs from the following muscles: first dorsal interosseous (1DI), abductor pollicis brevis (AbPB), flexor digitorum superficialis (FDS), extensor digitorum communis (EDC), brachioradialis (Brr) and anterior deltoid (Delt). Vertical lines relate the points in time of the TMS to the control trials; the timing of the TMS points was referenced on-line to kinematic recordings (point 1 to 3) and grip and load force recordings (point 4 to 8). The distance from the object was derived from the index finger and object markers (**B**), and the grasp aperture from the markers on the tips of the thumb and index finger (**B**). The channel that provided the trigger information is encircled at the relevant point. Data averaging was referenced to the 'touch' trigger. Adapted from [141].

take place directly, through the cortico-motoneuronal pathway, or indirectly, through spinal neural networks (e.g., [137]). Indeed, a significant number of neurons active in the hand area of the monkey primary motor cortex are modulated during lifting and restrain tasks similar to those discussed above [138, 139]. Many of these neurons alter their discharge frequency as a function of the object's weight and texture and in response to load perturbations. Afferent activity generated during grasping, particularly of cutaneous origin, exerts a direct influence on motor cortical cells; tactile inputs exert excitatory effects on some 60 % of monkey motor cortex neurons whose activity is related to hand movement [138, 140].

5.1.1. *Corticospinal control during reach, grasp and precision lift*

Using non-invasive transcranial magnetic brain stimulation (TMS) directed at the hand area of the motor cortex we recently attempted to assess the influence of the motor cortex on various hand and arm muscles while human subjects had to reach out, grasp and lift an object [141]. TMS can give rise to short-latency responses in most upper limb muscles [142], and the form and latency of the earliest component of these responses suggest that they are mediated by the direct, cortico-motoneuronal pathway [143-145]. Since the size of muscular responses evoked by TMS may to some extent be dependent on the level of cortical activity, TMS can be used to assess the level of corticospinal influence during the different phases of motor tasks (cf. [136, 146-149]).

In these experiments subjects reached forward about 25 cm to grip a small instrumented test object between the tips of the thumb and index finger, lift it from a table up to a height of ca. 3 cm, hold it in air for about 2.5 s and then replace it on the table. The trajectories of the subject's reach and grip were recorded by a reflex marker camera which tracked the position in the horizontal plane of four markers placed on the wrist, on the tips of the index finger and thumb, and on the upper surface of the object (Fig. 16B). Surface EMGs (root-mean-square (r.m.s.) processed) were recorded from two intrinsic hand muscles, first dorsal interosseous (1DI) and abductor pollicis brevis (AbPB); from two extrinsic hand muscles, extensor digitorum communis (EDC) and flexor digitorum superficialis (FDS) and from the arm muscles, brachioradialis (Brr) and anterior deltoid (Delt). Fig. 16C illustrates averaged data from control trials (n = 9) performed by one of the seven subjects investigated .

A magnetic stimulator was located at the optimal position for evoking a short-latency EMG response in the 1DI of the relaxed subject. Stimuli were delivered at eight different times relative to the task (see TMS, stimulation points in Fig. 16C); only one stimulus was given per test trial and no stimulation in control trials. The stimulus strength was usually 0.8 times the threshold intensity (T) required to evoke a response in the relaxed 1DI.

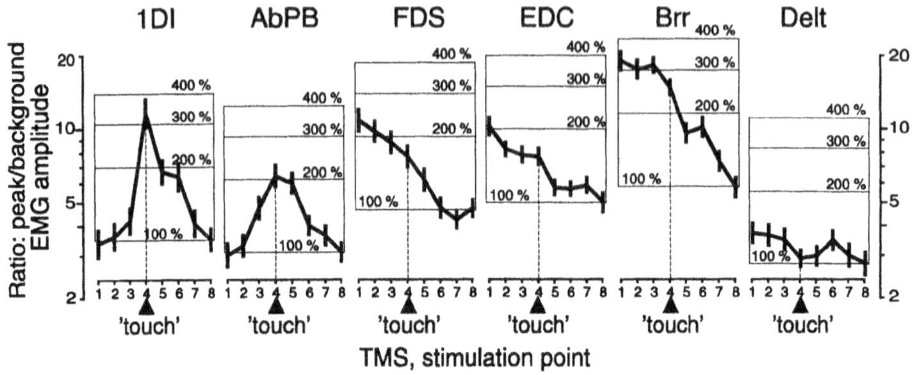

Figure 17. Pooled results from 7 subjects showing variation in responses to TMS during the task. Peak:background ratio for the eight TMS points (1—8) and all six muscles investigated. The measurements are joined by straight lines and the vertical bars indicate the mean ± 1 SEM. Dashed vertical lines are aligned with the first stimulus point following contact with the object ('touch'). Boxes superimposed on the ratio curves indicate the percentage changes in relation to the static values (100%). Note the logarithmic ordinates. For abbreviations see legend to Fig. 16. Adapted from [141].

Typical short-latency responses to TMS recorded from 1DI and Brr are shown in Fig. 16A. Responses in both muscles showed a similar and striking modulation in their amplitude, with an increase in response amplitude during the 'pre-touch' stimulation (point 3), maximum response for stimuli delivered at 'touch' (point 4), high response amplitudes during the 'load' (5) and 'transitional' (6) points, and a general decline through the succeeding 'lift' (7) point to a significantly smaller response during the 'static' phase (point 8).

Because each of the muscles also showed large changes in the background EMG activity in the different phases of the task (Fig. 16C, see also Fig. 16A), it is possible that the modulation in response amplitude simply reflected the changes in motoneuronal excitability across the task. An alternative or additional factor would be a change in corticospinal influence over some of the muscles investigated during different phases of the task, i.e., a phase-related increase in cortical excitability would result in larger descending corticospinal volleys producing larger EMG responses. To distinguish between these mechanisms we normalized the peak amplitude of the TMS evoked response to the level of background EMG activity by computing the ratio between the peak amplitude and the background activity (measured 10 ms before or after the TMS pulse). For each trial and muscle, this ratio was calculated by subtracting the logarithm of the background voltage from that of the peak voltage. All measurements were made from single trials. For further methodological details see [115] and [141].

Figure 17 shows that there was a marked modulation in the normalized amplitude of the EMG responses to TMS across the task; the boxes superimposed upon the ratio curves indicate the percentage changes relative to the static values (TMS-point 8). If the response amplitude during the different stimulation points had simply varied in proportion to the background EMG, these ratio plots would be represented as horizontal lines (note the logarithmic ordinates), which was not the case. Based on various evidence we conclude that the most likely explanation for the modulation in corticospinal influence during the task is a phase-related variation in cortical excitability that is probed by TMS (for a detailed discussion see [115] and [141]). For instance, the phase-related modulation in the peak:background ratio was very sensitive to the intensity of TMS. With the weak stimulus used (ca. 0.8T), the corticospinal volley changes in amplitude because the numbers of neurons activated by TMS across the task reflects, in part, their degree of excitability and therefore their susceptibility to discharge when TMS is delivered. Consequently, with stronger (1.0T) stimuli all responses were much larger, but there was little or no modulation of the peak:background ratio with phase of the task; the responses to TMS varied in proportion to the background EMG. This suggests that these strong stimuli activated corticospinal neurons irrespective of their excitatory state, and that the changes in the EMG responses to the resulting descending volley solely reflect changing levels of motoneuronal excitability during different phases of the task. Because the earliest component of the short-latency responses evoked by TMS is likely to be cortico-motoneuronal in origin [144, 150], any changes in the excitability of segmental interneuronal or propriospinal pathways [137, 151] would be unlikely to influence this component without a concomitant alteration in the level of target motoneuronal activity.

The pattern of modulation in responses to TMS across the task varied considerably from muscle to muscle (Fig. 17). For instance, Brr showed the highest ratios during the reach (points 1 - 3), despite being less active during the reach than during manipulation (cf. Fig. 16C). FDS also showed high ratios during the reach, as did EDC. In contrast both of the intrinsic hand muscles, 1DI and AbPB, exhibited maximum ratios at 'touch' (point 4). Note that despite the similarity in their response ratios, the pattern of EMG activity in 1DI and AbPB during the task was quite different (see Fig. 16C). The increase in AbPB activity from mid-reach (point 1) to late-reach (point 2) was probably related to the opening of the grasp aperture during the reach.

Taken together, these observations suggest that the intrinsic hand muscles receive their strongest cortical drive as the digits closed around the object, and during the early dynamic phases of the actual manipulation. Because tactile inputs are known to be essential for appropriate coordination of this task and are particularly intense during these phases [32, 79] it is possible that this strong drive partly relates to the central effects of these inputs (see below). In contrast, the extrinsic hand muscles (flexors and extensors), which act to orient the palm and finger tips, appear to be strongly influenced by the motor cortex throughout the reach. Once grip is established after the initial touch of the object, the excitatory drive of these muscles by the motor cortex falls rapidly during the subsequent phases of the lift. The brachioradialis and anterior deltoid muscles, which contribute to transport of the hand, show a similar pattern. Indeed, many studies in monkeys have shown that neurons in the primary motor cortex, whose activity relate to proximal muscles discharge during reaching [152, 153], are recruited earlier than are neurons with more distal involvement [154, 155].

The cortical drive to arm and hand muscles during the static phase while an object is held still in the air appears much reduced compared to during the dynamic phases of the lift. One possible explanation is that the parameterization of the motor output to suit the physical properties of the object (friction, shape and weight etc.) is carried out under tight cortical control during the early phases of manipulation (see above). But once adequately parameterized, the cortical influence over the motor output would not be needed and the function of maintaining a stable grasp may thereafter be taken over by subcortical mechanisms.

The weak TMS intensities used in this study [141] did not impair subjects' ability to carry out the task. However, TMS delivered late during the reach could significantly delaye the onset of the load phase, i.e., the isometric parallel increase in load and grip forces necessary to lift the object. During the late phases of reach cortical control may undergo a critical transition from reach to grasp-related programs which would be particularly vulnerable to disruption by TMS at this time. TMS might also interfere with the use of tactile information by the cortex, i.e., the initial afferent contact responses confirming that the grip has been established [79]. The observation that TMS can produce delays in the execution of a voluntary task was first made by Day *et al.* [156].

5.1.2. *Corticospinal control during a restraint tasks*

We have used transcranial magnetic brain stimulation (TMS) directed at the hand area of the human motor cortex also to assess the influences by the motor cortex over various hand and arm muscles during the restrain task discussed in previous sections [115]. In this experiment, subjects gripped an object that was subjected to an unexpected brisk pulling load of 1N, delivered at a high rate (ca. 80 N/s) (Fig. 18). This 'step' load increase triggered an increase in EMG activity in the gripping muscles at a minimum onset latency around 51 ms (see 'background' EMG signals referring to control trials in Fig. 18) followed shortly afterwards (10-20 ms) by an increase in grip force. Because the applied load generates a distinct cutaneous input ('Tactile afferent signal' in Fig. 18), the timing of its interaction with TMS can be investigated in considerable detail. In our experiments, the timing of the TMS was randomized to occur at intervals from ca. 20 ms before the onset of the unexpected load to 140 ms after it, or during the steady intertrial periods of grip ('TMS probes' in Fig. 18A).

Figure 18 shows, for different hand muscles, the variations in peak amplitude of the EMG responses evoked by the TMS (solid line with error bars) by depicting their amplitudes in percent of the amplitude of the intertrial response on a logarithmic scale. In 1DI, the TMS response to the 80 ms probe (see Fig. 18) increased on average more than eight times compared to the intertrial responses, and in several other muscles (AbPB, FDS and FCU) the change was more than two times. Again, these marked changes in response amplitude should not be taken to simply reflect the level of motoneuronal activity during the grip response. First, in most muscles, changes in amplitude of the TMS evoked responses were substantially larger than could be explained by a proportional relationship between a TMS evoked constant descending corticospinal volley and a time varying modulation in the excitability of the motoneuron pool (Fig. 19; note the logarithmic ordinates), and far larger than could be explained by an algebraic summation.

Second, modulation of response amplitude did not necessarily covary with background EMG. Third, the optimum modulation was obtained with intensities (0.8T) of TMS below the threshold (T) for exciting the relaxed 1DI muscle, and was not present with stronger intensities (also see [141]). Therefore, we infer an additional contribution to the modulation of the TMS evoked responses due to changes in the corticospinal volley elicited by TMS.

Indeed, we suggest that the response modulation occurring beyond that observed in the background EMGs reflected variation in the excitability of cortical neurons during the course of the load trial, and that it was these changes in excitability that were probed by low intensity TMS. Thus, a particularly large corticospinal volley would be generated if corticospinal cells were themselves active at the time of delivery of TMS, thereby increasing their probability of discharge by magnetic stimulation (see previous section). Hence, the results are consistent with a central interaction between the effects of tactile afferent inputs and those of TMS. A likely site of interaction is the primary motor cortex. Interestingly, in monkeys performing a similar restraint task an equal proportion of neurons in the motor cortex responsive to the perturbations had proprioceptive and cutaneous receptive fields, although those with cutaneous fields showed stronger excitation [139].

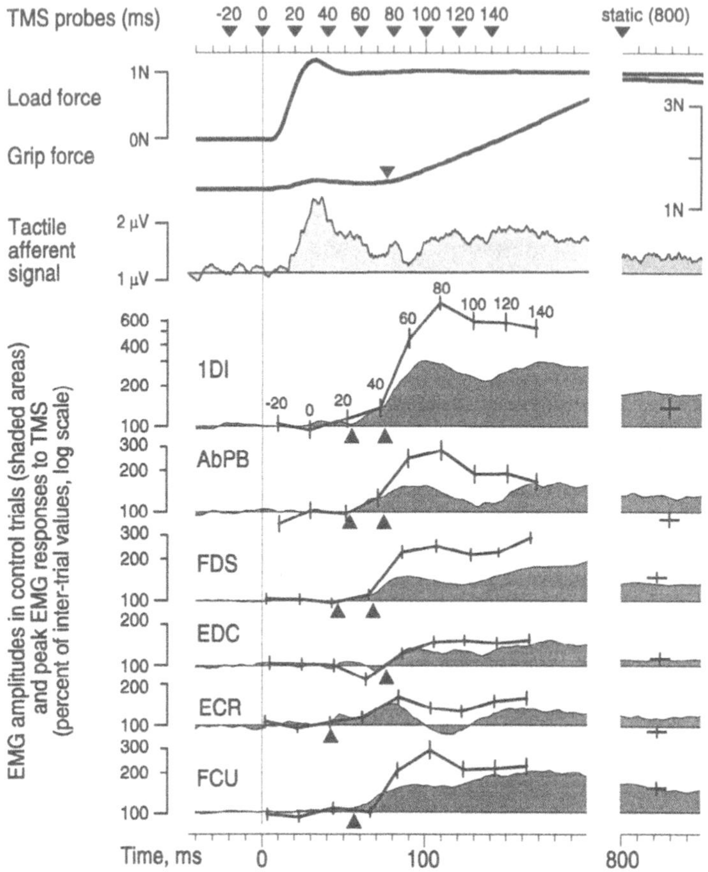

Figure 18. Grip force and EMG changes produced by an unexpected pulling force of 1N (with a force ramp of ca. 80 N/s) generated away from the subject. In control trials 'background' EMG was recorded (r.m.s. processed) from first dorsal interosseous (1DI), abductor pollicis brevis (AbPB), flexor digitorum superficialis (FDS), extensor digitorum communis (EDC), extensor carpi radiales (ECR) and flexor carpi ulnaris (FCU) muscles. The 'tactile afferent signal' refers to rectified multi-unit nerve activity recorded from cutaneous fascicles in the median nerve at the midlevel of the upper arm. On separate trials, TMS was delivered at different times before and after the command to generate the force pulse (see top line) referenced to the computer-generated command to the force servo, which actually preceded the start of the load increase by ca. 5 ms. The arrow on the grip force trace indicates the onset of the reflex grip response during trials in which no TMS was given. Onset of reflex EMG changes indicated by arrows; 1DI, AbPB and FDS showed a small early response, followed by a second, larger response (second arrow). The figure also shows the variation in the size of the peak amplitude of the TMS evoked EMG responses in the different muscles (line with vertical bars, which represents the mean ± SEM). The amplitude of this response and 'background' EMG is given in percent of the amplitude of the intertrial values (note the logarithmic scale). The peaks are represented on the time scale by the time of their occurrence rather than by the point in time of the corresponding TMS probe which occurred 21-28 ms earlier, depending on the muscle; the numbers above the vertical bars of the curve referring to 1DI indicate the time of delivery of these TMS probes. Adapted from [115].

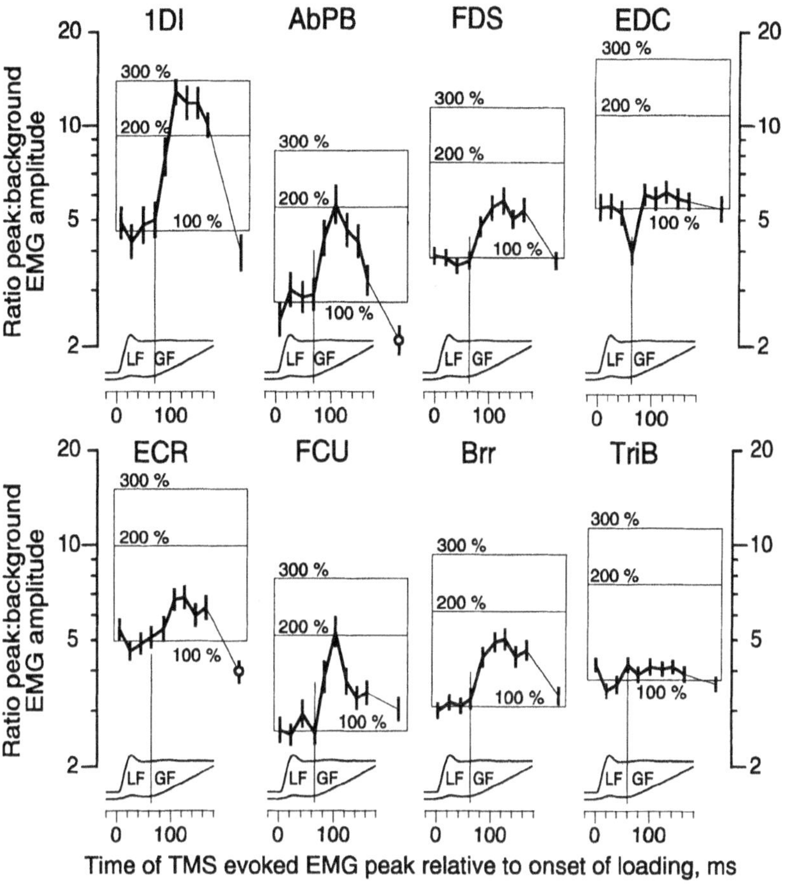

Figure 19. Ratio peak:background amplitude for TMS probes during a restraint task. The measurements are joined by straight lines and the vertical bars indicate the data points (mean +/- 1 SEM). These data points are represented on the time scale by the mean time of the occurrence of the peak of the TMS evoked EMG responses, except for data referring to the static TMS probe which is represented by the right-most data point. Thin vertical lines are aligned with the occurrence of the response to the 40 ms probe. The LF and GF curves below these boxes represent the time course of the load and grip forces in relation to the data points. Note the logarithmic ordinates. EMG was recorded from first dorsal interosseous (1DI), abductor pollicis brevis (AbPB), flexor digitorum superficialis (FDS), extensor digitorum communis (EDC), flexor carpi ulnaris (FCU), extensor carpi radiales (ECR), brachioradialis (Brr) and triceps brachii (TriB) muscles. Adapted from [115].

The very earliest increase in the TMS evoked EMG response over that of the background EMG was observed for the 60 ms TMS probe (see Figs. 18 and

19). This is more than 20 ms after the predicted first cortical arrival of tactile signals based of somatosensory evoked potential measurements [116]. Thus, cortex would have received and probably somehow processed most of the initial components of the afferent volley before emitting a descending command.

These finding together with various observations on response latencies etc. [114] suggests a subcortical but supraspinal origin of response initiation. However, motor cortical mechanisms would powerfully contribute to the regulation of the already initiated response. Similar conclusions have been drawn from recordings of electroencephalographic activity over the scalp overlying the motor area of the hand during the same task [116]. Furthermore, there is evidence that fast spinal regulatory mechanisms also contribute to regulation of the grip force response after initiated [114]. Thus, in parallel with the cortical mechanisms spinal networks provide response regulation after being adequately preset to the task by supraspinal mechanisms. Indeed, there is evidence in man that different sensory-motor processes may account for initiation of motor actions versus specification of response parameters [114, 157]and that these mechanisms may be organized in parallel [158].

5.2. SOME GENERAL ASPECTS ON CNS MECHANISMS IN MANIPULATION

Primary somatosensory cortex (SI) and individuated finger movements. SI receives and processes somatosensory information used in the regulation of fine grip forces during precision grip, but does not participate in force initiation per se in a visual step-tracking paradigm [159]. Integration of somatic sensory information proceeds within the postcentral gyrus in a hierarchical manner [160]. Many of the caudal neurons that integrate signals from cutaneous and deep afferent sources of the hand respond to specific types of stimulation related to manipulative actions rather than simple somatic stimulation. Indeed, pharmacological inactivation of SI neurons in area 2 causes deficits in precision grasping but not in visually guided reaching or hand shaping [161].

The independence of adjustment of fingertip force coordination in providing grasp stability noted earlier is complemented by parallel sensory information processing for the separate digits by the CNS. For example, adjacent digits of the hand are represented discontinuously in multiple sensory maps in the primary somatosensory cortex, but only as long as independent finger actions are possible (cf. artificial syndactyly; [162]). In contrast, similar ordered maps comprising digit-specific representations do not seem to exist in brain areas executing motor commands such as the primary motor cortex [163]. Indeed, given anatomical constraints (e.g., muscle-joint organization) and mechanical constraints (e.g., coupling of finger-tip forces through the object) it seems inconceivable that the CNS would achieve independent digit control for the force coordination of stable grasp by operating on digit-specific muscles [164] (see also [165, 166]). Instead, several digit and wrist/forearm movement representation areas of the motor cortex interdigitate in mosaic like patterns (e.g., [167]). It remains to be determined how digit specific information in sensory representations may be interfaced with the level of control related to fractionated finger actions.

Cerebellum and cortical association areas in anticipatory parameter control. As with the primary motor cortex neurons, when monkeys lift an object and hold it in the air the discharge rate in many cerebellar Purkinje cells and unidentified neurons reflects the object's weight and/or friction, as well as the occurrence of load perturbations tending to cause slips (see above) [168, 169]. For cerebellar neurons, the majority of Purkinje and unidentified cerebellar neurons that responded had cutaneous receptive fields, although some had proprioceptive fields.

The discharge modulation of these neurons related to objects weight and friction occurred not only during actual manipulation, but also prior to the application of grip and load forces suggesting a role of cerebellum in anticipatory parameter control pertaining to object properties. Besides, several theories have proposed that the cerebellum may form the internal predictive (memory) models of the body and its sensorimotor apparatus (and of external objects) to achieve the predictive representations necessary for feed-forward sensorimotor control policies [65]. However, it is not clear which CNS areas might store or represent weight- and frictional related somatosensory information acquired during previous lifts. In general terms it is believed that the neural computations underlying the use of implicit memory information probably are distributed in CNS areas that are involved in the processing of specific attributes of sensorimotor information (e.g., [170, 171]).

Patients with degenerative cerebellar lesions performing lifting tasks by the precision grip of thumb and index finger are able to adapt their grip force levels to the different object loads [172]. But they are less efficient than healthy controls in using sensorimotor memory about object weight for anticipatory control. Accordingly, the temporal profile of grip force rate of change featured an irregular pattern characteristic of a lack of sufficient anticipatory parameterization [5]. These patients also show prolonged latencies between contact with the object and onset of lift force, and the level of grip force at the start of lift force is elevated, i.e., impairments similar to those observed during cutaneous anesthesia [3].

Although the basic structure of a motor program may not be generated within the human cerebellum, it certainly plays an important role for the temporal processing and the distribution of the synergetic activatory inputs across muscles; the cerebellar efferents from the cerebellar nuclei strongly influences the primary motor cortex via the ventrolateral thalamus (for recent reviews see [173, 174]; also see [175]). Synergies which are initiated by external sensory signals as in the restrain task discussed above are may be supported by inputs from the dentate and interpositus nuclei and signals transmitted to parietal associative areas 5 and 7. The intermediate cerebellum appears specialized for the control of object-oriented manual actions such as those involving reaching to grasp (rather than reach per se) [176]. Synergies induced by internal signals or motivational states is likely to involve the dentate nucleus and prefrontal area 9 as well as the supplementary motor and premotor areas.

Indeed, animal as well as human clinical data implicate the posterior parietal and premotor cortical areas in control of object-oriented manual actions (for revent reviews see [8, 177-179]). However, most CNS studies have so far been concerned with visuomotor mechanisms in the control of hand transport, shaping and orientation rather than mechanisms involved in the adaptation of the finger-tip forces to intrinsic object properties during actual manipulation.

Interestingly, in lifting tasks there is a dissociation between perceiving objects and the operation of mechanisms for anticipatory control of finger forces. When healthy adults lift objects equal in weight but not in size, they reliably report smaller objects to be heavier [180], but still they use a larger force output for larger objects [41]. Hence, the operation of relating the target object to the information stored in memory needed for parameterizing the motor commands most likely represents an automatic process dissociated from conscious perception. This agrees with the notion that anticipatory control of actions based on visual identification of intrinsic physical properties of objects and the conscious perception of objects viewed involve different neural pathways, i.e., the 'dorsal stream' projecting from the striate cortex to the posterior parietal region and the 'ventral stream' of projections from the striate cortex to the inferotemporal cortex, respectively [177].

6. **Conclusions**

We have examined two prototypical prehensile tasks in dexterous manipulation: (1) lifting an object from a support surface, holding it in the air and then replacing it, and (2) restraining an object subjected to unpredictable load forces. Grasp stability is achieved by a process best described as predictive, feedforward sensory control rather than continuous closed-loop feedback control. At the heart of this is the need to carry out sensible motor behaviors despite sensory and effector systems subject to large neural conduction and processing delays [181]. Before the relevant physical properties of an object are definitely known (typically via somatosensory mechanisms) action has to be taken on the basis of what can be anticipated. For instance, the minimum possible time from the onset of an error episode to the onset of a corrective action expressed as a mechanical (force) response at the fingertips is ca. 45 ms (for ref. see [114]). A continuous closed-loop feedback control system with such long delays would be ineffective for manipulative actions with frequencies above ca. 1 Hz [182]; that is, such control would not work for the frequency range characteristic of many manipulative actions [183].

The sensorimotor mechanisms employed in precision grip lifting to adapt force output to frictional condition and to object weight operate according to a predictive feed-forward sensory control policy running on at least two time scales. Contextual, cognitive and movement phase dependent interpretations of multi-sensory input seem to be vital features of the underlying sensorimotor transformations. On an extended time scale, previous experience with the object at hand (or similar objects) is used to adjust the motor commands parametrically in advance of the movement. This ability to parameterize default motor commands has been called 'anticipatory parameter control' [46, 69, 184, 185]. A motor program is thus pre-shaped on the basis of memory systems pertaining to physical properties of objects gained in previous manipulative experiences and previous tasks of the same or similar nature. Through vision, for instance, common objects can be identified in terms of necessary initial fingertip forces.

While the task evolves, somatosensory as well as other sources of information may then modify the ongoing behavior. This takes place on a shorter time scale: During the actual manipulation discrete mechanical events in the digit-objet interface are encoded in the spatio-temporal pattern of signals in parallel sensory channels, particularly tactile. This information intervenes intermittently according to what has been called a 'discrete event, sensory driven' control policy [46, 184, 185]. In this control policy, which is distinguished from continuous feed-back or other continuous regulation, task progress is monitored to inform the CNS about completion of the goal for each of subsequent action phases of the task, and for triggering commands for the task's sequential phases. Moreover, disturbances in task execution due to erroneous anticipatory settings of the motor commands are reflected by discrete mechanical events that occur while not expected, or alternatively, that do not occur while expected. Accordingly, a set of predicted afferent signals are considered to be generated by the neural controller in conjunction with the efferent signals, and are compared to the actual afferent signals. Information about, more-or-less expected, mismatches immediately trigger pre-programmed corrective or compensatory actions that are appropriate for the task and the current phase of the task. This information would also serve as a signal to update the sensorimotor memories supporting the anticipatory parameter control policy.

Using the 'anticipatory parameter control' and the 'discrete event, sensory driven' control policies, grasp stability (avoidance of slips but also excessive grip forces) is obtained at each digit by independent control mechanisms. Force coordination across digits is partly an emergent property of these local control mechanisms. Apparently, the neural processes underlying the task related sensorimotor transformations employed in manipulation engage most CNS areas known to be involved in sensorimotor processes pertaining to limb actions. For instance, we have observed a phasic modulation of the corticospinal influence over the various hand and arm while humans reach out, grasp and lift an object and the pattern of modulation varies from muscle to muscle. Accordingly, coordinated prehension in humans develops relatively late during ontogeny.

While restraining an object subjected to unpredictable destabilizing load forces, subjects use sensory information related to load changes to automatically and reactively generate grip responses via a task specific 'load-to-grip' sensorimotor transformation. Tactile receptors of the skin in contact with the object are the only type of mechanoreceptors capable of triggering and scaling appropriate grip responses, but muscle and joint afferents may provide information related to the reactive forces produced by the subject. The employed sensorimotor transform appears to exploit short term predictions of the rate of load changes. Furthermore, the friction in the object-digit interface modifies the gain of the entire load-to-grip force transform based on frictional information obtained during initial skin-object contact. The initiation as well as the regulation of the motor responses seem to depend on supraspinal control mechanisms. Evidence is provided that the human motor cortex is involved in the regulation of already initiated grip responses, but that their initiation most likely depend on brain stem networks. Afferent inflow may also contribute in the regulation of already initiated responses via fast spinal networks.

Acknowledgments. This work was supported by the Swedish Medical Research Council (project 08667) and the Office of Naval Research, Arlington, VA (grant no N00014-92-J-1919).

7. References

1. MacKenzie, C.L and Iberall, T. (1994) The grasping hand, Elsevier Science Publishers, Amsterdam.
2. Johansson, R.S., Riso, R., Häger, C. and Bäckström, L. (1992) Somatosensory control of precision grip during unpredictable pulling loads. I. Changes in load force amplitude, *Experimental Brain Research* **89**, 181-191.
3. Johansson, R.S. and Westling, G. (1984) Roles of glabrous skin receptors and sensorimotor memory in automatic control of precision grip when lifting rougher or more slippery objects, *Experimental Brain Research* **56**, 550-564.

4. Westling, G. and Johansson, R.S. (1984) Factors influencing the force control during precision grip, *Experimental Brain Research* **53**, 277-284.
5. Johansson, R.S. and Westling, G. (1988) Coordinated isometric muscle commands adequately and erroneously programmed for the weight during lifting task with precision grip, *Experimental Brain Research* **71**, 59-71.
6. Jeannerod, M. (1984) The timing of natural prehension movements, *Journal of Motor Behavior* **16**, 235-254.
7. Jeannerod, M. (1986) The formation of finger grip during prehension. A cortically mediated visuomotor pattern, *Behavioural Brain Research* **19**, 99-116.
8. Jeannerod, M., Arbib, M.A., Rizzolatti, G. and Sakata, H. (1995) Grasping objects: the cortical mechanisms of visuomotor transformation, *Trends in Neurosciences* **18**, 314-320.
9. Gentilucci, M., Castiello, U., Corradini, M.L., Scarpa, M., Umilta, C. and Rizzolatti, G. (1991) Influence of Different Types of Grasping on the Transport Component of Prehension Movements, *Neuropsychologia* **29**, 361-378.
10. Jakobson, L.S. and Goodale, M.A. (1991) Factors affecting higher-order movement planning - A kinematic analysis of human prehension, *Experimental Brain Research* **86**, 199-208.
11. Paulignan, Y., Jeannerod, M., Mackenzie, C. and Marteniuk, R. (1991) Selective Perturbation of Visual Input During Prehension Movements. 2. The Effects of Changing Object Size, *Experimental Brain Research* **87**, 407-420.
12. Rosenbaum, D.A., Marchak, F., Barnes, H.J., Vaughan, J., Slotta, J.D. and Jorgensen, M.J. Constraints for action selection: Overhand versus underhand grips, in M. Jeannerod (ed.), *Attention and Performance.* Lawrence Erlbaum, Hillsdale, NJ, pp. 321-342.
13. Stelmach, G.E., Castiello, U. and Jeannerod, M. (1994) Orienting the Finger Opposition Space During Prehension Movements, *Journal of Motor Behavior* **26**, 178-186.
14. Servos, P. and Goodale, M.A. (1994) Binocular Vision and the Online Control of Human Prehension, *Experimental Brain Research* **98**, 119-127.
15. Jeannerod, M. Intersegmental coordination during reaching at natural visual objects, in J. Long and A. Baddeley (eds.), *Attention and Performance.* Erlbaum, Hillsdale, NJ, pp. 153-168.
16. Goodale, M.A., Jakobson, L.S. and Keillor, J.M. (1994) Differences in the Visual Control of Pantomimed and Natural Grasping Movements, *Neuropsychologia* **32**, 1159-1178.
17. Jackson, S.R., Jackson, G.M. and Rosicky, J. (1995) Are Non-Relevant Objects Represented in Working-Memory - The Effect of Nontarget Objects on Reach and Grasp Kinematics, *Experimental Brain Research* **102**, 519-530.
18. Johansson, R.S. and Westling, G. Afferent signals during manipulative tasks in man, in O. Franzen and J. Westman (eds.), *Somatosensory Mechanisms.* Macmillan Press, London, pp. 25-48.
19. Edin, B.B., Westling, G. and Johansson, R.S. (1992) Independent control of fingertip forces at individual digits during precision lifting in humans, *Journal of Physiology* **450**, 547-564.
20. Flanagan, J.R., Tresilian, J. and Wing, A.M. (1993) Coupling of Grip Force and Load Force During Arm Movements with Grasped Objects, *Neuroscience Letters* **152**, 53-56.
21. Flanagan, J.R. and Wing, A.M. (1993) Modulation of Grip Force with Load Force During Point-to-Point Arm Movements, *Experimental Brain Research* **95**, 131-143.
22. Kinoshita, H., Kawai, S., Ikuta, K. and Teraoka, T. (1995) Individual finger forces acting on a grasped object during shaking actions, *Ergonomics,* (In press).
23. Kinoshita, H., Ikuta, K., Kawai, S. and Udo, M. (1993) Effects of lifting speed and height on the regulation of forces during lifting tasks using a precision grip, *Journal of Human Movement Studies* **25**, 151-175.

24. Flanagan, J.R. and Tresilian, J.R. (1994) Grip Load Force Coupling - A General Control Strategy for Transporting Objects, *Journal of Experimental Psychology: Human Perception and Performance* **20**, 944-957.
25. Burstedt, M., Westling, G., Johansson, R.S. and Edin, B.B. Co-ordinated manipulative behavior can emerge by independent neural systems, *ENA-satellite symposium: Neural Control of Eye, Head and Limb Movements.* Ohlstadt, München, Germany, Sept 10-13.
26. Kinoshita, H., Kawai, S. and Ikuta, K. (1995) Contributions and coordination of individual fingers in multiple finger prehension, *Ergonomics* **38**, 1212-1230.
27. Johansson, R.S. and Westling, G. (1988) Programmed and triggered actions to rapid load changes during precision grip, *Experimental Brain Research* **71**, 72-86.
28. Johansson, R.S. and Westling, G. Tactile afferent input influencing motor coordination during precision grip, in A. Struppler and A. Weindl (eds.), *Clinical Aspects of Sensory Motor Integration,*. Springer, Berlin, pp. 3-13.
29. Bernstein, N. (1967) *The co-ordination and regulation of movements*, Pergamon Press, Oxford, England.
30. Turvey, M.T., Shaw, R.E. and Mace, W. Issues in the theory of action. Degrees of freedom, coordinative structures and coalitions, in J. Requin and N.J. Hillsdale (eds.), *Attention and performance VII.* Erlbaum, London, pp. 557-595.
31. Sporns, O. and Edelman, G.M. (1993) Solving Bernstein's problem: a proposal for the development of coordinated movement by selection, *Child Development* **664**, 960-969.
32. Johansson, R.S. and Westling, G. (1987) Signals in tactile afferents from the fingers eliciting adaptive motor responses during precision grip, *Experimental Brain Research* **66**, 141-154.
33. Cadoret, G. and Smith, A.M. (1995) Friction, not texture, dictates grip forces used during object manipulation, *Journal of Neurophysiology* (In press).
34. Cole, K.J. and Johansson, R.S. (1993) Friction in the digit-object interface scales the sensorimotor transformation for grip responses to pulling loads, *Experimental Brain Research* **95**, 523-532.
35. Forssberg, H., Eliasson, A.C., Kinoshita, H., Westling, G. and Johansson, R.S. (1995) Development of human precision grip IV. Tactile adaptation of isometric finger forces to the frictional condition. Exp. Brain Res. (In press), *Experimental Brain Research* **104**, 323-330.
36. Flanagan, J.R., Wing, A.M., Allison, S. and Spenceley, A. (1995) Effects of surface texture on weight perception when lifting objects with a precision grip, *Perception and Psychophysics* **57**, 282-290.
37. Johansson, R.S. and Westling, G. Influences of cutaneous sensory input on the motor coordination during precision manipulation, in C. von Euler, O. Franzen, U. Lindblom and D. Ottoson (eds.), *Somatosensory Mechanisms.* Macmillan Press, London, pp. 249-260.
38. Gordon, A.M., Westling, G., Cole, K.J. and Johansson, R.S. (1993) Memory representations underlying motor commands used during manipulation of common and novel objects, *Journal of Neurophysiology* **69**, 1789-1796.
39. Gordon, A.M., Forssberg, H., Johansson, R.S. and Westling, G. (1991) Integration of sensory information during the programming of precision grip: comments on the contributions of size cues, *Experimental Brain Research* **85**, 226-229.
40. Gordon, A.M., Forssberg, H. and Iwasaki, N. (1994) Formation and lateralization of internal representations underlying motor commands during precision grip, *Neuropsychologia* **32**, 555-568.
41. Gordon, A.M., Forssberg, H., Johansson, R.S. and Westling, G. (1991) Visual size cues in the programming of manipulative forces during precision grip, *Experimental Brain Research* **83**, 477-482.

42. Gordon, A.M., Forssberg, H., Johansson, R.S. and Westling, G. (1991) The integration of haptically acquired size information in the programming of precision grip, *Experimental Brain Research* **83**, 483-488.
43. Lacquaniti, F., Borghese, N.A. and Carrozzo, M. (1991) Transient Reversal of the Stretch Reflex in Human Arm Muscles, *Journal of Neurophysiology* **66**, 939-954.
44. Lacquaniti, F., Borghese, N.A. and Carrozzo, M. (1992) Internal models of limb geometry in the control of hand compliance, *Journal of Neuroscience* **12**, 1750-1762.
45. Ghez, C., Hening, W. and Gordon, J. (1991) Organization of voluntary movement, *Current Opinion in Neurobiology* **1**, 664-671.
46. Johansson, R.S. and Cole, K.J. (1992) Sensory-motor coordination during grasping and manipulative actions, *Current Opinion in Neurobiology* **2**, 815-823.
47. Lacquaniti, F. (1992) Automatic control of limb movement and posture, *Current Opinion in Neurobiology* **2**, 807-814.
48. Hugon, M., Massion, J. and Wiesendanger. (1982) Anticipatory postural changes induced by active unloading and comparison with passive unloading in man, *Pflugers Archiv European Journal of Physiology* **393**, 292-296.
49. Paulignan, Y., Dufossé, M., Hugon, M. and Massion, J. (1989) Acquisition of co-ordination between posture and movement in a bimanual task, *Experimental Brain Research* **77**, 337-348.
50. Massion, J. (1994) Postural Control-System, *Current Opinion in Neurobiology* **4**, 877-887.
51. Mott, F.W. and Sherrington, C.S. (1895) Experiments upon the influence of sensory nerves upon movement and nutrition of the limbs. Preliminary communication, *R Soc London Ser B Proc* **57**, 481-488.
52. Twitchell, T. (1954) Sensory factors in purposive movements, *Journal of Neurophysiology* **17**, 239-252.
53. Moberg, E. (1962) Criticism and study of methods for examining sensibility in the hand, *Neurology* **12**, 8-19.
54. Denny-Brown, D. (1966) *The cerebral control of movement*, The Liverpool University Press, Liverpool.
55. Caccia, M.R., McComas, A.J., Upton, A.R.M. and Blogg, T. (1973) Cutaneous reflexes in small muscles of the hand, *Journal of Neurology, Neurosurgery and Psychiatry* **36**, 960-977.
56. Garnett, R. and Stephens, J.A. (1980) The reflex responses of single motor units in human first dorsal interosseous muscle following cutaneous afferent stimulation, *Journal of Physiology* **303**, 351-364.
57. Jenner, J.R. and Stephens, J.A. (1982) Cutaneous reflex responses and their central nervous pathways studied in man, *Journal of Physiology* **333**, 405-419.
58. McCloskey, D.I., Gandevia, S., Potter, E.K. and Colebatch, J.G. Muscle sense and effort: motor commands and judgments about muscular contractions, in J.E. Desmedt (ed.), *Motor Control Mechanisms in Health and Disease.* Raven Press, New York. pp. 151-167.
59. Darton, K., Lippold, O.C.J., Shahani, M. and Shahani, U. (1985) Long-latency spinal reflexes in humans, *Journal of Neurophysiology* **53**, 1604-1618.
60. Marsden, C.D., Merton, P.A. and Morton, H.B. (1985) New observations on the human stretch reflex, *Journal of Physiology* **360**, 51P.
61. Evans, A.L., Harrison, L.M. and Stephens, J.A. (1989) Task-dependent changes in cutaneous reflexes recorded from various muscles controlling finger movement in man, *Journal of Physiology* **418**, 1-12.
62. Matthews, P.B.C. (1989) Long-latency stretch reflexes of two intrinsic muscles of the human hand analysed by cooling the arm, *Journal of Physiology* **419**, 519-538.

63. Cole, K.J. and Abbs, J.H. (1986) Coordination of three-joint digit movements for rapid finger-thumb grasp, *Journal of Neurophysiology* **55**, 1407-1423.
64. Cole, K.J. and Abbs, J.H. (1987) Kinematic and elctromyographic responses to perturbation of a rapid grasp, *Journal of Neurophysiology* **57**, 1-13.
65. Miall, R.C., Weir, D.J., Wolpert, D.M. and Stein, J.F. (1993) Is the Cerebellum a Smith Predictor, *Journal of Motor Behavior* **25**, 203-216.
66. Merfeld, D.M., Young, L.R., Oman, C.M. and Shelhamer, M.J. (1993) A Multidimensional model of the effect of gravity on the spatial orientation of the monkey, *J Vestib Res* **3**, 141-161.
67. Baev, K.V. and Shimansky, Y.P. (1992) Principles of organization of neural systems controlling automatic movements in animals, *Progress in Neurobiology* **39**, 45-112.
68. Prochazka, A. (1993) Comparison of Natural and Artificial Control of Movement, *IEEE Trans Rehabil Eng* **1**, 7-17.
69. Johansson, R.S. How is grasping modified by somatosensory input?, in D.R. Humphrey and H.-J. Freund (eds.), *Motor Control: Concepts and Issues. Dahlem Konferenzen.* John Wiley & Sons Ltd, Chichester, pp. 331-355.
70. Johansson, R.S. and Vallbo, Å.B. (1983) Tactile sensory coding in the glabrous skin of the human hand, *Trends in Neurosciences* **6**, 27-31.
71. Vallbo, A.B. and Johansson, R.S. (1984) Properties of cutaneous mechanoreceptors in the human hand related to touch sensation, *Human Neurobiology* **3**, 3-14.
72. Burgess, P.R. and Perl, E.R. Cutaneous mechanoreceptors and nociceptors, in A. Iggo (ed.), *Handbook of Sensory Physiology.* Springer, Berlin, pp. 30-78.
73. Iggo, A. Cutaneous receptors, in J.I. Hubbard (ed.), *The Peripheral Nervous System.* Plenum Press, New York, pp. 347-404.
74. Pubols, B.H.J. and Pubols, L.M. (1983) Tactile receptor discharge and mechanical properties of glabrous skin, *Federation Proceedings* **42**, 2528-2535.
75. Johnson, K.O. and Hsiao, S.S. (1992) Neural mechanisms of tactual form and texture perception, *Annual Review of Neuroscience* **15**, 227-250.
76. Johansson, R.S. (1978) Tactile sensibility in the human hand: receptive field characteristics of mechanoreceptive units in the glabrous skin area, *Journal of Physiology* **281**, 101-125.
77. Knibestöl, M. and Vallbo, Å. (1970) Single unit analysis of mechanoreceptor activity from the human glabrous skin, *Acta Physiologica Scandinavica* **80**, 178-195.
78. Knibestöl, M. (1975) Stimulus-response functions of slowly adapting mechanoreceptors in the human glabrous skin area, *Journal of Physiology* **245**, 63-80.
79. Westling, G. and Johansson, R.S. (1987) Responses in glabrous skin mechanoreceptors during precision grip in humans, *Experimental Brain Research* **66**, 128-140.
80. Macefield, V.G., Häger-Ross, C. and Johansson, R.S. (1995) Control of grip force during restraint of an object held between finger and thumb: responses of cutaneous afferents from the digits, *Experimental Brain Research,* (In press).
81. Marsden, C.D., Merton, P.A. and Morton, H.B. (1977) The sensory mechanism of servo action in human muscle, *Journal of Physiology* **265**, 521-535.
82. Gandevia, S.C. and McCloskey, D.I. (1977) Effects of related sensory inputs on motor performances in man studied through changes in perceived heaviness, *Journal of Physiology* **272**, 653-672.
83. Gandevia, S.C. and McCloskey, D.I. (1977) Changes in motor commands, as shown by changes in perceived heaviness, during partial curarization and peripheral anaesthesia in man, *Journal of Physiology* **272**, 673-689.
84. Garnett, R. and Stephens, J.A. (1981) Changes in the recruitment threshold of motor units produced by cutaneous stimulation in man, *Journal of Physiology* **311**, 463-473.

85. Day, B.L. and Marsden, C.D. (1982) Accurate repositioning of the human thumb against unpredictable dynamic loads is dependent upon peripheral feedback, *Journal of Physiology* **327**, 393-407.
86. Kanda, K. and Desmedt, J.E. Cutaneous facilitation of large motor units and motor control of human fingers in precision grip, in J.E. Desmedt (ed.), *Motor Control Mechanisms in Health and Disease.* Raven Press, New York, pp. 253-261.
87. Hulliger, M., Nordh, E., Thelin, A.-E. and Vallbo, Å.B. (1979) The responses of afferent fibres from the glabrous skin of the hand during voluntary finger movements in man, *Journal of Physiology* **291**, 233-249.
88. Edin, B.B. and Abbs, J.H. (1991) Finger movement responses of cutaneous mechanoreceptors in the dorsal skin of the human hand, *Journal of Neurophysiology* **65**, 657-670.
89. Edin, B.B. (1992) Quantitative analysis of static strain sensitivity in human mechanoreceptors from hairy skin, *Journal of Neurophysiology* **67**, 1105-1113.
90. Edin, B.B. and Johansson, N. (1995) Skin strain patterns provide kinaesthetic information to the human central nervous system, *Journal of Physiology* **487**, 243-251.
91. Matthews, P.B.C. (1988) Proprioceptors and their contribution to somatosensory mapping: complex messages require complex processing, *Canadian Journal of Physiology and Pharmacology* **66**, 430-438.
92. Edin, B.B. (1990) Finger joint movement sensitivity of non-cutaneous mechanoreceptor afferents in the human radial nerve, *Experimental Brain Research* **82**, 417-422.
93. Macefield, V.G. and Johansson, R.S. (1995) Control of grip force during restraint of an object held between finger and thumb: responses of muscle and joint afferents from the digits, *Experimental Brain Research,* (In press).
94. Darling, W.G., Cole, K.J. and Abbs, J.H. (1988) Kinematic variability of grasp movements as a function of practice and movement speed, *Experimental Brain Research* **73**, 225-235.
95. Evarts, E.V. (1981) Sherrington's concepts of proprioception, *Trends in Neurosciences* **4**, 44-46.
96. Vallbo, Å.B. Proprioceptive activity from human finger muscles, in W.J.P. Barnes and M.H. Gladden (eds.), *Feedback and motor control in invertebrates and vertebrates.* Croom Helm Ltd., London, pp. 411-430.
97. Rossignol, S., Lund, J.P. and Drew, T. The role of sensory inputs in regulating patterns of rhythmical movements in higher vertebrates, in A.H. Cohen, S. Rossignol and S. Grillner (eds.), *Neural control of rhythmic movements.* John Wiley and Sons, Inc., New York, pp. 201-283.
98. Cole, K. (1994) The effects of graded tactile anesthesia on the control of grip force, *Soc Neurosci Abstr 20, 4906* ,
99. Cole, K.J. (1991) Grasp force control in older adults, *Journal of Motor Behavior* **23**, 251-258.
100. Cole, K.J. and Beck, C.L. (1994) The Stability of Precision Grip Force in Older Adults, Journal of Motor Behavior 26, 171-177.
101. Gordon, A.M. (1994) Development of the Reach to Grasp Movement, *Advances in Psychology* **105**, 37-56.
102. Eyre, J.A., Miller, S. and Ramesh, V. (1991) Constancy of central conduction delays during development in man: investigation of motor and somatosensory pathways, *Journal of Physiology* **434**, 441-452.
103. Pehoski, C. Object manipulation in infants and children, in A. Henderson and C. Pehoski (eds.), *Hand function in the child, foundations for remediation.* Mosby, St Louis, pp. 136-153.

104. Touwen, B. Variability and stereotypy in normal and deviant development, in B. Touwen (ed.), *Neurological development in infancy.* SIMP and Heinemann Med Books, London, pp. 99-110.
105. Forssberg, H., Eliasson, A.C., Kinoshita, H., Johansson, R.S. and Westling, G. (1991) Development of human precision grip. I: Basic coordination of force, *Experimental Brain Research* **85**, 451-457.
106. Forssberg, H., Kinoshita, H., Eliasson, A.C., Johansson, R.S., Westling, G. and Gordon, A.M. (1992) Development of Human Precision Grip .2. Anticipatory Control of Isometric Forces Targeted for Objects Weight, *Experimental Brain Research* **90**, 393-398.
107. Gordon, A.M., Forssberg, H., Johansson, R.S., Eliasson, A.C. and Westling, G. (1992) Development of Human Precision Grip .3. Integration of Visual Size Cues During the Programming of Isometric Forces, *Experimental Brain Research* **90**, 399-403.
108. Eliasson, A., Forssberg, H., Ikuta, K., Apel, I., Westling, G. and Johansson, R.S. (1995) Development of human precision grip V. Anticipatory and triggered grip actions during sudden loading, *Experimental Brain Research,* (In press).
109. Eliasson, A.-C., Gordon, A.M. and Forssberg, H. (1991) Basic co-ordination of manipulative forces of children with Cerebral Palsy, *Developmental Medicine and Child Neurology* **33**, 661-670.
110. Eliasson, A.C., Gordon, A.M. and Forssberg, H. (1992) Impaired anticipatory control of isometric forces during grasping by children with cerebral palcy, *Developmental Medicine and Child Neurology* **34**, 216-225.
111. Eliasson, A.C., Gordon, A.M. and Forssberg, H. (1995) Tactile Control of Isometric Fingertip Forces During Grasping in Children with Cerebral-Palsy, *Developmental Medicine and Child Neurology* **37**, 72-84.
112. Cole, K.J. and Abbs, J.H. (1988) Grip force adjustments evoked by load force perturbations of a grasped object, *Journal of Neurophysiology* **60**, 1513-1522.
113. Johansson, R.S., Häger, C. and Bäckström, L. (1992) Somatosensory control of precision grip during unpredictable pulling loads. III. Impairments during digital anesthesia, *Experimental Brain Research* **89**, 204-213.
114. Johansson, R.S., Häger, C. and Riso, R. (1992) Somatosensory control of precision grip during unpredictable pulling loads. II. Changes in load force rate, *Experimental Brain Research* **89**, 192-203.
115. Johansson, R.S., Lemon, R.N. and Westling, G. (1994) Time varying enhancement of human cortical excitability mediated by cutaneous inputs during precision grip, *Journal of Physiology* **481**, 761-775.
116. Macefield, V.G. and Johansson, R.S. (1994) Electrical Signs of Cortical Involvement in the Automatic-Control of Grip Force, *Neuroreport* **5**, 2229-2232.
117. Häger-Ross, C., Cole, K.J. and Johansson, R.S. (1995) Grip force responses to unanticipated object loading: Load direction reveals body- and gravity-referenced intrinsic task variables, *Experimental Brain Research,* (In press).
118. Häger-Ross, C. and Johansson, R.S. (1995) Non-digital afferent input in reactive control of fingertip forces during precision grip, *Experimental Brain Research,* (In press).
119. Freund, H.-J., Büdingen, H. and J. (1978) The relationship between speed and amplitude of the fastest voluntary contractions of human arm muscles, *Experimental Brain Research* **31**, 1-12.
120. Ghez, C. and Vicario, D. (1978) The control of rapid limb movement in the cat. II. Scaling of isometric force adjustments, *Experimental Brain Research* **33**, 191-202.
121. Gordon, J. and Ghez, C. (1987) Trajectory control in targeted force impulses. II. Pulse height control, *Experimental Brain Research* **67**, 241-252.
122. Fuchs, A. (1967) Saccadic and smooth pursuit eye movements in the monkey, *Journal of Physiology* **191**, 609-631.

123. Lisberger, S.G., Morris, E.J. and Tychsen, L. (1987) Visual motor processing and sensory-motor integration for smooth pursuit eye movements, *Annual Review of Neuroscience* **10**, 97-129.
124. Robinson, D.A. The control of eye movements, in V.B. Brooks (ed.), *Handbook of Physiology: The Nervous System.* Am. Physiology Soc., Bethesda, MD, pp. 1275--1320.
125. Favilla, M., Gordon, J., Hening, W. and Ghez, C. (1990) Trajectory control in targeted force impulses. VII. Independent setting of amplitude and direction in response preparation, Experimental Brain Research 79, 530-538.
126. Vallbo, Å.B. and Hagbarth, K.E. (1968) Activity from skin mechanoreceptors recorded percutaneously in awake human subjects, *Experimental Neurology* **21**, 270-289.
127. Goodwin, A.W. and Morley, J.W. (1987) Sinusoidal movement of a grating across the monkey's fingerpad: Effect of contact angle and force of the grating on afferent fiber responses, *Journal of Neuroscience* **7**, 2192-2202.
128. Srinivasan, M.A., Whitehouse, J.M. and LaMotte, R.H. (1990) Tactile detection of slip: surface microgeometry and peripheral neural codes, *Journal of Neurophysiology* **63**, 1323-1332.
129. Edin, B.B., Essick, G.K., Trulsson, M. and Olsson, K.A. (1995) Receptor Encoding of Moving Tactile Stimuli in Humans .1. Temporal Pattern of Discharge of Individual Low-Threshold Mechanoreceptors, *Journal of Neuroscience* **15**, 830-847.
130. Johansson, R.S. and Vallbo, A.B. (1979) Tactile sensibility in the human hand: relative and absolute densities of four types of mechanoreceptive units in glabrous skin, *Journal of Physiology* **286**, 283-300.
131. Burke, D., Gandevia, S.C. and Macefield, G. (1988) Responses to passive movement of receptors in joint, skin and muscle of the human hand, *Journal of Physiology* **402**, 347-361.
132. Edin, B.B. (1990) Finger joint movement sensitivity of non-cutaneous mechanoreceptor afferents in the human radial nerve, *Experimental Brain Research* **82**, 417-422.
133. Lawrence, D., G., Kuypers, H., G., J. and M. (1968) The functional organization of the motor system in the monkey, *Brain* **91**, 1-14.
134. Kuypers, H.G.J.M. Anatomy of the descending pathways, in J.M. Brookhart and V.B. Mountcastle (eds.), *Handbook of Physiology. Sect. 1: The Nervous System.* American Physiological Society, Bethesda, Maryland, pp. 597-666.
135. Porter, R and Lemon, R.N. (1993) *Corticospinal Neurones and Voluntary Movement,* Oxford University Press, Oxford, UK.
136. Datta, A.K., Harrison, L.M. and Stephens, J.A. (1989) Task-dependent changes in the size of response to magnetic brain stimulation in human first dorsal interosseous muscle, *Journal of Physiology* **418**, 13-23.
137. Gracies, J.M., Meunier, S. and Pierrot-Deseilligny, E. (1994) Evidence for corticospinal excitation of presumed propriospinal neurons in man, *Journal of Physiology* **475**, 509-518.
138. Picard, N., Smith, A. and M. (1992) Primary motor cortical activity related to the weight and texture of grasped objects in the monkey, *Journal of Neurophysiology* **68**, 1867-1881.
139. Picard, N. and Smith, A.M. (1992) Primary motor cortical responses to perturbations of prehension in the monkey, *Journal of Neurophysiology* **68**, 1882-1894.
140. Lemon, R.N. (1981) Functional properties of monkey motor cortex neurones receiving afferent input from the hand and fingers, *Journal of Physiology* **311**, 497-519.
141. Lemon, R.N., Johansson, R.S. and Westling, G. (1995) Corticospinal control during reach, grasp and precision lift in man, *Journal of Neuroscience,* (In press).
142. Rothwell, J.C., Thompson, P.D., Day, B.L., Boyd, S. and Marsden, C.D. (1991) Stimulation of the Human Motor Cortex Through the Scalp, *Experimental Physiology* **76**, 159-200.

143. Day, B.L., Dressler, D., Maertens de Noordhout, A., Marsden, C.D., Nakashima, K., Rothwell, J.C. and Thompson, P.D. (1989) Electric and magnetic stimulation of human motor cortex: surface EMG and single motor unit responses, *Journal of Physiology* **412**, 449-473.
144. Edgley, S.A., Eyre, J.A., Lemon, R.N. and Miller, S. (1990) Excitation of the corticospinal tract by electromagnetic and electrical stimulation of the scalp in the Macaque monkey, *Journal of Physiology* **425**, 301-320.
145. Palmer, E. and Ashby, P. (1992) Corticospinal Projections to Upper Limb Motoneurones in Humans, *Journal of Physiology* **448**, 397-412.
146. Day, B.L., Riescher, H., Struppler, A., Rothwell, J.C. and Marsden, C.D. (1991) Changes in the response to magnetic and electrical stimulation of the motor cortex following muscle stretch in man, *Journal of Physiology* **433**, 41-57.
147. Palmer, E. and Ashby, P. (1992) Evidence that a long latency stretch reflex in humans is transcortical, *Journal of Physiology* **449**, 429-440.
148. Flament, D., Goldsmith, P., Buckley, C.J. and Lemon, R.N. (1993) Task dependence of responses in first dorsal interosseous muscle to magnetic brain stimulation in man, *Journal of Physiology* **464**, 361-378.
149. Baker, S.N.B., Olivier, E. and Lemon, R.N. (1994) Recording the pyramidal volley evoked by transcranial magnetic stimulation in consious monkey, *Experimental Brain Research* **99**, 529-533.
150. Baldissera, F. and Cavallari, P. (1995) Short latency subliminal effects of transcranial magnetic stimulation on forearm motoneurones, *Experimental Brain Research* **96**, 513-518.
151. Burke, D., Gracies, J.M., Mazevet, D., Meunier, S. and Pierrot-Deseilligny, E. (1994) Non-monosynaptic transmission of the cortical command for voluntary movement in man, *Journal of Physiology* **480**, 191-202.
152. Georgopoulos, A.P., Camaniti, R., Kalaska, J.F. and Massey, J.T. (1982) On the relations between the direction of two-dimensional movements and cell discharge in primate motor cortex, *Journal of Neuroscience* **2**, 1527-1537.
153. Fu, Q.G., Suarez, J.I. and Ebner, T.J. (1993) Neuronal specification of direction and distance during reaching movements in the superior precentral premotor area and primary motor cortex of monkeys, *Journal of Neurophysiology* **70**, 2097-2116.
154. Murphy, J.T., Kwan, H.C. and Wong, Y.C. (1985) Sequential activation of neurons in primate motor cortex during unrestrained forelimb movements, *Journal of Neurophysiology* **53**, 435-445.
155. Kwan, H.C., Murphy, J.T. and Wong, Y.C. (1987) Interactions between neurons in precentral cortical zones controlling different joints, *Brain Research* **400**, 259-269.
156. Day, B.L., Rothwell, J.C., Thompson, P.D., Maertens de Noordhout, A., Nakashima, K., Shannon, K. and Marsden, C.D. (1989) Delay in the execution of voluntary movement by electrical or magnetic brain stimulation in intact man. Evidence for the storage of motor programs in the brain, *Brain* **112**, 649-663.
157. Cordo, P.J. and Flanders, M. (1989) Sensory control of target acquisition, *Trends in Neurosciences* **12**, 110-116.
158. Ghez, C., Hening, W. and Favilla, M. Parallel interacting channels in the initiation and specification of motor response features, in M. Jeannerod (ed.), *Attention and Performance XIII*. Lawrence Erlbaum associates,, Hillsdale, NJ, pp. 265-293.
159. Wannier, T., Maier, M. and Hepp-Reymond, M. (1991) Contrasting properties of monkey somatosensory and motor cortex neurons activated during the control of force in precision grip, *Journal of Neurophysiology* **65**, 572-89.
160. Iwamura, Y. (1993) Dynamic and Hierarchical Processing in the Monkey Somatosensory Cortex, *Biomedical Research* **14**, 107-111.

161. Hikosaka, O., Tanaka, M., Sakamoto, M. and Iwamura, Y. (1985) Deficits in manipulative behaviors induced by local injectrions of muscimol in the first somatosensory cortex of the conscious monkey, *Brain Research* **325**, 375-380.
162. Allard, T., Clark, S.A., Jenkins, W.M. and Merzenich, M.M. (1991) Reorganization of somatosensory area 3b representations in adult owl monkeys after digital syndactyly, *Journal of Neurophysiology* **66**, 1048-1058.
163. Donoghue, J.P., Leibovic, S. and Sanes, J.N. (1992) Organization of the forelimb area in squirrel monkey motor cortex: representation of digit, wrist, and elbow muscles, *Experimental Brain Research* **89**, 1-19.
164. Schieber, M.H. (1995) Muscular production of individuated finger movements - the roles of extrinsic finger muscles, *Journal of Neuroscience* **15**, 284-297.
165. Maier, M.A. and Hepp-Reymond, M.C. (1995) EMG activation patterns during force production in precision grip .1. Contribution of 15 finger muscles to isometric force, *Experimental Brain Research* **103**, 108-122.
166. Maier, M.A. and Hepp-Reymond, M.C. (1995) EMG activation patterns during force production in precision grip .2. Muscular synergies in the spatial and temporal domain, *Experimental Brain Research* **103**, 123-136.
167. Nudo, R.J., Jenkins, W.M., Merzenich, M.M., Prejean, T. and Grenda, R. (1992) Neurophysiological correlates of hand preference in primary motor cortex of adult Squirrel-monkeys, *Journal of Neuroscience* **12**, 2918-2947.
168. Dugas, C. and Smith, A.M. (1992) Responses of cerebellar Purkinje cells to slip of a hand-held object, *Journal of Neurophysiology* **67**, 483-495.
169. Espinoza, E. and Smith, A.M. (1990) Purkinje cell simple spike activity during grasping and lifting objects of different textures and weights, *Journal of Neurophysiology* **64**, 698-714.
170. Seitz, R.J. and Roland, P.E. (1992) Learning of sequential finger movements in man - a combined kinematic and positron emission tomography (PET) Study, *European Journal of Neuroscience* **4**, X154-165.
171. Halsband, U. and Freund, H.J. (1993) Motor Learning, *Current Opinion in Neurobiology* **3**, 940-949.
172. Müller, F. and Dichgans, J. (1994) Dyscoordination of pinch and lift forces during grasp in patients with cerebellar lesions, *Experimental Brain Research* **101**, 485-492.
173. Rispalpadel, L. (1993) Contribution of cerebellar efferents to the organization of motor synergies, *Revue Neurologique* **149**, 716-727.
174. Dichgans, J. and Fetter, M. (1993) Compartmentalized cerebellar functions upon the stabilization of body posture, *Revue Neurologique* **149**, 654-664.
175. Holsapple, J.W., Preston, J.B. and Strick, P.L. (1991) The origin of thalamic inputs to the hand representation in the primary motor cortex, *Journal of Neuroscience* **11**, 2644-2654.
176. van Kan, P.L.E., Horn, K.M. and Gibson, A.R. (1994) The importance of hand use to discharge of interpositus neurons of the monkey, *Journal of Physiology* **480**, 171-190.
177. Goodale, M.A. and Milner, A.D. (1992) Separate visual pathways for perception and action, *Trends in Neurosciences* **15**, 20-25.
178. Jeannerod, M. (1994) The representing brain - neural correlates of motor intention and imagery, *Behavioral and Brain Sciences* **17**, 187-202.
179. Sakata, H. and Taira, M. (1994) Parietal control of hand action, *Current Opinion in Neurobiology* **4**, 847-856.
180. Charpentier, A. (1891) Analyse experimentale de quelgues elements de la sensation de poids, *Arch Physiol Norm Pathol* **3**, 122-135.
181. Rack, P.M.H. Limitations of somatosensory feedback in control of posture and movement, in J.M. Brookhart and V.B. Mountcastle (eds.), *Handbook of Physiology. Sect. 1: The Nervous System.* American Physiological Society, Bethesda, Maryland, pp. 229-256.

182. Hogan, N., Bizzi, E., Mussa-Ivaldi, F.A. and Flash, T. (1987) Controlling multijoint motor behavior, *Exercise and Sport Sciences Reviews* **15**, 153-190.
183. Kunesch, E., Binkofski, F. and Freund, H.-J. (1989) Invariant temporal characteristics of manipulative hand movements, *Experimental Brain Research* **78**, 539-546.
184. Johansson, R.S. and Edin, B.B. (1993) Predictive Feedforward Sensory Control During Grasping and Manipulation in Man, *Biomedical Research* **14**, 95-106.
185. Johansson, R.S. and Cole, K.J. (1994) Grasp Stability During Manipulative Actions, *Canadian Journal of Physiology and Pharmacology* **72**, 511-524.

THE TIMING OF HUMAN MOVEMENTS

H. HEUER

Institut für Arbeitsphysiologie

an der Universität Dortmund

Ardeystraße 67

D-44139 Dortmund

1. Introduction

For many tasks it is not only important what one does, but also when one does it. Catching is an example. A grasp that is too early or too late will result in a failure; the timing is an integral and critical part of the action. Timing is controlled, and the time window in which the action can be performed successfully is specified by environmental variables. Thus the timing is extrinsic in addition to being controlled.

In other tasks like playing music, precise timing is also a major task requirement and likely to be controlled, but there is no extrinsic signal that specifies the proper timing. Intrinsic timing is mostly ascribed to some internal clock, a kind of periodic process that can serve a clock-like function. Finally there are tasks which do not require a particular timing. Nevertheless movements necessarily have a temporal structure, and for tasks which do not require a particular timing this is likely to emerge as a by-product of other movement characteristics.

The distinction between controlled and emergent timing is not always drawn clearly. It refers primarily to task characteristics rather than to how the task is actually performed. With respect to control processes the distinction is somewhat unclear,

F. Lacquaniti and P. Viviani (eds.), Neural Bases of Motor Behaviour, 261–314.

because even when the task requires a certain timing, this can be achieved by varying other movement characteristics so that the proper timing comes about as a by-product. Clearer is the distinction between extrinsic and intrinsic timing. Extrinsic timing is guided by environmental information, while intrinsic timing is independent of the environment and can be controlled or emergent, which may be hard to distinguish.

In this chapter I shall discuss some results on extrinsic timing, followed by a discussion of some constraints for voluntary timing. Finally I shall describe a two-level framework that is useful for the formulation of specific models of intrinsic timing and which, in principle, can also be applied to extrinsic timing tasks.

2. Extrinsic Timing

A typical requirement of extrinsic timing tasks is that a certain state of a movement has to coincide temporally with a certain environmental state. For example, a grasp has to be synchronized with the arrival of the ball for a sucessful catch, the other side of the road has to be reached before arrival of the approaching car, or the movement of the bat has to cross the path of a ball at a certain moment. One of the questions that are raised with respect to performance in such tasks is concerned with the nature of the extrinsic timing information and how this information is used by processes of motor control. Another question which is hardly discussed at all but has a certain practical importance is concerned with individual differences in the accuracy of extrinsic timing. I shall address these two questions in turn.

2.1. EXTRINSIC TIMING INFORMATION

The information that is used in the temporal adjustment of human movements to the environment seems to be mostly visual. Since the seminal work of Lee (1976) a particular hypothesis on the nature of this visual timing information has received a great deal of attention. Basically this hypothesis holds that time - more precisely: time to contact with an approaching object - can be perceived directly on the basis of certain characteristics of the optical flow. I shall introduce this hypothesis by way of example and then present it in a somewhat more general way than is typical.

Lee, Young, Reddish, Lough, and Clayton (1983) studied the preparatory bending and stretching of the knee when subjects had to jump toward a falling ball and hit it with the fist. In different conditions the drop height of the ball was 3, 5 or 7.2 m. The question was to which perceivable characteristic of the falling ball the motor act of bending the knees and jumping was temporally coupled. Figure 1 presents three candidate variables.

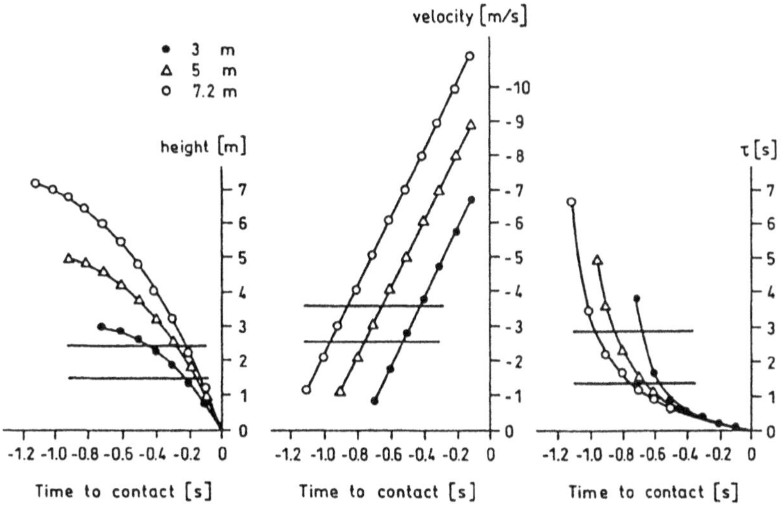

Figure 1: Height, velocity, and τ as a function of time to contact with the ground for objects falling from different heights (after Heuer, 1990)

The first candidate is the height of the ball. In the left graph of Figure 1 the height of the ball is shown as a function of time to contact with the ground. The two horizontal lines mark two landmarks of the knee movement like the start and the minimal knee angle. If the time course of the knee movement were linked to the height of the ball, the movement would start earlier with the small drop height than with the large one; the upper horizontal line marks the height at which knee bending could be initiated.

The time difference would be reduced at the minimal knee angle (lower horizontal line), and it would vanish during the stretching of the knee. The second candidate is the velocity of the ball, shown as a function of time to contact in the middle graph of Figure 1. If ball velocity provided the environmental timing information, bending of the knee should start earlier with the larger drop height than with the smaller one, and this difference should be the same at the second landmark. Finally, the right graph shows a variable called τ as a function of time to contact with the ground. It is equivalent to distance (or height in the particular instance) divided by velocity. Thus, at each moment in time, it specifies the time to contact under the assumption that velocity remained constant. If the knee movement were guided by this variable, it should start earlier with the larger drop height, and the effect of drop height should gradually vanish in the course of the movement. This is in fact what Lee et al. (1983) observed, and the differences between the three drop-height conditions essentially disappeared when the knee angles were plotted as a function of τ rather than of time of contact.

There is a good deal of support for the hypothesis that τ variables play an important role in extrinsic timing (e.g. Lee & Young, 1986). In recent years several detailed theoretical analyses of this type of variable have appeared, most notably perhaps a series of papers by Tresilian (e.g. 1994a). As compared with these analyses, my presentation here is somewhat simplified. Its main purpose is to convey the message that in assessing time to contact the visual system works as seems to be typical for it - it picks up all sorts of information that are available.

A τ ratio can be defined in general terms as the ratio of a variable x and its first derivative with respect to time, $\dot{x}$ (cf. Lee, 1992):

$$\tau_x = x / \dot{x} \tag{1}$$

Thus, τ is a time variable. When its value is negative (e.g., x is positive and $\dot{x}$ is negative), it indicates the time needed to reach $x = 0$ under the assumption that $\dot{x}$ is constant. This is a well-studied case. When τ takes a positive value, it indicates the time that has passed since $x = 0$, again under the assumption of $\dot{x}$ being constant; as far as I know this case has never been studied.

Tau ratios can be defined for variables in the external world. For example, when an object approaches another one, the ratio of the current distance and velocity can be computed to estimate the time of collision of the two objects. The same computations can be performed for egocentric distance, that is, distance of an object from an observer. In all cases the τ ratios are distal variables, variables that exist in the outside world and that are not immediately accessible to the visual system of an observer.

Distal variables give rise to proximal variables that can be sensed directly. For example, the distance between two objects is mapped on a distance between their images on the retina, or the distance of an object from an observer gives rise to a certain angle of convergence of the eyes and a certain retinal disparity. In many cases the relation between distal variables and proximal variables is not unique. The best-known example for the ambiguity of the relation between distal and proximal variables is perhaps the joint dependency of retinal size on the size and the distance of an object. The disambiguation of retinal size requires additional information about distance, and only to the extent that this becomes available becomes perception of the size of an object independent of its distance; the classic findings on size constancy (e.g. Holway & Boning, 1941) are reported in essentially all textbooks of visual perception.

Which proximal variables are used to assess distal τ? An answer to any question about how distal variables are assessed requires two kinds of analysis (cf. Brunswick, 1952). The first one is theoretical and deals with the "ecological validity" of proximal variables: which ones are systematically related to the distal variable and could be used in principle by the visual system? The second and following analysis is experimental and deals with the "functional validity" of the proximal variables that are ecologically valid. It requires an isolated variation of proximal variables - without the normal covariation of the distal variables of interest - to determine whether perception of the distal variable varies as hypothesized under the assumption that the proximal variable is functionally valid. By this the actual use of the proximal variable can be shown. Of course, it is not perception per se that is examined, but some type of response that indicates what is perceived. There are different types of response that can be used, and functional validity for one type of response does not necessarily imply functional validity for other types of response because different response systems can make different use of proximal variables (e.g. Bridgeman, Kirch, & Sperling, 1981).

Suppose that x is the distance of an object from an observer that approaches the point between the two eyes along the median plane of the head on a straight path. This is the situation that I shall discuss in some detail. For other situations like two objects that approach each other in a frontoparallel plane, other proximal variables may be relevant. For objects that have another direction of approach toward the observer or that will pass the observer like balls normally do, the proximal variables described below can be treated as approximations that, however, will not always be satisfactory (cf. Tresilian, 1990).

According to Lee's (1986) original hypothesis, the proximal variable for the perception of time to contact with an object that approches the observer or that is approached (like the red traffic light by a car) is the inverse relative rate of dilation of the retinal image. The size of the retinal image is generally indicated by the visual angle α. Up to about 10° the visual angle, measured in radians, can be approximated by

$$\alpha(t)[\mathrm{rad}] = \frac{S}{D(t)} \tag{2}$$

with S as object size and D as distance of the object. This approximation holds because for small angles $S/D \approx \tan(S/D)$. When the size S of the object is constant and the distance changes with velocity $D'(t)$, the rate of change of the visual angle is

$$\alpha'(t) = -\frac{SD'(t)}{D^2(t)} \tag{3}$$

For the ratio one obtains

$$\frac{\alpha(t)}{\alpha'(t)} = -\frac{SD^2(t)}{D(t)SD'(t)} = -\frac{D(t)}{D'(t)} \quad (4)$$

There are straightforward procedures for the examination of the functional validity of the proximal variable $\tau_\alpha(t) = \alpha(t)/\alpha'(t)$ for the perception of time to contact, and there can be hardly any doubt that it has an effect on estimates of time to contact (e.g. Schiff & Detwiler, 1979; Todd, 1981) as well as on behavioral adjustments (e.g. Savelsbergh, Whiting, & Bootsma, 1991). Although some authors hold explicitly or implicitly that τ_α is the only proximal variable relevant for the perception of time to contact, this is certainly not true. First, Tresilian (1990, 1991) has shown that there are conditions in which τ_α does not provide adequate information, in particular when the object does not move directly toward the eye of the observer and when a non-spherical object rotates during its approach. In fact, there are other ecologically valid proximal variables, and most likely these different variables are picked up and integrated (Tresilian, 1994a, b).

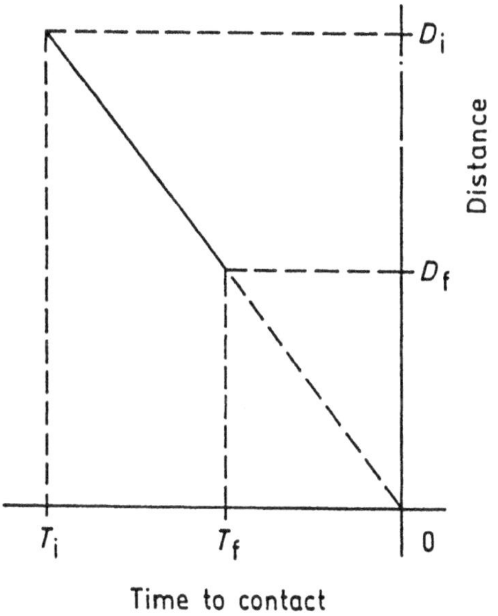

Figure 2: Distance-time plot of the object-motion simulated in the experiments of Heuer (1993)

Typically the perception of time to contact is discussed only with respect to retinal variables or optic-flow variables that give rise to them (e.g. Bruce & Green, 1985, and other textbooks on visual perception). Thus it is an intriguing question whether extraretinal information plays a role as well. Some results of Heuer (1993) indicate that it does so. Figure 2 and Table 1 describe the distal situation that was simulated optically: it was a constant-velocity approach of a circle on a straight path that hit the midpoint between the two eyes. The approach started in distance D_i and took time T_i. During the first part of its approach the circle was visible (proportion p of total duration and distance of the approach). The subjects had to extrapolate the motion of the circle and to press a key at the time the circle would have hit their head.

Table 1: Parameters of simulated constant-velocity approaches in Heuer's (1993) experiments (T_i, D_i, T_f, D_f : initial and final times to contact and distances; T_v, D_v : duration and distance of visible approach with proportion p of total duration and distance, v: velocity).

Distal event	T_i (s)	D_i (cm)	p	T_f (s)	D_f (cm)	T_v (s)	D_v (cm)	v (cm/s)
1	2	200	0.5	1.0	100	1.0	100	100
2	2	200	0.7	0.6	60	1.4	140	100
3	2	100	0.5	1.0	50	1.0	50	50
4	2	100	0.7	0.6	30	1.4	70	50
5	4	200	0.5	2.0	100	2.0	100	50
6	4	200	0.7	1.2	60	2.8	140	50
7	4	100	0.5	2.0	50	2.0	50	25
8	4	100	0.7	1.2	30	2.8	70	25
9	8	200	0.5	4.0	100	4.0	100	25
10	8	200	0.7	2.4	60	5.6	140	25
11	8	100	0.5	4.0	50	4.0	50	12.5
12	8	100	0.7	2.4	30	5.6	70	12.5

The approach was optically simulated by means of two variables, (retinal) size and target vergence. Subjects looked into a large wooden box so that only the circles were visible in a completely dark surround. They were presented on an oscilloscope 20 cm in front of the eyes. Subject looked through 4-D lenses, so that their accommodation was fixed at 1 m. Two circles were presented, one to the left and one to the right eye by way of appropriately placed polarizing filters. Retinal size was varied by means of changing the radius of the circles, and target vergence was varied by means of changing the distance between their midpoints. Changing size and changing target vergence represented the two stimuli in this experiment.

The changing-size stimulus gives rise to changing retinal size as a proximal variable. When the size is increased, one perceives an approach of the circle together with an increase of its apparent size. Changing target vergence can be mapped on two different proximal variables. With a perfect fixation of the circle, target vergence is mapped on ocular vergence; with less than perfect fixation a part of the target-vergence stimulus is also mapped on (absolute) retinal disparity. In the experiment the target vergence changed up to several degrees. Thus, the fact that the circles remained fused indicates that a major part of the target-vergence changes was mapped on changes of ocular vergence. When target vergence is increased (lateral distance between the circles is reduced) one perceives an approach of the circle together with a decrease of its apparent size. The covariation of apparent size with target vergence is produced by the size-constancy mechanism mentioned above (in normal life visual angles that arise from objects at different distances are constant only when the closer objects are smaller). This is what one perceives with circles (Heuer, 1987, 1993) or lines (Swanston & Gogel, 1986, Exp. 2), but with random-dot stereograms no apparent motion in depth has been found when target vergence was changed, although ocular vergence changed according to the stimulus (Erkelens & Collewijn, 1985a, b). The reasons for this difference seem to be unknown.

In Figure 3 the mean estimates of time to contact are shown for vergence stimuli (size was constant and indicated a stationary object), size stimuli (target vergence was constant and indicated a stationary object) and congruent stimuli (both stimuli simulated the same distal object motion). The estimates are given as time since the start of the apparent approach, and they are means across different initial distances and different visible proportions. These latter two factors affected the estimates, but these effects will be skipped here. The shortest estimates were given when the stimuli were congruent; the estimates were the largest when only target vergence indicated the approach but size a stationary object, and they were shorter when the approach was indicated by changing size (and target vergence indicated a stationary object). According to these results estimates of time to contact depend on both changing size and changing target vergence, although the former variable seems to be more powerful than the latter.

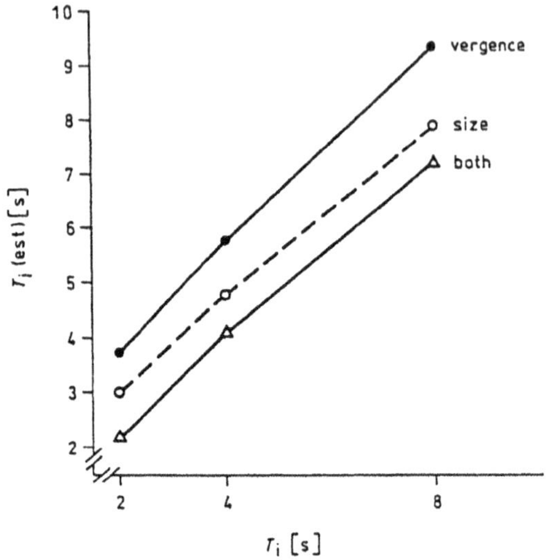

Figure 3: Estimated time to contact as a function of time to contact for object approaches simulated by target vergence, size, and both stimuli (after Heuer, 1993)

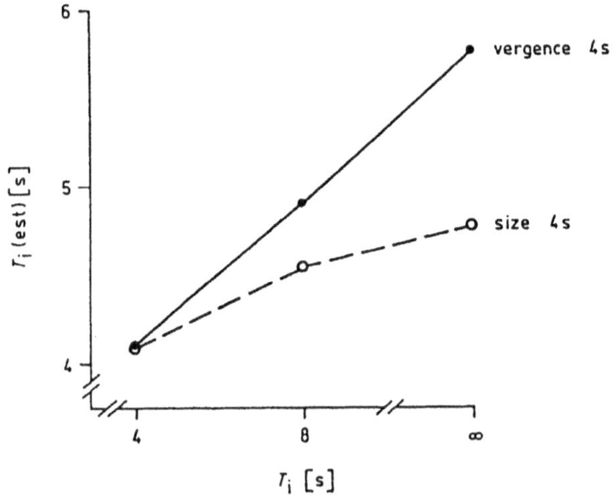

Figure 4: Estimated time to contact as a function of time to contact indicated by changing size (target vergence set to T_i = 4 s) or changing target vergence (size set to T_i = 4 s) (after Heuer, 1993)

The same conclusion can be drawn from the findings shown in Figure 4. Some additional incongruent stimuli were used in which the one stimulus indicated a time to contact of 4 s, but the other one a slower approach with a time to contact of 8 s. When the changing-size stimulus was set to 4 s time to contact and the changing-target-vergence stimulus to 4 s, 8 s, and stationary object, estimates of time to contact increased somewhat. But they increased much stronger when the role of the stimuli was reversed so that changing target vergence was set to 4 s and changing size to 4 s, 8 s, and stationary object.

How general is the finding that changing size has more powerful effects on estimates of time to contact than changing target vergence? Consider objects that are progressively reduced in size. When very small objects approach an observer, changing size may become hard to detect, and the changing-size stimulus is likely to become less powerful. Figure 5 shows that this is indeed the case. In the second experiment of Heuer (1993) the size of the simulated circle was varied; its radius was set to 0.35, 0.70, and 1.40 cm. With the largest radius the results are very similar to those of the first experiment in which the radius was 1.0 cm. When the radius was reduced, estimates based on congruent stimuli remained the shortest ones, but estimates based on single stimuli showed a reversal. With the smallest object size they were largest when changing size indicated the approach and target vergence a stationary object, and they were shorter when the approach was indicated by target vergence.

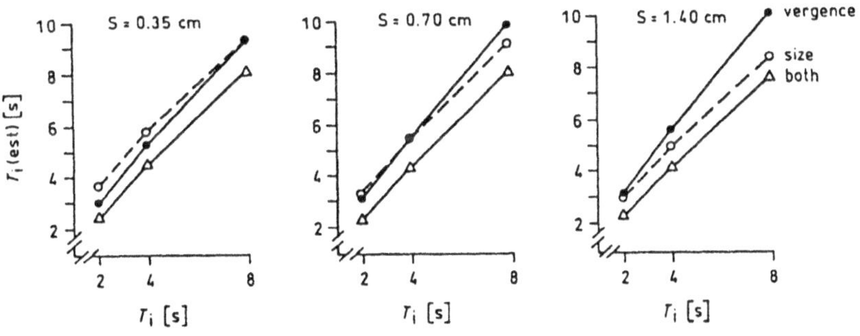

Figure 5: Estimated time to contact as a function of time to contact for object approaches simulated by target vergence, size, and both stimuli and for different object sizes S (after Heuer, 1993)

The estimates of time to contact reflect both variations of changing-size and changing-target-vergence stimuli. There seems to be nothing special about the inverse relative dilation of the retinal image. As is generally the case in the perception of space, the visual system picks up whatever information is available and combines this information in some way, the specifics of which may depend on the sensitivity or validity of the different proximal variables with respect to their distal referents. The combination of changing target vergence and changing size can also be evidenced from judgements of the perceived amplitude of optically simulated oscillatory motions in depth, as long as the two stimuli are in phase. When they are anti-phase so that the one indicates an approach of the object and the other one a simultaneous recession, the combination can break down. In a certain range of amplitudes of simulated anti-phase motions bistable perceptions can result that are determined either by the one or by the other stimulus (Heuer, 1987).

As mentioned above, the functional validity of proximal variables may depend on the task. The findings that indicate the combination of different extrinsic timing variables were obtained with highly "cognitive" tasks: subjects perceived the moving object, had to extrapolate the motion and then to press a single key, or they had to provide a well-considered judgement of the apparent amplitude of an oscillatory motion. In these tasks the situation is different from tasks like catching for which one could suppose a tighter perceptual-motor linkage. An extensive discussion of task differences can be found in Tresilian (1995). For example, the accuracy in hitting an approaching ball is much higher than the accuracy of estimates of time to contact - and there are good reasons for this difference. Nevertheless, I am not aware of evidence according to which the use of different proximal variables is really different for these tasks. In the experiment with oscillatory motion in depth (Heuer, 1987) the main findings were not only observed with verbal judgements of the perceived amplitude, but also with pointing movements to the last position of the circle that disappeared at one or the other extremum of its path. Thus it is likely that extrinsic timing is based on whatever information is useful - and this exploitation of redundancy seems to be typical for the visual system.

2.2. EXTRINSIC TIMING INFORMATION AND MOTOR CONTROL

How is extrinsic timing information linked to processes of motor control? Most likely the answer to this question will depend on the kind of task and the kind of information. One task that has received some attention in the literature is that of Lee et al. (1993). Although originally the proximal variable thought relevant was the inverse relative dilation of the retinal image, other proximal variables may have been used in that task. However, for the question of how the extrinsic timing information is integrated with processes of motor control the precise nature of the extrinsic information can perhaps be neglected.

One way to conceptualize the integration of extrinsic timing information into processes of motor control is in terms of a generalized motor program that can be characterized by way of a prototypical function (e.g. Meyer, Smith, & Wright, 1982; Heuer, 1984). Upon execution the prototypical function $f(\vartheta)$ is transformed into real time. This can be done by reading a spatially organized representation of the movement trajectory (cf. Lashley, 1951). In this case ϑ is replaced by an internally generated timing signal. As suggested by Cruse, Dean, Heuer, and Schmidt (1991), ϑ could also be replaced by an extrinsic timing signal τ. Although this is a simple motor-program model, it makes continuous use of the extrinsic timing signal.

Tresilian (1994b) discussed other models for the particular task. For a continuous model he assumed that the extrinsic timing variable determines the equilibrium position of the joint. With certain ancillary assumptions this model reproduced the experimental data of Lee et al. (1983) fairly adequately. However, Tresilian (1994b) did also fit a model without continuous use of extrinsic timing information. Thus, fundamentally different models for the available data are possible, and there is yet no way to decide upon how information about time to contact is really integrated with processes of motor control.

Information about time to contact is anticipatory. It specifies the time that remains until a certain event will occur and thereby allows the proper temporal adjustment of anticipatory movements. However, there are timing tasks in which such anticipatory information is not available and in which anticipatory motor control is achieved in other ways. The best-studied task of this kind is to tap in synchrony with some external pacing signal. This is possible only with a periodic pacing signal (or perhaps a more complex regularity than periodicity). Models for such tasks typically assume the adjustment of an internal timing mechanism to the period and phase of the pacing signal (e.g. Mates, 1994a, b; Schöner & Kelso, 1988), however different the models are in other respects.

The synchronization task reveals some complications in the extrinsic timing of motor patterns that are not evident from the studies of time-to-contact information. When subjects have to tap in synchrony with external pacing signals like simple clicks, they exhibit a so-called negative asynchrony that has been observed consistently since the original finding of Dunlap (1910): the taps lead the clicks by about 20 to 100 ms. According to the available evidence (Aschersleben, 1994; Aschersleben & Prinz, 1995) the lead of the tap results from the fact that its feedback is synchronized with the click rather than the (physical) tap itself. For example, because of longer nerve conduction times the feedback delay is longer for the foot than for the hand, and so is the negative asynchrony. For auditory feedback the delay is shorter than for tactile and/or kinesthetic feedback; the negative asynchrony is reduced correspondingly when auditory feedback is added. When auditory feedback is artificially delayed, negative asynchrony increases again. Thus, what is synchronized is the perception of the movement with the perception of the click.

For catching a ball or hitting a baseball, extrinsic timing with reference to sensory feedback rather than to the physical movement would most likely be disastrous for performance. As pointed out by Aschersleben (1994) and Aschersleben and Prinz (1995), these tasks have a feature that the synchronization task does not have: there is knowledge of results - e.g. catch or failure - that relates the physical movement to the timing signal. Thus, anticipatory motor control can be adjusted accordingly. When the processing of time-to-contact information is analyzed in the light of these considerations, it becomes obvious that one of the key problems is the choice of the proper anticipation time. This aspect of the problem has thus far been neglected in attempts to formally model the integration of continuous extrinsic timing information into motor control.

2.3. INDIVIDUAL DIFFERENCES IN EXTRINSIC TIMING

One of the everyday tasks that require extrinsic timing is crossing a road. The observation that boys at about 7 years of age are particularly at risk triggered Ehrmann (1988) to explore potential reasons for this. Specifically he was interested in how the behavior is adjusted to the time that remains until a car reaches the place where the road is crossed. The task of crossing a road when faced with an approaching car can be considered as a particular type of coincident-timing task.

Coincident-timing tasks require a certain limb position at a certain time like in hitting a ball. In most experimental variants of the task the temporal target is presented with preview so that it can be anticipated. An example is a falling ball (e.g. Bootsma, 1989). Schmidt (1972; Schmidt & Russell, 1972) introduced a certain method of analyzing the timing in such tasks that was only slightly modified by Ehrmann (1989). The start time is defined as the time the movement begins relative to the temporal target. Movement time is the time from the start of the movement until it reaches the target position. The error is the difference of movement time minus start time, which is negative when the target position is reached too early and positive when the target position is reached too late, after the arrival of the target object. Schmidt (1972) was particularly interested in the correlation between start time and error. The reason for this interest was the following consideration: To the extent that the movement is started and runs off thereafter without any processing of extrinsic timing information, the error should be largely determined by the start time. If, however, the movement is perfectly adjusted to the extrinsic timing information, the error should always be close to zero. Consistent with this rationale the correlation decreased when movement time was increased from about 150 ms to about 750 ms (Schmidt & Russell, 1972). This suggests that, as total movement time increases, it becomes better adjusted to the time the movement was initiated and thus to the time that remains until the target object will reach the target position.

In crossing a road the temporal errors are associated with different costs. A positive error (being too late) is costly - injury or death in the real situation - while a negative error (being too early) is associated with low costs - the costs of running at most. Therefore one should expect a constant negative error even in a simulated road-crossing task. But otherwise the task is suited for the kind of analyses that have been applied to coincident-timing tasks.

Figure 6 shows the experimental situation that was used by Ehrmann (1988). The extrinsic timing information was presented in a quite unnatural, but simple way: A marker on a video screen approached a vertical line with a constant speed. With this display the time to contact is given by the retinal distance divided by its rate of change. The movement time of the cursor varied between about 7 and 11 s for the shorter walking distance and between about 15 and 22 s for the longer walking distance; the times correspond to those that a car would need at a speed of 40 to 60 km/h to pass a distance of 125 or 250 m.

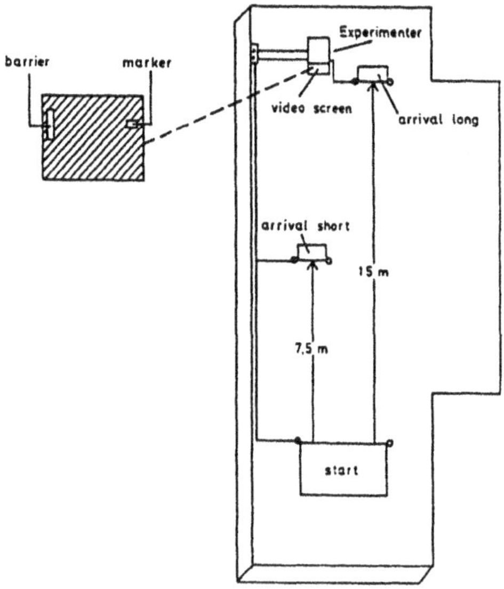

Figure 6: The experimental setup of Ehrmann (1988)

After the start of the marker the subjects could start to cross the road at a freely chosen time. They were requested to walk a distance of 7.5 m or 15 m, which correspond to typical German 2-lane and 4-lane roads. Start and end of the walk were timed automatically by means of a photoelectric make-and-breaks at each end. The instruction required the subjects to start as late as possible, but to arrive at the safe side of the road before the car without having to run (except in the case of danger). Four groups of subjects participated, 18 boys (mean age: 7;4), 18 girls (7;2), 18 adult males (30;10), and 18 females (30;5).

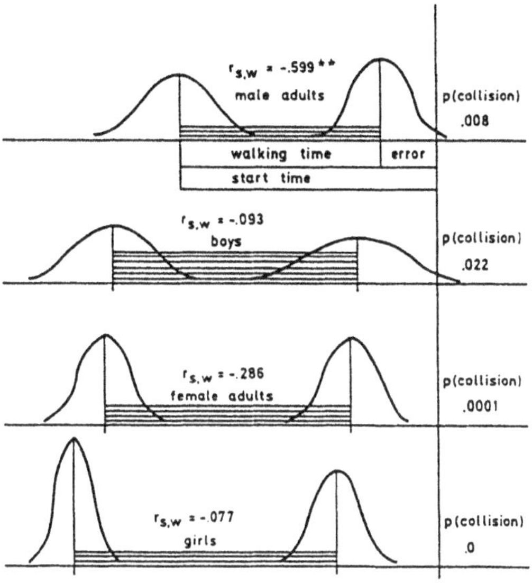

Figure 7: Summary of results of Ehrmann (1988); $r_{s,w}$ is the correlation between start time and walking time

The results are summarized in Figure 7. The width of the distributions reflects the mean individual standard deviations; the mean standard deviation of walking times is indicated by the number of horizontal lines. From the means and standard deviations of arrival times the probabilities of "collisions", that is, of positive errors, were estimated (under the assumption of normal distributions) and are given in Figure 7. The ranking of the groups according to these probabilies was the same as the ranking according to the number of collisions actually observed in each group (6, 12, 1, and 0 for males, boys, females, and girls, respectively). Thus the particular risk of the boys, that seems to exist in the real world, did also exist in the artificial task.

What is special about the boys? They were similar to adult females with respect to the mean start time and mean arrival time (or error); girls started earlier and arrived earlier on the average, while males started later and arrived later. Boys were similar to adult males with respect to the variability of start times; females and particularly girls exhibited a smaller variability. Finally, they were similar to girls in that their walking time was essentially independent of their start time, while in females and particularly in males a later start time was compensated by a shorter walking time. Thus, no single characteristic of behavior was unique for the boys, but the combination of two characteristics caused a high scatter of arrival times and, among the four groups, the highest number of collisions: high variability of start times and a lack of compensation by the appropriate covariation of walking times. Males compensated their start-time variability, and girls didn't need such compensation because their start times were more consistent. These findings suggest that the high accident risk of 7-year-old boys might be related to their insufficient use of extrinsic timing information together with a high variability in choosing the times for crossing a road in face of an approaching car; the insufficient use of extrinsic timing information was also found in the 7-year-old girls, but there it was inconsequential because of the low variability of start times and the earlier mean arrival times.

3. Timing constraints

Although the timing of human movements is quite flexible, there are limitations. Some physical timing constraints are obvious. For example, temporal patterns cannot be produced when they require forces that exceed human capabilities. But there are other and more subtle constraints that seem to be related to the exploitation of the characteristics of the passive muscle-bone systems and to characteristics of the neural machinery that is involved in motor control. I shall consider a couple of such constraints.

3.1. THE RELATION BETWEEN CURVATURE AND VELOCITY

Figure 8 shows several recordings of movements of a finger in the air. They are taken from Derwort (1938) and were produced by means of a stroboscopic light that was attached to the finger tip. The density of the dots indicates velocity, with a higher density corresponding to a smaller velocity. In Figure 8a examples for freely drawn shapes are shown. Without particular instructions on velocity there is a clearcut covariation between velocity and curvature: velocity is small for high curvature, and it is high for small curvature. This relation apparently cannot be overcome voluntarily. The movement shown in Figure 8b was produced with a constant-velocity instruction, but nevertheless exhibits the covariation. More recent data show that the covariation even survives the tracking of a constant-velocity target (Viviani, Campadelli, & Mounoud, 1987). Finally, it is not only curvature that determines (relative) velocity, but voluntary variations of velocity do also produce covariations of curvature. The left circle in Figure 8c was drawn with a constant-velocity instruction, but the right circle with a variable-velocity instruction, which led to a shape distortion.

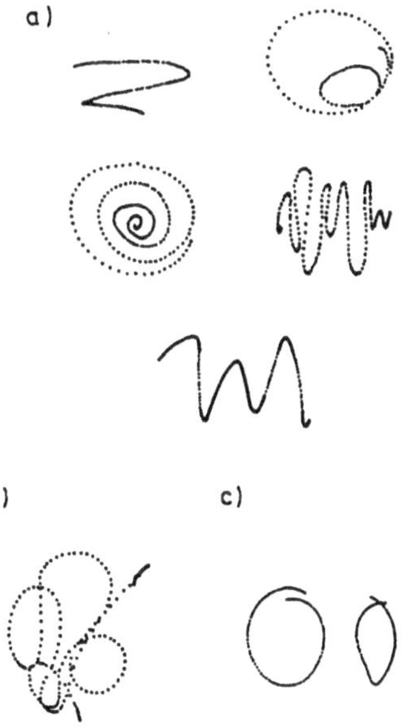

Figure 8. Traces of finger movements in the air; higher dot density indicates smaller velocity. (a) Shapes with spontaneously chosen velocities; (b) Shape produced with a constant-velocity instruction; (c) Circle drawn with constant- (left) and variable-(right) velocity instruction (after Derwort, 1938)

The relation between curvature and velocity has been systematically explored by Viviani and coworkers in a number of studies (e.g. Lacquaniti, Terzuolo, & Viviani, 1983; Viviani & Terzuolo, 1982; Viviani & Schneider, 1991). It has become known as the "two-thirds power law"

$$A(t) = kC(t)^{2/3} \tag{5}$$

with A as angular velocity and C as curvature. When expressed as relation between radius of curvature R and tangential velocity V, with C=1/R and A=V/(2πR), the two-thirds power law becomes a "one-third power law":

$$V(t) = K \cdot R(t)^{1/3} \tag{6}$$

These two relations are meaningful only for $|C|>0$ or $|R|<\infty$. They have been found to hold for the drawing of circles and ellipses, but also for amorphous scribbles, at least as reasonably good approximations. Stronger violations have been found in handwriting (Thomassen & Teulings, 1985).

One perspective on the one-third power law is that it is an empirical relation similar to Stevens' law in psychophysics. Under this perspective it is a reasonable procedure to fit the linear equation log V = log K + n log R to data and thereby to determine the constant K and to examine whether n deviates from 1/3 (e.g. Wann, Nimmo-Smith, & Wing, 1988). Another perspective is to consider the one-third power law as a truism, except for one its features, the constancy of K. In my view this is a useful perspective, so I shall describe it in more detail.

A movement in a plane can be described by the pair of equations x = x(t), y = y (t). For example, with $x(t) = A_x \cos\omega t$, $y(t) = A_y \sin\omega t$ one obtains an elliptical movement or, with $A_x = A_y$, a circular movement. With temporal derivatives in the dot notation, the following two definitions hold:

$$V = \left(\dot{x}^2 + \dot{y}^2\right)^{1/2} \tag{7}$$

$$R = \frac{\left(\dot{x}^2 + \dot{y}^2\right)^{3/2}}{\dot{x}\ddot{y} - \dot{y}\ddot{x}} \tag{8}$$

From these definitions one obtains immediately:

$$R = \frac{1}{\dot{x}\ddot{y} - \dot{y}\ddot{x}} V^3$$
$$V = \left(\dot{x}\ddot{y} - \dot{y}\ddot{x}\right)^{1/3} R^{1/3} \tag{9}$$

This is the one-third power law of Eq. (6) with $K = (\dot{x}\ddot{y} - \dot{y}\ddot{x})^{1/3}$ as a time-varying quantity. Thus, the factorization $V = K \cdot R^{1/3}$ follows from the definitions of tangential velocity and radius of curvature, and the only substantial feature of the one-third power law that goes beyond the mathematical truism is the constancy of K over time.

When, under the second perspective, a power function is fitted to data and the exponent n is not equal 1/3, this indicates that K depends on the curvature. With n = 1/3 + m one obtains:

$$V = bR^{1/3+m} = bR^m R^{1/3} = KR^{1/3} \tag{10}$$

with $K = bR^m$. Thus, with positive m K will gradually increase as the radius of curvature increases and the path becomes straighter, while with negative m K will gradually decline.

The second perspective does also suggest to explore in more detail those movement characteristics that will lead to a constant value of K. The critical characteristic is

$$\frac{d\left(\dot{x}\,\ddot{y} - \dot{y}\,\ddot{x}\right)}{dt} = \dot{x}\,\dddot{y} - \dot{y}\,\dddot{x} = 0 \quad \text{or} \qquad \frac{\dot{x}}{\dot{y}} = \frac{\dddot{x}}{\dddot{y}} \tag{11}$$

This condition is satisfied for $x(t) = A_x \cos\omega t$, $y(t) = A_y \sin\omega t$ with $\dot{x}/\dot{y} = -(A_x \sin\omega t)/(A_y \cos\omega t) = \dddot{x}/\dddot{y}$, that is, for all ellipses that are drawn at a reasonable speed so that the movements along both axes of the plane look harmonic. This is a sufficient condition for the one-third power law to hold, but perhaps not a necessary condition.

More generally, the necessary condition for the one-third power law that is given by Eq. (11) is satisfied when the jerk ratio is identical to the velocity ratio. The velocity ratio $\dot{x}/\dot{y}$ is the tangent of the angle of the velocity vector with the y axis, while the jerk ratio $\dddot{x}/\dddot{y}$ is the tangent of the same angle of the jerk vector. Thus, the condition of Eq. (11) is satisfied and the one-third power law will hold whenever the current jerk vector is aligned with the current axis of movement, pointing in the same or opposite direction. This seems to happen, for example, in scribbling-like movements for certain stretches of time (cf. Viviani, 1986). Bearing in mind the proportionality of acceleration and force, one might suspect that there are certain advantages in keeping the axis of the force change aligned with the axis of movement, so that this is a way of producing movements that is chosen spontaneously and hard to overcome when required by the task.

3.2. TIMING CONSTRAINTS FOR UNI-ARTICULAR MOVEMENTS

For movements of a single joint there is no meaningful relation between curvature and velocity. Nevertheless, the timing is not arbitrary. To illustrate the existence of timing constraints even in such simple tasks, I shall use an example from Heuer and Schmidt (1988). The task of the subjects was to reproduce a certain movement of a cursor on a screen by way of flexions and extensions of the elbow joint in a horizontal plane. In each trial the target pattern was presented, reproduced by the subject, and feedback was presented thereafter. The position-time curves of the two target patterns are shown in Figure 9. The "harmonic" pattern was generated by way of double integration of an amplitude-modulated sine function. From this pattern the "non-harmonic" target pattern was derived by a nonlinear transformation of the time axis so that the middle two quarters were temporally expanded and the initial and final quarters were compressed. Thus, the sequence of positions was exactly the same in the two target patterns, only the arrangement of these positions on the time axis was different.

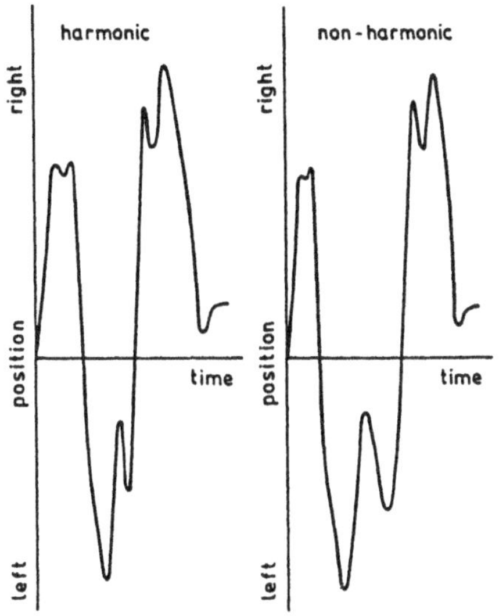

Figure 9: Position-time curves of the target patterns used by Heuer and Schmidt (1988)

Reproduction of the two target patterns was practiced by two groups of subjects; the details of practice are described by Heuer and Schmidt (1988) and not really important here. Each movement of the subjects was linearly compressed or expanded in time so that its duration was the same as that of the target pattern, and the cross-correlation R which the target pattern was computed. These cross-correlations are sensitive to timing errors, in particular to errors in the timing of the large and rapid position changes. In Figure 10 (upper graph) their means are shown. They increase with practice, and the non-harmonic pattern is reproduced less accurately than the harmonic one.

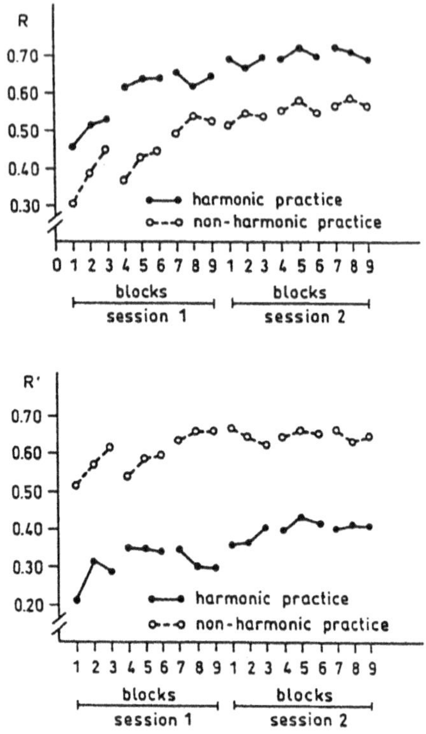

Figure 10: Upper graph: Cross-correlation of reproductions with practiced pattern (R). Lower graph: Cross-correlation of reproductions with target patterns practiced by the other group (R') (after Heuer and Schmidt, 1988)

In addition to the cross-correlations R of the reproductions with the target pattern practiced, the cross-correlations R' with the target pattern practiced by the other group were computed. As Figure 10 (lower graph) shows, this correlation reached a value of about 0.4 for the subjects who practiced the harmonic pattern, while their correlation with the target pattern that was actually practiced reached 0.7. However, for the subjects who reproduced and practiced the non-harmonic pattern, the reproduction had a higher correlation with the harmonic pattern than with the non-harmonic one. Thus, the reproductions of the non-harmonic pattern had a considerable bias toward the harmonic target pattern, while the reproductions of the harmonic pattern apparently did not have a bias toward the non-harmonic one (the correlation of the reproductions of the harmonic pattern with the non-harmonic target pattern did most likely result from the correlation of 0.68 between the two target patterns).

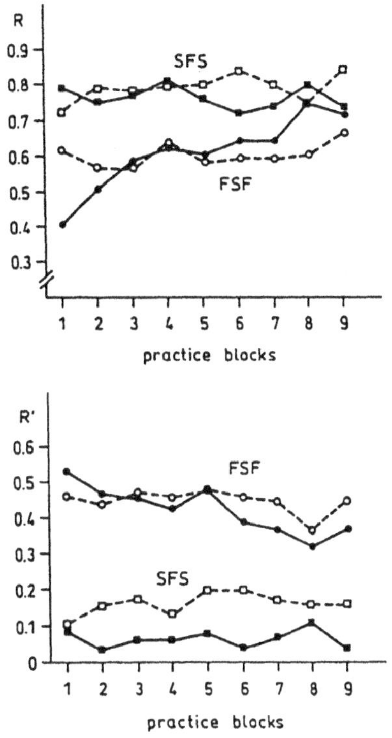

Figure 11: Upper graph: Cross-correlation of reproductions with practiced target patterns (R); each of the two target patterns SFS and FSF was practiced by two groups. Lower graph: Cross-correlations of reproductions with target patterns practiced by the other groups (R') (after Heuer, 1988a)

Figure 11 shows equivalent results from an experiment in which the cross-correlation between the two target patterns was close to zero (Heuer, 1988a). Both were derived from a harmonic pattern by nonlinear transformations of the time axis; the harmonic pattern was similar to the one shown in Figure 9. In the one transformation (fast-slow-fast, FSF) the initial and final quarters were compressed, while the middle two were expanded; in the other transformation (slow-fast-slow, SFS) the initial and final quarters were expanded and the middle two compressed. In the two groups who practiced the SFS pattern the reproductions were highly correlated with that pattern from the very start, while in the two groups who practized the FSF pattern the correlations with the target pattern were smaller initially and only gradually reached a higher level (Figure 11, upper graph).

Reproductions of the SFS pattern were only little correlated with the FSF target pattern (Figure 11, lower graph), while reproductions of the FSF pattern had higher correlations with the SFS target pattern, initially as high as the correlations with the FSF pattern. In the course of practice, as the correlations with the FSF target pattern increased, those with the SFS target pattern declined. Thus there was a clear bias toward the SFS pattern that was somewhat reduced in the course of practice. Heuer and Schmidt (1988) called the patterns towards which the reproductions are biased "natural patterns". The existence of natural patterns is perhaps the reason for the failure of subjects to achieve a certain instructed timing pattern, as has been observed occasionally (e.g. Carter & Shapiro, 1984). Whether or not a timing pattern is natural cannot only be inferred from biases in performance, but performers seem to experience it immediately.

What is it what makes a certain temporal pattern, given a certain sequence of spatial positions, a natural pattern or a less natural one? I am not aware of a concise formulation of the timing constraints in one-dimensional movements comparable to the one-third power law for movements in a plane. One may suspect that there are at least two sources for such constraints: The one are the intrinsic mechanical characteristics of the moving limb, the other are some characteristics of human rhythm production.

3.3. CHARACTERISTICS OF PASSIVE LIMBS

Typically a movement results from an interplay of active and passive forces. When muscular forces are used to drive a limb through a certain spatio-temporal pattern, they must be adjusted to the passive characteristics of the muscle-bone system. The passive characteristics of a joint can be approximated by a mechanical system with a mass between two opposing damped springs, that is, by a second-order system. Once set into motion, such a system tends toward oscillatory behavior. Because of the damping the amplitude shrinks over time, with the frequency being determined by the parameters. This frequency can be observed when a passive limb is pushed by the experimenter and the subject actively continues the originally passive oscillation.

How is such a mechanical system driven at different frequencies by way of changing the length-tension characteristics of the antagonistic springs, that is, by way of contracting the antagonistic muscles? Wachholder and Altenburger (1926) recorded the EMG from the extensor carpi radialis and the flexor radialis during hand oscillations with different frequencies. Figure 12 summarizes their findings. With slow movements, flexor activity starts during flexion and tends to continue into extension, while extensor activity starts during extension and tends to continue into flexion. Thus the reversal of the direction of the movement is passive. With the medium tempo, the flexor starts at the beginning of flexion and ends before the reversal of direction; similarly the extensor becomes active at the beginning of extension and ends before the reversal. Finally, with a high tempo, the flexor starts while the wrist is still in extension, and the extensor starts while the wrist is still in flexion. Not shown in the diagram is the case that the flexor is active during extension and the extensor during flexion, which can happen at very high frequencies. Thus, the phase relation between EMG and movement depends in a systematic manner on oscillation frequency.

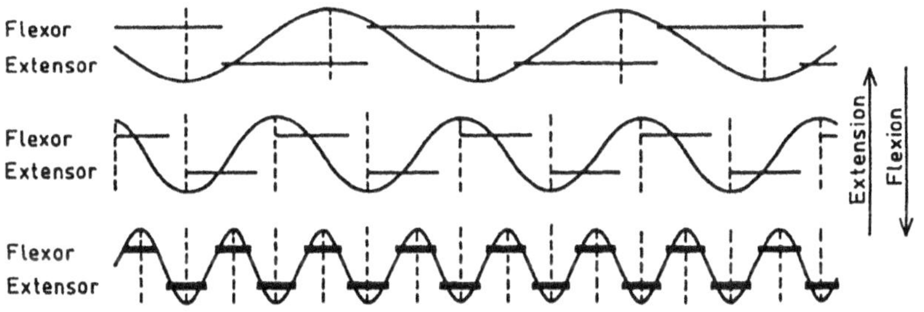

Figure 12: Phase relations between EMG bursts (horizontal lines) and oscillatory hand movements at different frequencies (after Wachholder & Altenburger, 1926)

In considering the interplay of active and passive forces, Wachholder and Altenburger (1926) concluded that there is an intermediate tempo at which the passive forces are movement until reversal, and in which no active forces are needed to decelerate the limb and to reverse the direction of movement. At this optimal tempo the agonist becomes active as soon as the movement is in the corresponding direction, and is active only to the extent that the damping is compensated.

The observation that oscillations of a certain frequency are associated with a certain phase relation between EMG and movement leads to the question of how these two are coordinated and what will happen when there are small errors in timing the EMG relative to the movement. Of course, the timing of the EMG is likely to be particularly delicate when movements are very rapid as in skilled typing. In rapid-tapping tasks so-called blockings can be observed in which the inter-tap interval is about twice as long as the normal interval or even longer (e.g. Warren & Clark, 1937). It is possible that such blockings result from an improper timing of EMG bursts, and some preliminary findings indicate that this is so.

Figure 13 shows in more detail what happens during a blocking; it covers a period of about 1 s in very rapid tapping. The topmost trace shows the contact of the index finger with the platform on which the tapping was performed, the trace below the position-time curve of the finger (angle at the metacarpo-phalangeal joint). What happens during the blocking is that the finger moves downward, but not far enough to make contact with the ground.

The two lowest traces show the EMG recordings: the extensor burst is more or less coincident with the contact of the finger with the ground, while the flexor burst is located in the extension phase of the joint movement. In this particular example of a blocking the one extensor burst seems to have ocurred too early so that the downward movement was reversed before the finger reached the platform. A smaller than usual interval between two extensor bursts (A and B) and also between the flexor burst in-between (C) and the second of the extensor bursts (B) was characteristic for one of the subjects from whom we have taken such recordings; in another subject the interval between successive extensor bursts was shorter than usual as well, but here the shortening occurred between the first of the extensor bursts and the next flexor burst.

Thus, according to these somewhat preliminary observations, blocking can result from an imprecise timing of the EMG relative to the movement.

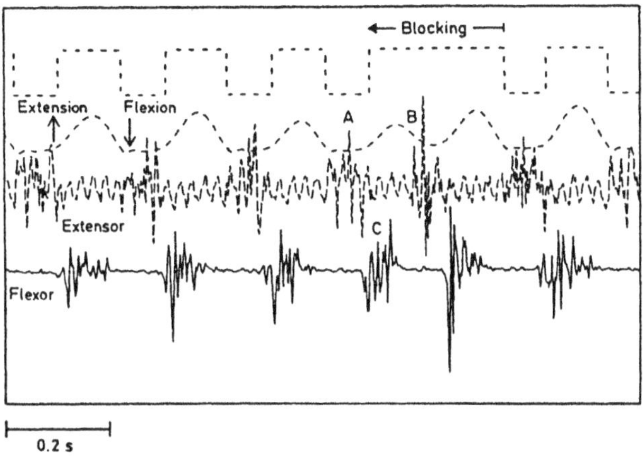

Figure 13: Recording of a blocking in very rapid tapping. Traces are (from top to bottom): contact (low) vs. no-contact (high) with the tapping surface, angle of metacarpo-phalangeal joint of index finger, extensor EMG, flexor EMG

The findings of Wachholder and Altenburger (1926) do not only point to the importance of a precise temporal relation between muscle activity and overt movement which is frequency dependent, but also to the existence of an optimal relation that determines an optimal frequency at which passive forces are exploited best. Wachholder and Altenburger (1926) did also observe that the "Elastizitätstempo", the passive frequency of a limb, depended on amplitude. More recent data (Rosenbaum, Slotta, Vaughan, & Plamondon, 1991) show in a more systematic manner that the preferred frequency of periodic movements increases with smaller instructed amplitude, and also that the preferred amplitude increases when the instructed frequency is reduced. Thus, there seems to be not only a preferred and perhaps most economic frequency-and-amplitude combination, but also a certain preferred frequency for a certain given amplitude and vice versa. Such relations could be one of the underlying factors for natural temporal patterns, although other factors might contribute when the position-time curves of uni-articular movements become more complex.

3.4. RHYTHMS

Rhythms seem to exist in essentially all sequences of human movements. Even in such simple sequences as in oscillatory movements of the hand humans tend to emphasize either flexion or extension so that there are stressed and unstressed elements in the sequence. Wachholder and Altenburger (1926) observed that stress is reflected in the pattern of muscular activity; in particular the interval between the bursts of the stressed and unstressed muscle was longer than the interval between unstressed and the following stressed muscle. According to subjective experience a rhythm is also associated with the production of more irregular patterns, for example in the reproductions of the target patterns shown in Figure 9.

Rhythms constitute constraints for the timing of movements. These constraints have been studied in sequences of simple key presses, and this is a task in which the rhythmic constraints on motor timing can perhaps be studied best in that most likely there are no constraints that result from the interplay of active and passive forces. Keele (1986) has summarized the constraints on timing in these task by way of three principles which define "natural" temporal patterns. When other patterns are required, the task becomes more difficult and most likely biases will be observed so that the temporal pattern produced becomes more similar to a natural pattern.

The first principle according to Keele (1986) is that of beats. At one level of a hierarchical temporal structure there must be repeated intervals that are marked by (sensory or motor) events. For example, the patterns 250-250-500 or 250-250-250-750 obey this principle with beat intervals 500 and 750, but the pattern 300-300-400 does not. The second principle is that of equal subdivision at a certain level of the hierarchy (except that some intervals might not be subdivided at all). The pattern 200-200-200-200-400-400 is consistent with this principle because the beat interval 400 is subdivided into two parts or not at all, but the pattern 400-400-400-600-600 is not because the beat interval 1200 is subdivided into three and two parts on the next level of the hierarchy. Finally, at least for alternating intervals of different durations, there is a preference for a 1:2 ratio; bias in the intervals produced will be toward this ratio whenever other ratios are required, as in 200-300-200-300 ....

3.5. TEMPORAL COUPLING

The final constraint on motor timing that I shall touch upon is effective whenever two or more limbs are moved simultaneously. In bimanual movements, for example, timing of the limbs is not independent, but there is a strong temporal coupling that is hard to overcome. For example, aiming movements with the right and left arm that have different movement times in isolation because of different amplitude and accuracy requirements tend to be synchronized in bimanual performance (Kelso, Southard, & Goodman, 1979; Marteniuk, MacKenzie, & Baba, 1984). It is hard to overcome this synchronization, but possible (Spijkers, Tachmatzidis, Debus, Fischer, & Kausche, 1994).

The task for which temporal coupling has been studied most extensively is the production of bimanual sequences of discrete finger taps. Temporal accuracy in such tasks is greatly reduced when the rhythms of the two hands are non-harmonic, that is, their ratio is not an integer (Klapp, 1979). There is considerable evidence that timing control in such tasks is integrated rather than independent for the two hands, and that the timing accuracy declines with an increasing complexity of the timing structure required by the task (e.g. Deutsch, 1983; Klapp, Hill, Tyler, Martin, Jagacinski, & Jones, 1985; Jagacinski, Marshburn, Klapp, & Jones, 1988). Below a specific model for integrated timing of bimanual discrete movements will be discussed in some detail.

4. Two-level models of motor timing

Two-levels models represent a certain type of model of motor timing among others, e.g. oscillator models. Here I shall describe the two-level framework and its application to two specific problems. The major ingredients of this type of model are

1. The distinction between a central level of timing control and a peripheral level at which the observations are made.
2. A specification of timing at the central level that is specific to the kind of problem to which the framework is applied.

I shall present the basic distinction between levels and describe the seminal model that is the origin of the more general two-level framework. In this seminal model timing at the central level is conceived as a sequence of clock pulses with identically distributed intervals. Thereafter I shall focus on the phenomenon of invariant relative timing and two-level multiplicative-rate models, in which timing at the central level is conceived as being proportional to total movement duration. I shall not discuss other specifications of central timing; an extensive discussion of such models is provided by Vorberg and Wing (1994), who also describe models for the adjustment of motor acts to environmental signals. A more comprehensive model for such synchronization tasks has been developed by Mates (1994a, b).

4.1 THE TWO-LEVEL FRAMEWORK

The basic scheme for the two-level framework is shown in Figure 14. At the central level of control there is a sequence of events with intervals $y_{i,1}, y_{i,2}, y_{i,3} \ldots$, where $i = 1,\ldots,n$ refers to different replications of the motor pattern. These events are mostly conceived as commands that are delivered by a central timer. Observations are made at the peripheral level. Again there is a sequence of events like key taps or some landmarks that are identified in a continuous recording of a movement. The central events give rise to the events that can be observed at the peripheral level with variable delays $m_{i,0}, m_{i,1} \ldots$, $i = 1,\ldots,n$. Thus the intervals between the events that are observed at the peripheral level are:

$$x_{i,j} = y_{i,j} + m_{i,j} - m_{i,j-1} \tag{12}$$

with $j = 1,\ldots,m$. The $x_{i,j}, y_{i,j}$, and $m_{i,j}$ are samples from random variables X_j, Y_j, M_j that are drawn for each repetition of the movement pattern. Thus Eq. (12) can be re-written as

$$X_j = Y_j + M_j - M_{j-1} \tag{13}$$

In addition to the distinction between levels, two-level models typically involve two independence assumptions. The first assumption is independence between levels so that $\mathrm{cov}(Y_j, M_k) = 0$. The second assumption is independence between motor delays so that $\mathrm{cov}(M_j, M_k) = 0$ for $j \neq k$. Both these assumptions are non-trivial and can be violated. For example, when faster performances are associated with stronger forces, motor delays could covary with intervals between central commands. When peripheral events are close together, successive motor delays could be correlated because the fluctuation of motor delay over time during the execution of a movement will have a limited bandwidth.

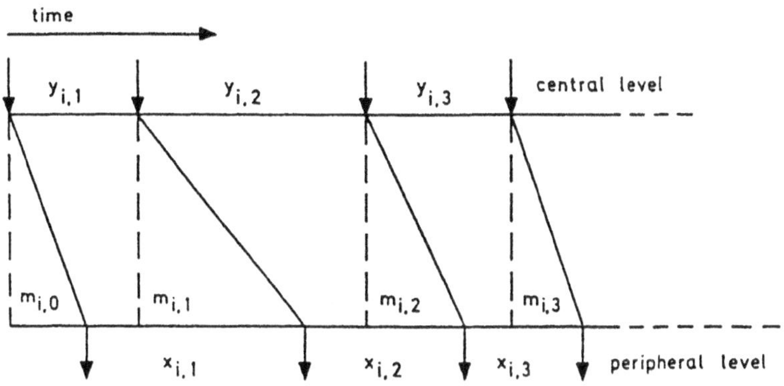

Figure 14: Schematic representation of the two-level framework

Vorberg and Wing (1994) mention another assumption that is implied by most applications of the two-level framework, the assumption of order preservation. According to this assumption the order of peripheral events is the same as the order of central events. In principle a reversal is possible when the motor delays have a high variability and the central intervals are short, so that it can happen that $M_{j-1} > M_j + Y_j$ (cf. Figure 14). In most tasks that is unlikely to occur because central intervals are much longer than the estimated standard deviation of motor delays. In models for other tasks the assumption of order preservation may not be implied because the events can be identified by other features than their order (for example in bimanual tasks where hand is one such feature).

The analysis of two-level models makes extensive use of covariances. One important characteristic of covariances that is used in the derivation of predictions is their distributivity (cf. Vorberg & Wing, 1994):

$$\mathrm{cov}(aW + bX,\ cY + dZ) = ac \cdot \mathrm{cov}(W, Y) + ad \cdot \mathrm{cov}(W, Z) + bc \cdot \mathrm{cov}(X, Y) + bd \cdot \mathrm{cov}(X, Z)$$

Application of this rule to the covariances of observed intervals results in:

$$\begin{aligned}\mathrm{cov}(X_j, X_k) &= \mathrm{cov}(Y_j + M_j - M_{j-1}, Y_k + M_k - M_{k-1}) \\ &= \mathrm{cov}(Y_j, Y_k) \\ &\quad + \mathrm{cov}(Y_j, M_k) - \mathrm{cov}(Y_j, M_{k-1}) + \mathrm{cov}(Y_k, M_j) - \mathrm{cov}(Y_k, M_{j-1}) \\ &\quad + \mathrm{cov}(M_j, M_k) - \mathrm{cov}(M_j, M_{k-1}) - \mathrm{cov}(M_{j-1}, M_k) + \mathrm{cov}(M_{j-1}, M_{k-1})\end{aligned} \tag{14}$$

With the two independence assumptions and thus $\mathrm{cov}(Y_j, M_k) = 0$ and $\mathrm{cov}(M_j, M_k) = 0$ for $j \neq k$ this expression can be simplified to

$$\mathrm{cov}(X_j, X_k) = \begin{cases} \mathrm{var}(Y_j) + \mathrm{var}(M_j) + \mathrm{var}(M_{j-1}) & \text{for } j = k \\ \mathrm{cov}(Y_j, Y_{j+1}) - \mathrm{var}(M_j) & \text{for } k = j+1 \\ \mathrm{cov}(Y_j, Y_k) & \text{for } k > j+1 \end{cases} \tag{15}$$

These equations are valid for two-level models in general. For applications to specific tasks additional specifications of the characteristics of the central intervals Y_j are needed.

4.2. REGULAR TAPPING

Two-level models have originally been developed for regular-tapping tasks (Wing & Kristofferson, 1973a, b). The seminal model is often referred to as the Wing-

Kristofferson model. More specifically, the task is a continuation task. Each trial consists of a synchronization phase and a continuation phase. During the synchronization phase pacing stimuli are presented. The subjects start tapping in synchrony with the pacing stimuli, and continue the tapping in the continuation phase without the pacing stimuli. The essential requirement in this task is the precise timing of taps, and this requires the repetition of one and the same basic interval. In the model that is applied to this task, the Y_j are treated as independent and identically distributed random variables, so that $\operatorname{cov}(Y_j, Y_k) = 0$ for $j \neq k$ and $\operatorname{var}(Y_j) = \operatorname{var}(Y_k) = \operatorname{var}(Y)$. Further, since one and the same motor response is repeated, the M_j are also taken as identically distributed random variables so that $\operatorname{var}(M_j) = \operatorname{var}(M_k) = \operatorname{var}(M)$. By these assumptions Eq. (15) becomes:

$$\operatorname{cov}(X_j, X_{j+l}) = \begin{cases} \operatorname{var}(Y) + 2\operatorname{var}(M) & \text{for } l = 0 \\ -\operatorname{var}(M) & \text{for } l = 1 \\ 0 & \text{for } l > 1 \end{cases} \tag{16}$$

The $\operatorname{cov}(X_j, X_{j+l})$ can be estimated in different ways from data. Assume that a single subject performs i=1,...,n trials with j=1,...,m inter-tap intervals in each trial's continuation phase. Generally the covariances can be estimated as the covariances of interval j and interval j + l computed across the set of n trials. However, in the two-level model for regular tapping the covariances are independent of j. Therefore, Eq. (16) is also true for the autocovariance (acv) with lag l, and in the analysis of continuation tasks it is typically the autocovariance with lag 1 that is estimated from single trials as:

$$\hat{\operatorname{acv}}(1) = \frac{1}{m-1} \sum_{j=1}^{m-1} \left[(x_j - \bar{x})(x_{j+1} - \bar{x}) \right] \quad \text{with} \quad \bar{x} = \frac{1}{m} \sum_{j=1}^{m} x_i$$

Vorberg and Wing (1994) discuss a number of problems with these estimates and ways to solve them. Such problems result mainly from the fact that the estimates are biased, but also from trends in the series of inter-tap intervals: during a continuation phase the tapping frequency can gradually increase or decline. Such trends imply a violation of the assumption that successive intervals are uncorrelated. It is almost impossible to remove them post hoc. Of course, short continuation phases can serve to eliminate or at least to reduce the trends, but on the other hand they increase the bias of the autocovariance estimates. Thus, there is not much that one can do except for training subjects to avoid trends or otherwise to reject trials from the analysis in which trends are clearly present.

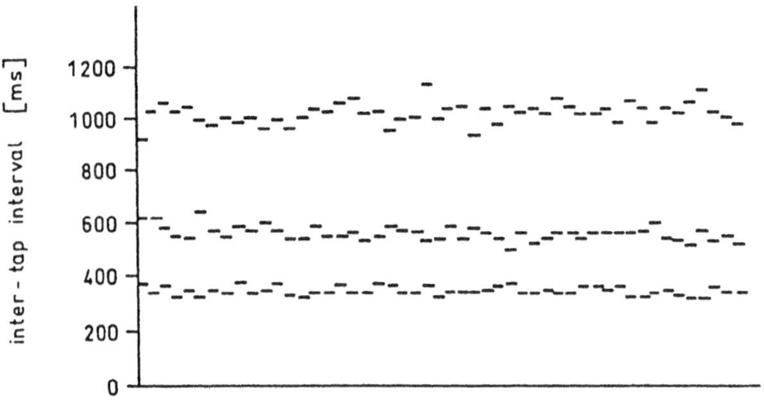

Figure 15: Inter-tap intervals in regular tapping at different frequencies (after Vorberg & Wing, 1994)

In Figure 15 some classic tapping data are shown (Stevens, 1886). These are sequences of inter-tap intervals that were produced by a professional musician. The first thing to note is the increase of the variability with the mean interval. For intervals with a mean length above about 300 ms, this is a quite regular result (e.g. Peters, 1989). Thus the production of regular tapping intervals is one of the few tasks that are performed more accurately at higher tempos (cf. Hancock & Newell, 1985). Another noticable characteristic of the intervals is the zig-zag pattern: a longer interval is more often followed by a shorter one than by an even longer interval, and a shorter interval is more often followed by a longer one than by an even shorter interval. This amounts to a negative correlation between successive intervals, which Stevens attributed to corrections of too long and too short intervals. From Eq. (16) it is apparent that such a negative correlation is predicted by the Wing-Kristofferson model without any correction being involved. It results from the fact that a long motor delay at the end of an interval increases the duration of that interval and simultaneously reduces the duration of the immediately following interval.

There are two major predictions of the Wing-Kristofferson model. The first one is that autocorrelations or autocovariances with a lag larger than 1 should be zero; non-adjacent intervals should be uncorrelated according to Eq. (16). The second prediction is the one already mentioned: the autocorrelation or autocovariance with lag 1 should be negative, and the autocorrelation should be larger than -0.5. (It will be -0.5 when var(Y) = 0.) When these predictions are met, the central variability and the motor-delay variability can be estimated as

$$\begin{aligned} \hat{\text{var}}(M) &= -\hat{\text{acv}}(1) \\ \hat{\text{var}}(Y) &= \hat{\text{acv}}(0) - 2\,\hat{\text{acv}}(1) \end{aligned} \tag{17}$$

where $\hat{\text{acv}}(0)$ and $\hat{\text{acv}}(1)$ are equivalent to $\text{cov}(X_j, X_j)$ and $\text{cov}(X_j, X_{j+1})$ in Eq. (16).

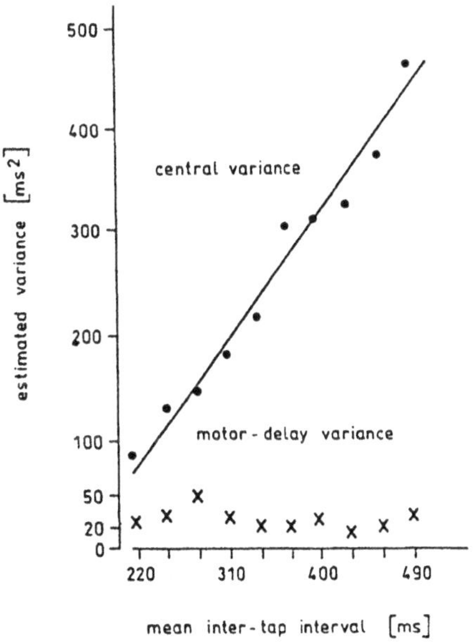

Figure 16: Estimates of central variance and motor-delay variance as a function of mean inter-tap interval (after Wing, 1980)

Figure 16 presents such estimates that were obtained by Wing (1980). The linear increase of the central variance with the mean tapping interval and the constancy of the motor-delay variance are highly plausible and give credit to the validity of the model. Other supportive evidence has been reported by Ivry, Keele, and Diener (1989) who found selective influences of different cerebellar lesions on either the central variability or the motor-delay variability.

Before I turn to another application of the two-level framework, I shall briefly discuss two different perspectives on the Wing-Kristofferson model. Under the first perspective it is a model of regular tapping that can be right or wrong. For example, if the major predictions are not met by a sufficient number of trials, one would reject it for a certain subject, and if it would be rejected for a sufficient number of subjects, it would be rejected at all. Under the second perspective the model is similar to a statistical model. Statistical models start with certain assumptions about probability distributions, characteristics of covariance matrices etc., and if these are met by the data to a reasonable degree, one can make use of the models. If the data do not meet the assumptions, this does not imply that the model is wrong, but only that it is not applicable to the particular data.

In my view the Wing-Kristofferson model is very much like a statistical model. It is based on certain assumptions that can be wrong for various reasons, e.g. because of non-stationary series of inter-tap intervals in a number of trials. Whether or not the assumptions are met can be tested, and when they are met, the model allows a useful partitioning of the variability of the inter-tap intervals. This partitioning is statistical in nature in that it rests on the independence assumptions. Attempts to identify, for example, the motor delay physiologically appear futile. If this view is correct, the question whether the model fits some data or not is mainly a methodological question - and it would be legitimate to fit the test situation to the model. The findings of substantial interest would have to do with how the variance components are affected by various other factors, like the different types of lesions studied by Ivry et al. (1989). Such results give meaning to the components beyond their statistical separability.

4.3. RELATIVE-TIMING INVARIANCE

The second application of the two-level framework that I shall discuss deals with the phenomenon of relative-timing invariance, that is, with the invariance of the relative durations of segments of a motor pattern when the overall speed or some other task characteristic is varied. The experimental basis has been reviewed several times (e.g. Schmidt, 1980, 1985; Gentner, 1987), so there is no need for another review. Before I describe the application of the two-level framework to the phenomenon, I shall briefly comment on two major issues that are associated with it. The first issue is whether relative timing is really invariant, and the second is the origin of the invariance (cf. Heuer, 1991).

The invariance of relative timing has been questioned several times, and there are clear instances of a lack of it. For example, Marteniuk, MacKenzie, Jeannerod, Athenes, and Dugas (1987) found that the relative timing of reaching movements depended on the nature of the target, and MacKenzie, Marteniuk, Dugas, Liske, and Eickmeier (1987) reported an effect of the target size on the relative timing of aiming movements. On the other hand Howarth, Beggs, and Bowden (1971) found that when movement time was varied, the distance that was covered in certain proportions of movement time remained constant. Thus, it seems that relative timing may remain invariant when the variation of task characteristics is of a certain type, but may not remain constant with other types of task variation.

In certain types of movement gross violations of relative-timing invariance can be observed when total duration is varied (e.g. Roth, 1987; Young & Schmidt, 1990; Schneider & Schmidt, 1995). In particular throwing-like or hitting-like movements seem to consist of two parts with relative-timing invariance within each part (e.g. backswing and forward swing), but not across parts. Thus, relative-timing invariance cannot be expected to be found in all types of movement. Most likely there are movements that exhibit this invariance, others that exhibit it during certain parts, and finally there will perhaps also be motor patterns without any indications of invariant relative timing.

In addition to gross deviations from relative-timing invariance, there will be small deviations that can be detected by powerful statistical tests, but will escape less powerful ones. Gentner (1987) provides some examples for this. Here one may ask how the statistically significant deviation from invariance should be evaluated. After all, the tendency toward invariance may be much more conspicuous than the remaining deviation from a perfect invariance. In fact, one can argue that there is really no reason to expect a perfect invariance, so the detection of deviations is only a matter of statistical power. If this is true, one can accept the small deviations as resulting from the use of conceptual simplifications, and there is actually no reason to expect a perfect fit between some natural phenomenon and human conceptualizations of it.

Summing up, there are gross violations of relative-timing invariance - some movement patterns do not have that characteristic when total duration is varied - and there are minor deviations that can be accepted as discrepancies between nature and human conceptualizations of it. Even when approximate invariance exists, however, it is another question how it comes about.

The invariance of relative timing has been attributed to the characteristics of a template or motor engram (Terzulo & Viviani, 1980), to the characteristics of a coordinative structure (Kelso, 1981) or to the essential features of a generalized motor program (Schmidt, 1980, 1985). Such accounts seem to imply that, for a certain type of movement, relative-timing invariance is a mandatory feature that cannot easily be changed (cf. Heuer. 1991). Thus, after practive of a certain temporal pattern transfer to a new pattern should be poor. Contrary to this expectation, Heuer and Schmidt (1988) found essentially perfect transfer to a motor pattern with a new relative timing as compared to one for which only overall duration was changed. Similar findings of an easy transfer to a new temporal pattern have been reported by Langley and Zelaznik (1984), Wulf and Schmidt (1988), and Carnahan and Lee (1989). Thus, it seems that the invariance of relative timing is more a strategical than a mandatory phenomenon that is likely to be related to the constraints on timing discussed above. No matter whether this view is correct or not, the examination of relative-timing invariance requires adequate test procedures.

4.4. A TWO-LEVEL MULTIPLICATIVE-RATE MODEL

In the literature several methods for testing the invariance of relative timing can be found (e.g Gentner, 1987). Here I shall describe methods that can be derived from the two-level framework. In addition, I shall examine some other tests, that have been proposed, from this perspective. These analyses start from a single equation: Under the two-level framework relative-timing invariance means that the central intervals Y_j can be factorized as $Y_j = R\omega_j$ with R as a multiplicative rate parameter and ω_j as the prototypical duration of interval j between two central events j and j-1. Replacing Y_j in Eq. (12) by $R\omega_j$ results in:

$$X_j = R\omega_j + M_j - M_{j-1} \tag{18}$$

The two-level model has two important features that distinguish it from other specifications of relative-timing invariance. First, it is based on time intervals between discrete events, not on the temporal scaling of a continuous prototypical function (e.g. Meyer, Smith, & Wright, 1982). The continuous model implies the discrete one, but not vice versa. There can be invariance of the relative timing of certain events, while other events in-between do not obey the multiplicative scaling rule; grace notes in playing a piece of music on the piano can be an example (cf. Repp, 1994). The second feature is that relative-timing invariance is posited for a central level of control and may not appear at the peripheral level; at the peripheral level central invariance can be masked by systematic differences in motor delays. This feature has the consequence that almost all of the available tests of relative-timing invariance, which are tests of invariance at the peripheral level of observation, are no longer adequate from the perspective of a two-level model.

As an example, consider the constant-proportion test which has been proposed and used extensively by Gentner (1987). This seems to be a very powerful test of relative-timing invariance that only very few data sets survive. Basically it is the linear regression of the relative duration of a certain time interval within a motor pattern on the total duration computed across a number of repetitions of the pattern with different overall durations. A slope that deviates from zero indicates a violation of the invariance. Slopes can be computed for all intervals within a motor pattern, and the estimates are not independent. This poses a problem for the evaluation of a certain number of significant deviations from zero; however, by way of simulations Gentner (1987) provided some estimates of how many significant deviations can be expected by chance.

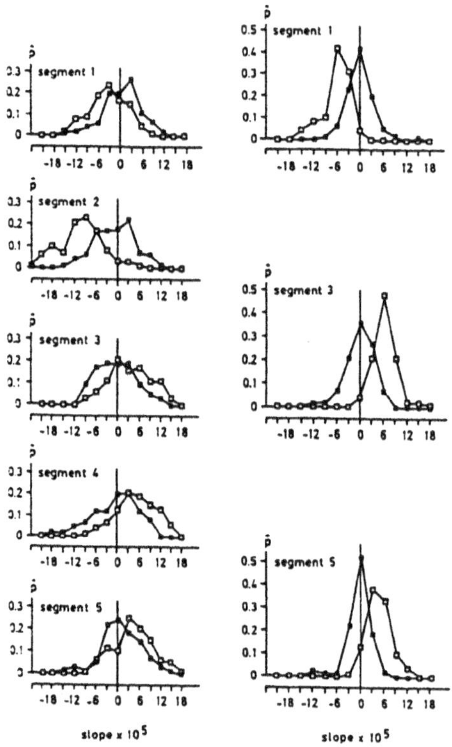

Figure 17: Distributions of slopes of the constant-proportion test with data simulated according to Eq. (18). Rate-parameter variance was smaller (left culumn) or larger (right culumn), mean motor delays $E(M_j)$ were the same for all j (filled squares) or different (open squares) (after Heuer, 1988b)

Figure 17 presents some simulation results with the two-level model of Eq. (18) (Heuer, 1988b). What is shown are the distributions of slopes in the constant-proportion test. There were five segments within each pattern, and the parameters of the model were varied across different simulation runs. The important variation was whether the mean motor delays were equal or different within each motor pattern. With different mean motor delays the central invariance of relative timing will be distorted at the peripheral level. As is evident from Figure 17, with all mean motor delays being equal the slopes of the constant-proportion test have a more or less symmetrical distribution around zero. In contrast, when the mean motor delays are different, the distributions are no longer centered at zero and the slopes will be systematically positive or negative. Thus, even when a data set does not pass the constant-proportion test, this does not imply a lack of relative-timing invariance at a central level of control. Instead , although it can indicate a lack of invariance, it can easily be caused by a systematic variation of motor delays in the course of the movement and thus by a distortion of the central invariance at the peripheral level of observation.

As far as I am aware (cf. Heuer, 1988b), there is only one regression procedure that is valid under the two-level model. This is the regression of mean interval durations on mean total duration computed across different experimental conditions which differ in the instructed overall tempo. For the validity of the procedure it is important that the assignment of the movements to classes with different total durations is a priori and not based on actual movement durations; a posteriori classifications introduce errors because they are affected by motor delays. The regression of mean interval duration on mean total duration should be linear:

$$E_c(X_j) = aE_c(X) + b$$

$$\text{with} \quad a = \frac{\omega_j}{\sum_{s=1}^{m} \omega_s}, \quad b = \left[E(M_j) - E(M_{j-1})\right] - \frac{\omega_j}{\sum_{s=1}^{m} \omega_s}\left[E(M_m) - E(M_o)\right] \qquad (19)$$

In this equation $E_c(X_j)$ and $E_c(X)$ indicate the expected values in condition c of the duration X_j of interval j and the total duration X. The slope a equals the relative prototypical duration of interval j, and the two terms in the intercept b are the differences between the motor delays at the end and start of interval j and those at the end and start of the whole movement, the latter difference being multiplied by the relative prototypical duration of interval j. Since $b \neq 0$ in general, it is evident that this regression equation does not imply a proportional scaling of observed interval durations.

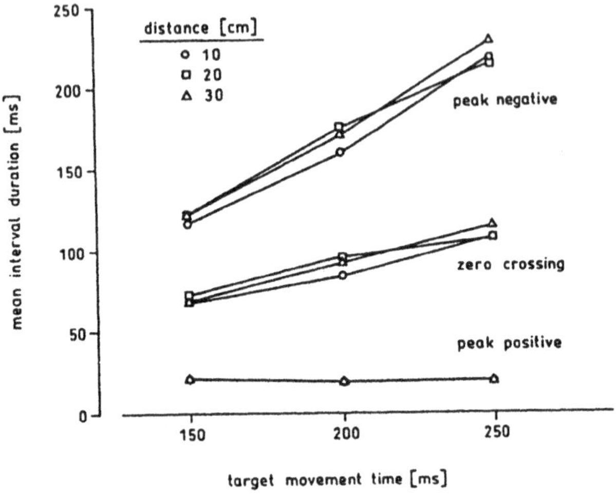

Figure 18: Mean durations of segments of rapid movements as a function of target movement time (after Zelaznik, Schmidt, & Gielen, 1986)

Figure 18 presents some results of Zelaznik, Schmidt, and Gielen (1986) to which Eq. (19) is applicable. The data are the mean durations until peak acceleration, zero crossing, and peak deceleration in rapid movements with different amplitudes and different target movement times (which replace the mean total durations in Eq. (19), but should be similar to them). Originally these data have been interpreted as evidence against relative-timing invariance because the time until peak acceleration is essentially independent of total duration. However, in the light of Eq. (19) these data indicate that ω_1, the prototypical duration of the interval until peak acceleration, is zero; then the intercept of about 20 ms estimates the difference between the expected motor delays $E(M_1) - E(M_0)$. According to this interpretation the temporal placement of peak acceleration is not a centrally controlled event, but the temporal separation between the start of the movement and peak acceleration is the result of different motor delays. Thus, under the two-level model these data are consistent with the notion of multiplicate-rate control at a central level.

4.5. TETRAD TESTS

Typically the analysis of two-level models of motor timing is based on variances and covariances (cf. Vorberg & Wing, 1994). Therefore the application of the two-level framework to the notion of relative-timing invariance suggests new tests. Perhaps the most interesting ones are tetrad tests that have a history in factor analysis (e.g. Spearman, 1927). They can be derived quite easily from Eq. (15) with $Y_j = R\omega_j$. With this assumption of multiplicative-rate control the covariances between non-adjacent intervals becomes $\mathrm{cov}(X_j, X_k) = \omega_j \omega_k \mathrm{var}(R)$.

Consider a set of four intervals j, k, s, and t, arranged in that order, and two sets of four covariances between non-adjacent intervals for which the following two ratios can be computed:

$$\delta_1 = \frac{\mathrm{cov}(X_j, X_k)\mathrm{cov}(X_s, X_t)}{\mathrm{cov}(X_j, X_t)\mathrm{cov}(X_k, X_s)} = \frac{\omega_j \omega_k \mathrm{var}(R) \omega_s \omega_t \mathrm{var}(R)}{\omega_j \omega_t \mathrm{var}(R) \omega_k \omega_s \mathrm{var}(R)} = 1$$

$$\delta_2 = \frac{\mathrm{cov}(X_j, X_s)\mathrm{cov}(X_k, X_t)}{\mathrm{cov}(X_j, X_t)\mathrm{cov}(X_k, X_s)} = \frac{\omega_j \omega_s \mathrm{var}(R) \omega_k \omega_t \mathrm{var}(R)}{\omega_j \omega_t \mathrm{var}(R) \omega_k \omega_s \mathrm{var}(R)} = 1 \qquad (20)$$

The computation of δ_1 requires at least 7 intervals in the motor pattern because all four intervals must be non-adjacent, while for the computation of δ_2 only 5 intervals are needed - intervals j, k and s, t can be adjacent because the covariances $\mathrm{cov}(X_j, X_k)$ and $\mathrm{cov}(X_s, X_t)$ are not included in the ratio. Instead of tetrad ratios tetrad differences can be used, which should be zero under the model. When there are more than 7 or 5 segments in a motor pattern, there will be several such ratios or differences. Heuer (1988b) reports some simulations that give information on the sampling distribution of tetrad differences.

Heuer, Schmidt, and Ghodsian (1995) have extended the two-level multiplicative-rate model to bimanual movements. In principle the extension can be straightforward in that the intervals that are analysed are not only defined by events in the movement of a single hand, but also across hands. Such between-hand intervals, however, are affected by systematic differences between the limbs, and these can result in violations of the assumption that motor delays are uncorrelated. Given the anatomical differences between the two hands and the differences in how they are connected to higher levels of motor control, it is likely that systematic differences in motor delays do exist. For example, Stucchi and Viviani (1993) found a consistent lead of the dominant hand in drawing-like movements. Thus, there can be systematic temporal offsets between the two hands in bimanual movements. Such offsets can be added to the model, and this leads to an interesting prediction for the covariances of between-hand intervals, which begin at a certain event or landmark in the movement of the one hand and end with a certain event or landmark in the movement of the other hand.

Consider an offset ΔC between Hand 1 and Hand 2. Depending on which hand leads the other by ΔC, the observed between-hand intervals, which under the model (and without a temporal offset) would be given by Eq. (18), are longer or shorter by ΔC. This has an effect on the covariances between non-adjacent between-hand intervals: they are increased or reduced by var (ΔC). They are increased, when both intervals begin in the same hand and both end in the other hand; they are reduced, when they begin in different hands. Thus the covariance matrix will show a systematic pattern of high covariances and low (or even negative) covariances; such a pattern was indeed observed by Heuer et al. (1995).

The existence of temporal offsets between the hands plagues the analysis of between-hand intervals. However, it is possible to derive tetrad ratios (or differences) for within-hand intervals that are based on covariances between intervals from different hands. Consider non-adjacent intervals j, k in Hand 1 and non-adjacent interval s, t in Hand 2. With a common rate parameter R for the two hands the following tetrad ratios should equal 1 under the model:

$$\vartheta_1 = \frac{\mathrm{cov}(X_j,X_s)\mathrm{cov}(X_k,X_t)}{\mathrm{cov}(X_j,X_k)\mathrm{cov}(X_s,X_t)} = \frac{\omega_j\omega_s \mathrm{var}(R)\omega_k\omega_t \mathrm{var}(R)}{\omega_j\omega_k \mathrm{var}(R)\omega_s\omega_t \mathrm{var}(R)} = 1$$

$$\vartheta_2 = \frac{\mathrm{cov}(X_j,X_t)\mathrm{cov}(X_k,X_s)}{\mathrm{cov}(X_j,X_k)\mathrm{cov}(X_s,X_t)} = \frac{\omega_j\omega_t \mathrm{var}(R)\omega_k\omega_s \mathrm{var}(R)}{\omega_j\omega_k \mathrm{var}(R)\omega_s\omega_t \mathrm{var}(R)} = 1 \tag{21}$$

Heuer et al. (1995) provide some information on the sampling distribution of these tetrad-ratios under the two-level multiplicative-rate model which can be fairly closely approximated by a log-normal distribution. Although the tetrad ratios are ratios of products of covariances, their random fluctuation is astonishingly small. In the simulations and in the experimental task of Heuer et al. (1995) their scatter was in the range of about 0.65 - 1.45.

Tetrad ratios test whether all covariation between intervals can be attributed to the single common rate parameter. When there is additional covariation between intervals, the ratios will deviate from 1. Such deviations, however, do not only indicate a failure of the multiplicative-rate model, but they can be informative on the nature of this failure. There is a straightforward way to analyse how tetrad ratios respond to violations of the assumption of a single rate parameter R. Even in the general case that there is no common rate parameter at all, each central interval Y_j can be decomposed into a prototypical duration ω_j and an interval-specific rate parameter R_j. Then the tetrad ratios become:

$$\vartheta_1 = \frac{\mathrm{cov}(R_j,R_s)\mathrm{cov}(R_k,R_t)}{\mathrm{cov}(R_j,R_k)\mathrm{cov}(R_s,R_t)} \qquad \vartheta_2 = \frac{\mathrm{cov}(R_j,R_t)\mathrm{cov}(R_k,R_s)}{\mathrm{cov}(R_j,R_k)\mathrm{cov}(R_s,R_t)} \tag{22}$$

From these equations the effects of different sets of equal and different rate parameters R_j, R_k, R_s, and R_t can easily be determined. Some examples are described by Heuer et al. (1995). A particularly interesting deviation from the model is that $R_j = R_k = R_1$ and $R_j = R_t = R_2$, so that the rate parameters in the two hands are different. In this case the tetrad ratios give the squared correlation of the rate parameters:

$$\vartheta_1 = \vartheta_2 = \frac{\mathrm{cov}^2(R_1, R_2)}{\mathrm{var}(R_1)\,\mathrm{var}(R_2)} = r^2_{R_1R_2} \tag{23}$$

Thus far tetrad ratios have hardly been used in the analysis of relative timing. There are statistical problems associated with them, but that seems to be the case with almost all procedures that have been employed as tests of relative-timing invariance. In particular, from a certain data set a potentially large number of tetrad ratios can be computed that are not independent. However, the deviations that I have seen in real data tended to be conspicuous with tetrad ratios lying outside the range of 0.5 - 2.0, and this is a range that will hardly be exceeded by chance with about 100 to 150 repetitions of the motor pattern and some voluntary variation of the overall tempo so that the variability of the rate parameter - more precisely: the variance of the central intervals - will be larger than the motor-delay variability (cf. Heuer et al., 1995, for a more detailed discussion of means to reduce the random fluctuation of tetrad ratios).

5. Concluding remarks

In this chapter I have touched upon a few topics related to the timing of human movements. The selection was heavily biased by personal interests and is rather non-systematic. Also, I feel that there are many open ends. To list just a few questions: How is extrinsic timing information integrated with processes of motor control? Most likely there will be several answers depending on the particular kind of motor skill, being different e.g. for a reach to a target and a series of somersaults in high diving. How is timing constrained by spatial characteristics of a motor pattern, and what is the basis for such constraints? What produces relative-timing invariance in some skills, and what are the conditions under which this invariance breaks down? These are just three questions that are related to the major sections of this chapter; there are many more unanswered questions. Given that one can end this chapter with a question for each of the major topics that I have discussed rather than with an answer, the chapter is perhaps primarily an invitation to dive into these problems: it doesn't give answers, but it may contain some hints as to where answers could be found and provide some equipment that might be needed on the way.

6. References

Aschersleben, G. (1994). *Afferente Informationen und die Synchronisation von Ereignissen.* Frankfurt: Lang

Aschersleben, G. & Prinz, W. (1995). Synchronizing actions with events: The role of sensory information. *Perception & Psychophysics* (in press).

Bootsma, R.J. (1989). Accuracy of perceptual processes subserving different perception-action systems. *Quarterly Journal of Experimental Psychology*, **41A**, 489-500.

Bridgeman, B., Kirch, M. & Sperling, A. (1981). Segregation of cognitive and motor aspects of visual function using induced motion. *Perception & Psychophysics*, **29**, 336-342.

Bruce, V. & Green, P.R. (1985). *Visual perception. Physiology, Psychology and Ecology.* London: Erlbaum.

Brunswik, E. (1952). *The conceptual framework of psychology.* Chicago: University of Chicago Press.

Carnahan, H. & Lee, T.D. (1989). Training for transfer of a movement timing skill. *Journal of Motor Behavior*, **21**, 48-59.

Carter, M.C. & Shapiro, D.C. (1984). Control of sequential movements: Evidence for generalized motor programs. *Journal of Neurophysiology*, **52**, 787-796.

Cruse, H., Dean, J., Heuer, H., & Schmidt, R.A. (1990). Utilization of sensory information for motor control. In: O. Neumann & W. Prinz (Eds), *Relationships between perception and action. Current approaches.* Berlin: Springer, pp. 43-79.

Derwort, A. (1938). Untersuchungen über den Zeitablauf figurierter Bewegungen beim Menschen. *Pflügers Archiv für die gesamte Physiologie*, **240**, 661-675.

Deutsch, D. (1983). The generation of two isochronous sequences in parallel. *Perception & Psychophysics*, **34**, 331-337.

Dunlap, K. (1910). Reactions on rhythmic stimuli, with attempt to synchronize. *Psychological Review*, **17**, 399-416.

Ehrmann, V. (1988). Gefahrenrisiko und Unfallgefährdung von Fußgängern im Straßenverkehr: Simulationsexperiment zum Überqueren von Fahrbahnen im fließenden Verkehr. Unveröffentlichte Diplomarbeit, Universität Bielefeld.

Erkelens, C.J. & Collewijn, H. (1985a). Motion perception during dichoptic viewing of moving random-dot stereograms. *Vision Research*, **25**, 583-588.

Erkelens, C.J. & Collewijn, H. (1985b). Eye movements and stereopsis during dichoptic viewing of moving random-dot stereograms. *Vision Research*, **25**, 1689-1700.

Gentner, D.R. (1987). Timing of skilled motor performance: tests of the proportional duration model. *Psychological Review*, **94**, 255-276.

Hancock, P.A. & Newell, K.M. (1985). The movement speed-accuracy relationship in space-time. In: H. Heuer, U. Kleinbeck, & K.-H. Schmidt (Eds.), *Motor behavior. Programming, control, and acquisition.* Berlin: Springer, pp. 153-188.

Heuer, H. (1984). On re-scaleability of force and time in aiming movements. *Psychological Research*, **46**, 73-86.

Heuer, H. (1987). Apparent motion in depth resulting from changing size and changing vergence. *Perception*, **16**, 337-350.

Heuer, H. (1988a) Adjustment and re-adjustment of the relative timing of a motor pattern. *Psychological Research*, **50**, 83-93.

Heuer, H. (1988b). Testing the invariance of relative timing. Comment on Gentner (1987). *Psychological Review*, **95**, 552-557.

Heuer, H. (1990). Psychomotorrik. In: H. Spada (Ed.), *Lehrbuch Allgemeines Psychologie.* Bern: Huber, pp. 495-559

Heuer, H. (1991). Invariant relative timing in motor-program theory. In: J. Fagard & P.H. Wolff (Eds.), *The development of timing control and temporal organization in coordinated action. Invariant relative timing, rhythms, and coordination.* Amsterdam: North-Holland, pp. 37-68.

Heuer, H. (1993). Estimates of time to contact based on changing size and changing target vergence. *Perception*, **22**, 549-563.

Heuer, H. & Schmidt, R.A. (1988). Transfer of learning among motor patterns with different relative timing. *Journal of Experimental Psychology: Human Perception and Performance*, **14**, 241-252.

Heuer, H., Schmidt, R.A., & Ghodsian, D. (1995). Generalized motor programs for rapid bimanual tasks: A two-level multiplicative-rate model. (submitted).

Holway, A.H. & Boring, E.G. (1941). Determinants of apparent visual size with distance variant. *American Journal of Psychology*, **54**, 21-37.

Howarth, C.I., Beggs, W.D.A., & Bowden, J.M. (1971). The relationship between speed and accuracy of movement aimed at a target. *Acta Psychologica*, **95**, 207-218.

Ivry, R.I., Keele, S.W. & Diener, H.C. (1988). Dissociation of the lateral and medial cerebellum in movement timing and movement execution. *Experimental Brain Research*, **73**, 167-180.

Jagacinski, R.J., Marshburn, E., Klapp, S.T., & Jones, M.R. (1988). Tests of parallel versus integrated structure in polyrhythmic tapping. *Journal of Motor Behavior*, **20**, 416-442.

Keele, S.W. (1986). Motor control. In: K.R. Boff, L. Kaufman, & J.P. Thomas (Eds.), *Handbook of perception and performance. Vol.II. Cognitive processes and performance.* New York: Wiley 1986.

Kelso, J.A.S. (1981). Contrasting perspectives on order and regulation in movement. In: J. Long & A. Baddeley (Eds.), *Attention and performance IX*. Hillsdale,N.J.: Erlbaum, pp. 437-457.

Kelso, J.A.S., Southard, D.L., & Goodman, D. (1979). On the coordination of two-handed movements. *Journal of Experimental Psychology: Human Perception and Performance*, **5**, 229-238.

Klapp, S.T. (1979). Doing two things at once: the role of temporal compytibility. *Memory and Cognition*, **7**, 375-381.

Klapp, S.T., Hill, M., Tyler, J., Martin, Z., Jagacinski, R., & Jones, M. (1985). On marching to two different drummers: Perceptual aspects of the dificulties. *Journal of Experimental Psychology: Human Perception and Performance*, **11**, 814-828.

Lacquaniti, F., Terzuolo, C.A., & Viviani, P. (1983). The law relating kinematic and figural aspects of drawing movements. *Acta Psychologica*, **54**, 115-130.

Langley, D.J. & Zelaznik, H.N. (1984). The acquisition of time properties associated with a sequential motor skill. *Journal of Motor Behavior*, **16**, 275-301.

Lashley, K.S. (1951). The problem of serial order in behavior. In L.A. Jeffress (Ed.), *Cerebral mechanism in behavior*. New York: Wiley, pp. 112-131.

Lee, D.N. (1976). A theory of visual control of braking based on information about time-to-collision. *Perception*, **5**, 437-459.

Lee, D.N. (1992). Body-environment coupling. In: U. Neisser (Ed.), *Ecological and interpersonal knowledge of the self*. Cambridge: Cambridge University Press.

Lee, D.N. & Young, D.S. (1986). Gearing action to the environment. In: H. Heuer & C. Fromm (Eds), *Generation and modulation of action patterns*. Berlin: Springer, pp. 217-230.

Lee, D.N., Young, D.S., Reddish, P.E., Lough, S. & Clayton, T.M.H. (1983). Visual timing in hitting an accelerating ball. *Quarterly Journal of Experimental Psychology*, **35A**, 333-346.

MacKenzie, C.L., Marteniuk, R.G., Dugas, C., Liske, D. & Eickmeier, B. (1987). Three-dimensional movement trajectories in Fitts' task: Implications for control. *Quarterly Journal of Experimental Psychology*, **39A**, 629-647.

Marteniuk, R.G., MacKenzie, C.L., & Baba, D.M. (1984). Bimanual movement control: Information processing and interaction effects. *Quarterly Journal of Experimental Psychology*, **36A**, 335-365.

Marteniuk, R.G., MacKenzie, C.L., Jeannerod, M., Athenes, S., & Dugas, C. (1987). Constraints on human arm movement trajectories. *Canadian Journal of Psychology*, **41**, 365-378.

Mates, J. (1994a). A model of synchronization of motor acts to a stimulus sequence. I. Timing and error corrections. *Biological Cybernetics*, 70, 463-473.

Mates, J. (1994b). A model of synchronization of motor acts to a stimulus sequence. II. Stability analysis, error estimation and simulations. *Biological Cybernetics*, **70**, 475-484.

Meyer, D.E., Smith, J.E.K., & Wright, C.E. (1982). Models for the speed and accuracy of aimed movements. *Psychological Review*, **89**, 449-482.

Peters, M. (1989). The relationship between variability of intertap intervals and interval duration. *Psychological Research*, **51**, 38-42.

Repp, B.H. (1994). Relational invariance of expressive microstructure across global tempo changes in music performance: An exploratory study. *Psychological Research/Psychologische Forschung*, 56, 269-292.

Rosenbaum, D.A., Slotta, J.D., Vaughan, J., & Plamondon, R. (1991). Optimal movement selection. *Psychological Science*, **2**, 86-91.

Roth, K. (1989). *Taktik im Sportspiel. Zum Erklärungswert der Theorie generalisierter motorischer Programme für die Regulation komplexer Bewegungshandlungen*. Schorndorf: Hofmann.

Savelsbergh, G.J.P., Whiting, H.T.A. & Bootsma, R.J. (1991). Grasping tau. *Journal of Experimental Psychology: Human Perception and Performance*, **17**, 315-322.

Schiff, W. & Detwiler, M.L. (1979). Information used in judging impending collision. *Perception*, **8**, 647-658.

Schmidt, R.A. (1972). The Index of Preprogramming (IP): A statistical method for evaluating the role of feedback in simple movements. *Psychonomic Science*, **27**, 83-85.

Schmidt, R.A. (1980). On the theoretical status of time in motor-program representations. In: G.E. Stelmach & J. Requin (Eds.), *Tutorial in motor behavior*. Amsterdam: North-Holland, pp. 145-165.

Schmidt, R.A. (1985). The search for invariance in skilled movement behavior. *Research Quarterly for Exercise and Sport*, **56**, 188-200.

Schmidt, R.A. & Russell, D.G. (1972). Movement velocity and movement time as determinants of degree of preprogramming in simple movements. *Journal of Experimental Psychology*, **96**, 315-320.

Schneider, D.M. & Schmidt, R.A. (1995) Units of action in motor control: Role of response complexity and target speed. *Human Performance*, **8**, 27-49.

Schöner, G. & Kelso, J.A.S. (1988). A synergetic theory of environmentally-specified and learned patterns of movement coordination.I: Relative phase dynamics. *Biological Cybernetics*, **58**, 71-80.

Spearman, C. (1927). *The abilities of man*. New York: MacMillan.

Spijkers, W., Tachmatzidis, K., Debus, G., Fischer, M., & Kausche,I. (1994). Temporal coordination of alternative and simultaneous aiming movements of constrained timing structure. *Psychological Research/Psychologische Forschung*, **57**, 20-29.

Stevens, L.T. (1886). On the time sense. *Mind*, **11**, 393-404.

Stucchi, N. & Viviani, P. (1993). Cerebral dominance and asynchrony between bimanual 2-D movements. *Journal of Experimental Psychology: Human Perception and Performance*, **19**, 1200-1220.

Swanston, M.T. & Gogel, W.C. (1986). Perceived size and motion in depth from optical expansion. *Perception & Psychophysics*, **39**, 309-326.

Terzuolo, C.A. & Viviani, P. (1980). Determinants and characteristics of motor patterns used for typing. *Neuroscience*, 5, 1085-1103.

Thomassen, A.J.W.M., & Teulings, H.-L. (1985). Size and shape in handwriting: Exploring spatio-temporal relationships at different levels. In: J.A. Michon & J.L. Jackson (Eds.), *Time, mind, and behavior*. Berlin: Springer, pp. 253-263.

Todd, J.T. (1981). Visual information about moving objects. *Journal of Experimental Psychology: Human Perception and Performance*, **7**, 795-810.

Tresilian, J.R. (1990). Perceptual information for the timing of interceptive action. *Perception*, **19**, 223-239.

Tresilian, J.R. (1991). Empirical and theoretical issues in the perception of time to contact. *Journal of Experimental Psychology: Human Perception and Performance*, **17**, 865-876.

Tresilian, J.R. (1994a). Approximate information sources and perceptual variables in interceptive timing. *Journal of Experimental Psychology: Human Perception and Performance*, **20**, 154-173.

Tresilian, J.R. (1994b). Perceptual and motor processes in interceptive timing. *Human Movement Science*, **13**, 335-373.

Tresilian, J.R. (1995). Perceptual and cognitive processes in time-to-contact estimation: Analysis of prediction-motion and relative judgement tasks. *Perception & Psychophysics* (in press).

Viviani, P. (1986). Do units of motor action really exist? In: H. Heuer & C. Fromm (Eds.), *Generation and modulation of action patterns*. Berlin: Springer, pp. 201-216.

Viviani, P., Campadelli, P., & Mounoud, P. (1987). Visuo-manual pursuit tracking of human two-dimensional movements. *Journal of Experimental Psychology: Human Perception and Performance*, **13**, 62-78.

Viviani, P. & Schneider, R. (1991). A developmental study of the relation between geometry and kinematics in drawing movements. *Journal of Experimental Psychology: Human Perception and Performance*, **17**, 198-218.

Viviani, P. & Terzuolo, C.A. (1982). Trajectory determines movement dynamics. *Neuroscience*, **7**, 431-437.

Vorberg, D. & Wing, A.M. (1993). Modelle für Variabilität und Abhängigkeit bei der zeitlichen Steuerung. In: H. Heuer & S.W. Keele (Eds.), *Enzyklopädie der Psychologie C II 3: Psychomotorik*. Göttingen: Hogrefe, pp. 223-320.

Wachholder, K. & Altenburger, H. (1926). Beiträge zur Physiologie der willkürlichen Bewegung. IX. Mitteilung. Fortlaufende Hin- und Herbewegungen. *Pflügers Archiv für die gesamte Physiologie*, **214**, 625-641.

Wann, J., Nimmo-Smith, I, & Wing, A.M. (1988). Relation between velocity and curvature in movement: Equivalence and divergence between a power law and a minimum-jerk model. *Journal of Experimental Psychology: Human Perception and Performance*, **14**, 622-637.

Warren, N. & Clark, B. (1937). Blocking in mental and motor tasks during a 65-hour vigil. *Journal of Experimental Psychology*, **21**, 97-105.

Wing, A.M. (1980). The long and short of timing in response sequences. In: G.E. Stelmach & J. Requin (Eds.), *Tutorials in motor behavior*. Amsterdam: North-Holland, pp. 469-486.

Wing, A.M. & Kristofferson, A.B. (1973a). The timing of interresponse intervals. *Perception & Psychophysics*, **13**, 455-460.

Wing, A.M. & Kristofferson, A.B. (1973b). Response delays and the timing of discrete motor responses. *Perception & Psychophysics*, **14**, 5-12.

Wulf, G. & Schmidt, R.A. (1988). Variability in practice: Facilitation in retention and transfer through schema formation or context effects? *Journal of Motor Behavior*, **20**, 133-149.

Young, D.E. & Schmidt,R.A. (1990). Units of motor behavior: Modifications with practice and feedback. In: M. Jeannerod (Ed.), *Attention and performance XIII*. Hillsdale,N.J.: Erlbaum, pp. 763-795.

Zelaznik, H.N., Schmidt, R.A., & Gielen, C.C.A.M. (1986). Kinematic properties of rapid aimed hand movements. *Journal of Motor Behavior*, **18**, 353-372

ON MODELLING

P. VIVIANI
Department of Psychobiology, FAPSE
University of Geneva, Switzerland, and
Department of Cognitive Science
Fondazione San Raffaele, Milan, Italy

1. Introduction

There is little agreement - both across, and within disciplines - on what it takes to have a model. In psychology and, to some extent, in economics it is not uncommon to call models any doctrine with a debatable amount of empirical support. Understandably, predictions in these domains are not a major concern. Instead, the ability to derive empirical predictions from first principles has a paramount importance in physics where the concept of model is almost coextensive with that of post-Galilean natural science. Virtually all systems of explanation in physics, from Newtonian mechanics to string theory, fall under such an all-encompassing definition. Of course, also in physics predictive power is limited in more than one way. For instance, even well established models - such as quantum mechanics and relativity theory - are not parameter-free as one would like them to be. Moreover, we have to deal with the disturbing fact that sometimes first principles clash almost unbearably with our intuitions. Despite all this, however, there is a general consensus that there can be no natural science without models. Indeed, some physicists hold that modelling is not merely necessary but, ultimately, also sufficient for understanding natural phenomena, and that empirical evidence should be relegated to a confirmatory role.

Such an attitude is not at all acceptable in brain science. While no serious neuroscientist would question that brain functions are natural phenomena to be investigated within the epistemologic framework of the natural sciences, the complexity of brain processes exceeds - or at least appears to us as exceeding - that of all other natural processes to a degree that discourages, for the moment, any attempt to proceed deductively from first principles. Of course, some basic neural mechanisms - e.g. those involved in the generation and transmission of action potentials - are fairly well understood, and can be modelled with reasonable accuracy with the tools of chemistry and molecular biology.

F. Lacquaniti and P. Viviani (eds.), Neural Bases of Motor Behaviour, 315–334.

Moreover, certain forms of behavior - e.g. the stabilization of gaze by visual and vestibular inputs - can be modelled at the functional level, without taking into account all the subtleties of the intervening biochemical processes. Altogether, however, the picture in brain science is much more patchy and confused than in physics. Often, two models are supposed to describe the same mechanisms at different explanatory levels, and yet there are no criteria to guarantee that one description is compatible with the other. In fact, it is not even uncommon that different authors proceed to model one function from diverging principles. Moreover, for most of the truly complex brain functions, such as memory or speech processing, we quite simply lack the empirical and theoretical bases for generating comprehensive models.

A noticeable tension exists within the neuroscience community as a consequence of this state of affairs. On the one hand, it is obvious that, unlike physics, a lot of useful brain science is actually being produced without relaying upon formalized models. On the other hand, it is equally obvious that the enormous quantity of experimental findings that neuroscientists are accumulating in certain domains do not always arrange themselves into comprehensible patterns. Some argue that further progress is impossible unless we adopt guiding principles for accommodating these findings. Others are more skeptical, and contemplate with no anguish the possibility there is no more synthetic description of the functioning of the brain than the collection of empirical facts pertaining to the brain - just as there is no more synthetic description of the genome than the sequence of its base pairs. Ultimately, one cannot even rule out conclusively worst-case scenarios in which it would turn out that the logic underlying the main cerebral functions is too complex for us to grasp. At any rate, pending the resolution of these quasi-philosophical questions, we can expect the present tension to persist for a while, with an ever-increasing number of researchers trying to cast their findings in some more or less formal framework, and an ever-increasing demand for guidelines for this difficult exercise.

So far, no one has been able to provide a universal language for modelling the working of the brain, as Newton did for modelling the motion of the masses. Past attempts in this direction - such as cybernetics, information theory, or computational theory - have all failed in one way or another. Also the more modest task of laying out heuristic principles for choosing the formal framework adequate to a given class of functions seems to be beyond our present reach. What can be done, and may actually be of some use at this point, is to assess critically the limitations and difficulties that one encounters in modelling specific brain functions. This chapter is a contribution to such an enterprise. The illustrative examples are drawn from my own field of expertise, sensori-motor coordination, but I hope that some remarks turn out to be relevant to other domains as well.

2. What is a model that we can talk about ?

Models in brain science constitute a fairly fuzzy set. Even within restricted areas, what is referred to as a model varies from case to case, ranging across the dimensions of complexity, scope, and formalism. Establishing a taxonomy within such a zoo would be a scholastic exercise of very limited interest. For the purpose of this chapter, it is sufficient (and necessary) to identify to a first approximation the subset of models that I had in mind in formulating my remarks. Along the scope dimension, the target is somewhere in the middle range, excluding on the one hand things like models of the cell membrane, and, on the other hand, over-arching constructions such as those that neuropsychologists propose sometimes to account for language production. As for the degree of formalism, the target is instead close the high end of the range, where one typically finds models capable of testable, quantitative predictions. Finally, my remarks will mostly concern models that are committed to some amount of biological plausibility.

These criteria reflect a personal bias that I might just as well spell out as part and parcel of the views advocated here. The importance of highly detailed and quantitative models of elementary components of the brain can hardly be overrated. By applying rigorous formalisms to functionally autonomous subsystems, these models provide us with some amount of epistemologic encapsulation, a commodity that turn out to be quite useful to capture the bigger picture. Their unqualified success - think of the Hodgins-Huxley equations - is the very reason why I will not have much to say about them. At the other end of the spectrum, models that are too encompassing often pay the price of being underspecified to an extent that makes it difficult to evaluate their import. Unlike hypotheses, models that one can talk about should not contain too many degrees of freedom. Finally, a word about biological plausibility. Arguably, this is the most controversial criterion for assessing a model, one that is often invoked by neuroscientists to dismiss aryificial neural network, and equally often cited by the advocates of these models as evidence of the narrowmindedness of the neuroscientists. I will not take side on this particular controversy, at least not the ground that PDP models are biologically implausible. Rather, in selecting the target for my remarks, I implicitly endorsed the view that would-be modellists should remain conceptually sympathetic to biological reality. This is reflected in the fact that not much thought is given to models with strong cybernetic and AI flavor.

3. The problem of underdetermination

Despite the recent development of techniques for charting nervous activity using measures of cerebral blood flow, brain studies in humans still rely heavily on behavioral data in normal and pathological individuals. Overt behavior that one can measure in humans results from the coordinated action of many distinct functional modules, each one of which, if we had access to the brain, could be described at many different levels, from biochemistry to neural dynamics. Thus, whether or not modelling is involved, going beyond the mere description of a performance entails a serious problem of indeterminacy. Even in the case of complex behaviors described with high accuracy, the number of features in the data that must be accounted for in terms of underlying mechanisms, is intrinsically limited. On the other hand, the number of free parameters in the description of a module (i.e. the number of degrees of freedom in the problem) increases from the more global levels of description to the more analytic ones. In the abstract, a conservative strategy of system identification would prescribe to consider only the levels that afford a true synthesis, i.e. those that entail fewer parameters than features. Oftentimes, however, already at the most global level, one has more parameters than needed. This is the case, for example, of the mechanisms that preside over the control of hand movements; even the most oversimplified accounts of how this control is exerted cannot avoid the indeterminacy implicit in the fact that there are more articulations in the arm than it is strictly necessary for orienting the hand in space.

Even more frequent are those situations in which a conservative strategy can actually be pursued, but the result turns out to be uninteresting from the conceptual point of view. A typical instance is provided by the study of human movements with strong oscillatory components. Almost isochronic oscillations are present in a wide variety of gestures ranging from simple spontaneous finger tapping, to periodic tracking movements, to complex actions such as handwriting. The motor strategies involved in all these situations are likely to be quite different; yet, the are often modelled within the same formal framework, namely the theory of linear system. Specifically, the control mechanism is represented by a linear integro-differential system converting an input into an output. The input is either an external stimulus (e.g. the movement of visual targets, an external force applied to the body, an acceleration etc), or an internally generated blueprint of the intended action; in all cases the output is an overt motor behavior. The model is then tailored to the specific control mechanism under examination by identifying the order of the differential system and evaluating its parameters.

The success of this modelling technique can be assessed by the accuracy with which it predicts the result of changing some of the operating conditions, for instance, the variations of movement amplitude as a function of frequency in the case of forearm oscillations. A vast literature documents the most common departures of the predictions from the data (non-linearities, frequency-independent time delays, etc), as well as the various generalizations proposed to improve this modelling technique. For the purpose of our argument, we concentrate instead on those (not so infrequent) instances in which the predictions of the model are very accurate. The import of such a favorable outcome depends critically on the nature of the situation under examination. This is illustrated by considering two somewhat extreme cases.

Suppose first that one is interested in describing the dynamic control of a relatively simple biomechanical system such as the forearm. One way of doing so is to ask to a subject to hold the forearm at given angle with respect to the arm, and then apply a sudden pulse of force to the wrist. The time course of the response to the external perturbation - a dumped sequence of oscillations - depends on factors that are fairly well understood, i.e. the passive visco-elastic properties of the muscles, and the dynamic properties of the reflexes. In particular, it is known that the stretch reflex acts as a negative feed-back on the length of the muscles. Together, these factors result into a dynamic response similar to that of a highly viscous second-order linear system. Thus, the analysis of the oscillations permits one to estimate certain parameters of the system, such as the gain of the reflex loop, that are relevant to the dynamic control problem.

Consider now the case in which one is interested in the mechanisms that make visuo-manual tracking possible. Pursuing with the hand a visual target moving along a spatial trajectory is a challenging task for the nervous system, requiring the coordinated intervention of many diverse cognitive, perceptual and motor components. Yet, experiments have shown that the complexity of the transformation leading from the stimulus to the resulting action is not always reflected in the relationship between measurable quantities. For instance, in one study subjects were requested to track with a stylus two-dimensional movements of a light target. The stimuli were in fact extemporaneous scribblings recorded in other subjects. Thus, both their path and velocity were realistic instances of the kinematic parameters of biological movements. The results demonstrated that under these circumstances the path of the pursuit is a fairly faithful reproduction of that of the target, the variability from one trial to the next being also quite small. Moreover, as long as the velocity of the target covaries with the curvature of the path in a way that is compatible with the intrinsic properties of the motor control system (a condition that was obviously fulfilled in these experiments), also the kinematics of the pursuing stylus reproduces with good approximation that of the visual stimulus.

This sounds encouraging. Had the performance been erratic or overly complex, it would have been difficult to draw accurate predictions from a reasonably small set of assumptions. On second thought, however, it becomes evident that too much accuracy in the response sets equally severe limitations to what one can understand about the mechanisms involved in visuo-manual tracking. In the limit, if the response were identical to the stimulus, not much could be said concerning these mechanisms other than ruling out a number of hypotheses that, contrary to evidence, would predict one or another type of tracking error. It can be interesting therefore to consider how deep down below the purely descriptive level one is allowed to go with an amount of error and variability comparable to the one documented in the typical study illustrated above.

Keeping in mind that one should not introduce more concepts and hypotheses that are really necessary, we modeled visuo-motor tracking by a very simple mass-spring-dashpot system in which the visual stimulus plays the role of the applied force, and the resulting displacement is supposed to represent the tracking response. In order to account for anticipatory actions, we also postulated that the processing of the input leads temporally its application to the system by a constant amount. The predictions of this naive model proved to be surprisingly accurate. With a judicious choice of the parameters the model was able to predict acceleration traces that were almost indistinguishable from the experimental ones (notice that comparing accelerations affords a very severe test of a model for biological movements, and that even better results would have been obtained by the customary comparison of displacements.)

These results are satisfactory from the aesthetical point of view; yet, the very fact that the predictions are so accurate points to a sobering conclusion. When dealing with behavioral data, even in the presence of a rather complex pattern of results, it is not particularly difficult to devise an ad hoc set of hypotheses that accounts for the data so well that no further refinement is warranted. Then, what we can hope to understand through modelling depends mainly on the extent to which the various elements of the model can be put in correspondence with functional modules of biological relevance. The example shows that this extent can be quite limited indeed. The mass-spring system outlined before is so rudimentary that it does not even discriminate visual from motor processing; moreover, it encapsulates the ability to anticipate the future course of the target - a highly complex cognitive process - into a single parameter. Clearly, talking about biological relevance in this case would be preposterous. Most studies of one-dimensional tracking in adults are even less satisfactory from the modelling point of view than the one that I have illustrated.

One might object that drawing pessimistic conclusions from just one type of experiment is unwarranted, and that modelling would yield more meaningful insights if only one took the pain of devising more congenial paradigms. Examples can be cited to show that there is some truth in this rejoinder. In particular, it has been argued that the developmental approach to the study of sensori-motor skills is more promising than the study of adults because the age-dependent evolution toward adult performance exposes some of the complexities in the underlying mechanisms that become opaque when the skill is fully mastered.

Inasmuch as these complexities elude the predictions of shallow, system-oriented models that prove adequate for adults, one is justified to consider more articulated models. The trouble, however, is that what is peculiar about the behavior of developing children is often related to the complex structure of the representational system in charge of transforming perceptual stimulation into an appropriate input for the motor system, and that we simply do not have yet the appropriate conceptual and formal tools for modelling these cognitive processes (see later). In other words, while it is true that we can complexify the experimental situation to a point that behavior is no longer predicted by simple models, it is also true that in doing so we often end up dealing with processes for which we do not possess yet an appropriate theory.

4. The problem of the relevant information

Virtually all models, including those based on some metaphorical representation of actual neural networks, feature a finite number of distinct processing modules. Irrespective of whether they fall into distinct functional categories, these modules are connected by "wires" or channels through which "outputs" are fed into "inputs". The question to be discussed here is the nature of what is supposed to be transmitted, in other words the nature of what drives the action of the modules. This is no moot question: even a quick glance at the literature shows that, while the general context is often sufficient to infer what is supposed to be transferred between modules, the lack of explicit stipulations is equally often responsible for conceptual ambiguities. As I see it, the most serious danger comes from conflating the concepts of *signal* and *information* that should instead be distinguished carefully. A signal is any temporal or spatial modulation of some form of energy. Signals are ubiquitous in the nervous system, from the organs that provide the interface between the environment and us, to the modules that, further inward, are responsible for the more solipsistic phenomena of the mind: memory, will, consciousness. In materialistic brain science, nothing can happen without some signal being sent and received. However, it is important to acknowledge the variety of modes with which the receiving module responds to the incoming signal.

At one end of the range of possibilities we find the responses that are nomologically determined by the laws of physics, like the depolarization of the cell membrane by presynaptic excitatory action potentials. These responses are not different conceptually from the motion that, by Newton's Second Law, forces impress to masses. At the other end, are the responses determined primarily by knowledge and beliefs about the state of the world that the nervous system happens to entertain when the signal is received. The announcement that my flight is being boarded from gate 18 triggers a complex locomotor pattern which depends on the incoming acoustic energy in a way that has nothing to do with the laws of acoustics or with the physiology of the acoustic nerve. The same walking reaction would be triggered by displaying the message visually. What counts in this example is my belief that flight announcements are trustworthy, the knowledge of the location of the gate, the estimate of the time necessary to reach this location, and so on.

All sorts of intermediate possibilities exist between these two extremes situations. We would take an important step toward the formal representation of brain functions if we were able to parametrize this continuum of possibilities, i.e. if we were able, in each case, to evaluate the extent to which the changes induced by the signal result from some arbitrary pattern in the spatio-temporal distribution of the signal energy, and the extent to which they result instead from the laws of physics. Unfortunately, producing such a measure of the symbolic content of arbitrary signals is still well beyond our present capabilities; only a much more modest programme has been suggested and partially carried out under the general banner of "Information Processing Theory". The theory is inspired to the well-known measure of information developed by Weaver and Shannon in the late forties to deal with the technical problem of transmitting electronic messages through noisy channels. Leaving aside the mathematical developments, the framework for defining this measure is as follows.

Let us assume two separate entities, a sender S and a receiver R. Upon making an observation on what's going on where he is, S transmits this acquired knowledge to R by dispatching a message through a communication channel. At the other end of the channel the incoming message causes R to believe that a certain event has been observed by S. The communication is supposed to go on indefinitely, successive observations being made independently at a constant rate. For simplicity, we stipulate that the set of observations that the sender may be induced to communicate is finite, and so is the set of beliefs that the receiver may be induced to entertain. Since any physical channel is bound to introduce noise in the communication, the two sets do not necessarily coincide, so that the mapping of the sender's world onto the receiver's world is in general a one-to-many association.

According to Weaver and Shannon, in order to evaluate the effectiveness of the communication channel, three pieces of knowledge must be available: 1) the probability with which each possible event is observed by S, 2) the probability that R attaches a priori (i.e. before the communication is activated) to each one of the events that he construes as possible, and 3) the joint probability that each observation by S is matched with any one of its possible interpretations by R.

These probabilities were used by Weaver and Shannon to define three corresponding *entropies*, or global measures of uncertainty. The first, H(S), is relative to the uncertainty about which event is actually observed by S. Therefore H(S) depends on factors outside the scope of the communication channel. The second entropy, H(R), is relative to the expectations by R concerning the set of possible interpretations that might emerge in the course of the communication. Finally, the conditional entropy H(S,R) defines the statistical properties of the linking channel; it depends on the probability that messages are either transmitted correctly, or equivocated with any other among those that R expects. Then, the average quantity of information transmitted by the channel, computed over a suitably long sequences of messages, is defined as $I(S,R) = H(S) + H(R) - H(S,R)$. The intimate relation between the notions of information and expectation becomes particularly clear when one considers the ideal case in which the transmission of the messages is faultless. Then, it can be shown that $H(S,R) = H(S) = H(R)$ and therefore that $I(S,R) = H(R)$: the quantity of information transmitted is equal to the a priori uncertainty of R which, of course, the message eliminates completely. When the communication channel is noisy, equivocation is possible, and some residual uncertainty remains in the receiver. Even in this more realistic case, however, the quantity of information in a message is proportional to the decrease in uncertainty that is experienced by the receiver after decoding the message.

The framework outlined above was developed (and proved tremendously useful) to optimize the efficiency of artificial communication channels. In employing such terms as "interpretation" ,"expectation", and "belief" that make no sense for man-made devices, we adopted the optimistic attitude of those who advocate the possibility of extending the domain of validity of this framework to communication within the central nervous system. Indeed, there exist in the brain identifiable structures that may be given the role of sending and receiving stations in a communication system. Also, the trains of action potentials along the axons and dendrites connecting these structures are somewhat reminiscent of the coded messages exchanged between electronic devices. However, as far as the transmission of information is concerned, biological and artificial systems differ in many important respects.

In electronic communication, information theory is used to design the codes that map events into messages at the sender station, and messages back into events at the receiver station. The noise characteristics of a channel sets an upper bound to the average quantity of information that may be carried. It can be shown that this upper bound is reached when both entropies H(S) and H(R) are maximum, that is, when all the messages have the same probability. Since the frequency of occurrence of the events to be communicated is an independent datum that cannot be modified, the only way to maximize the average amount of information transmitted is to adapt the code to the properties of the channel. By associating to each event an appropriate string of elementary characters (code), one can in fact make the probability distribution of the *characters* as uniform as possible, irrespective of the probability of the events. In other words, information theory provides the conceptual framework for choosing the best way of coding messages, and it is useful only insofar as the choice is essentially free, as it is the case for all symbolic systems of communication.

The situation is altogether different when we attempt to apply the same framework to the analysis of the nervous system. Here, as I mentioned before, one cannot even be sure of the extent to which the communication is truly symbolic (i.e. based on arbitrary codes). Furthermore, even if we had evidence that a certain symbolic coding is being used, we would still be confronted with another more serious difficulty. While in both artificial and biological communication systems the noise in the channel can be specified, at least in principle, by a matrix of conditional probabilities (i.e. the probability that a given message is received, given that another message is sent), the status of the sender and receiver entropies is quite different in the two cases. In man-made systems, the set of possible messages, both at the sender and receiver end, is defined in advance, and so are the (objective) corresponding probabilities. This is not in general the case for the brain.

First of all, it is far from obvious in this case what counts as a symbol and what counts as a message. Consider, for instance the the visual system. Clearly, the retina is not equipped for segmenting the optic flow into a preset array of visual primitives; to the extent that such primitives exist, they must emerge somewhere along the pathway from the eyes to the higher visual centers where object identity and location becomes available to consciousness. However, nothing that we are aware of in the course of the integration of the afferent pattern of stimulation marks the boundary between elementary and complex units. Moreover, it is also difficult to determine a priori the set of messages expected at the receiving end of the channel because what counts as a relevant visual event depends on the context, and is bound to change continuously as a function of previous messages. Finally, even assuming that a priori likelihoods can be defined over the set of expected messages, these likelihoods would be essentially subjective estimates. As such, they too are bound to be influenced by the ongoing flow of messages. For these reasons, in all but the most trivial conditions, there seems to be little hope of determining the relevant entropies, and therefore of evaluating the quantity of information transmitted in a biological system.

Since the conditions under which information theory is relevant to brain mechanisms cannot be investigated within the theory itself, since we cannot produce reliable quantitative estimates of the information being exchanged between brain structures, and since, at any rate, the biological coding scheme - whatever it might be - cannot be adapted, the agenda for a theory of biological information processing must be altogether different and much more modest than the one followed so successfully in the case man-made systems.

A line of thinking advertised by several authors consists of *assuming* a symbolic coding, and use *qualitatively* the concept of information for framing questions of biological relevance. An example, drawn from a well known textbook, outlines this approach and, at the same time, illustrates the kind of fallacious conclusions that one may be driven to when the analogy between man-made and biological systems is pushed too far. Consider the following points:

1) Our retina contains many millions receptors, each coding independently the intensity and wavelength of the light at its location.

2) The optic nerve contains far less fibers, each integrating the outputs of many contiguous retinal cells.

3) Whenever two or more signals are mixed, the result contains less information that the sum of the information carried by the individual signals.

4) If were able to grasp the functional significance of such compression of the incoming information, we would stand a better chance of understanding the emergence of visual primitives.

Now, points 1 and 2 state uncontroversial facts. Point 4 is logically valid; provided that point 3) holds, it is also factually true. Point 3 sounds plausible. It is instead plainly false, at least as long as one preserves the standard definition of information. This is shown by a simple counter-example. Let A and B be two independent channels, each carrying an infinitely long random sequence of just two symbols, 1 and 2 in A, and 2 and 3 in B. It is easy to show that if the probability of occurrence of each symbols is 0.5 in both channels, A and B together carry an average quantity of information of 2 bits per symbol. Suppose now that A and B are funnelled into a third channel C which sums the *numerical values* of their respective contribution. Then C carries a sequence that contains three different symbols: 3, 4, and 5 with probability 0.25, 0.5 and 0.25, respectively. Does the convergence of A and B results into a compression of the information ? It all depends on what we mean by the term information. On the one side, it is indeed the case that, by looking at A and B, we could discriminate the case in which A is carrying 1 and B 3, from the case in which both A and B are carrying 2. This is no longer possible after their convergence into C which in both cases would carry the symbol 4. However, the average quantity of information à la Shannon carried by C is still 2 bits per symbol: if the receiver computes expectations according to the formal rules of probability (something that cannot be taken for granted in a biological system, even assuming that the system cares about expectations) nothing has been lost or compressed. It certainly legitimate to assume that the relevant interpretation as far as neurobiology of vision is concerned is the first one. However, this amounts to make precisely the kind of hypotheses about visual processing that, according to the line of argument outlined above, should have been suggested and motivated by the theory of biological information processing. If we stick instead to the original Shannonian definition, point 4) above is a *non sequitur*, and the strategy suggested by this application of information theory leads nowhere.

More generally, it cannot be ruled out that in the future we will be able to devise other definitions of information that conport more naturally with biological communication, nor that point 4 above may hold true under one such definition. However, the motivation for a new definition must be found in some independent insight in what the correct rules of biological communication are. Till then, it should be feared that the loose usage of the term information, with the inevitable Shannonian overtone that it carries nowadays, may induce precisely the error exemplified above of expecting an insight from a formal definition.

Why is such a delusion so much widespread ? The reason, I think, is to be found in the failure to appreciate one profound difference that sets apart the logic of biological phenomena from that of physics. Leaving aside some recent rather wild cosmological speculations, such as Coleman's hypothesis that the our universe is just a minute, self-contained part of a much bigger picture, and that other parts may be ruled by other laws, it is a fact that the basic laws of physics are not historical accidents that could have been otherwise. The laws of physics are immanent to the universe and are valid at any time and place. As such, they constitute by necessity a coherent whole held together by the equally immanent rules of logic and mathematics. By contrast, the vast majority of physical phenomena, and life is one of them, are indeed historical accidents contingent upon circumstances that have no necessity. After all, even the human brain as we know it may have not existed if the saures had not been brought to extinction by the very accidental impact of the Earth with a huge asteroid.

Of course, everything that happens, including brain processes, is ultimately compatible with quantum mechanics and gravitation laws. Moreover, the distinction between necessary and accidental is not clear-cut, as demonstrated by the existence of phenomena that lie somewhere at the borderline between these two categories. For instance, chaotic systems are entirely deterministic. Yet, no matter how accurately one specifies the initial conditions, in the long run their state always diverge from our predictions and ends up being a contingent accident (on the other hand, no matter how remote in the future is the state that we want to predict, is it always possible to determine the accuracy with which one has to specify the initial conditions in order for prediction to come true). In spite of all this, however, it should be obvious that in the realm of necessity formal definitions and concepts may be expected to have not only heuristic power, but in fact substantive import. In the history of physics there are plenty of examples of completely abstract constructions - think of group theory - that provided the key to the solution of major problems decades, or even centuries after they had been developed, often for the sheer pleasure of it. The effectiveness of mathematics in explaining the physical world may appear perplexing and even unreasonable to some, but cannot be denied. Nothing similar can be expected in the realm of historical accidents, certainly not for most complex of all accidents that we are know of, the human brain.

True, mathematics proved equally effective in explaining some basic electrophysiological facts, such as the generation and propagation of action potentials, that partake the necessary nature of non-biological physical events. But when it comes to processes that are more remote from physical necessity, it is dubious that mathematics - or, for that matter, other kind of formal reasoning - may do more than casting in rigorous form intuitions already derived from empirical data. If so, it is not surprising that borrowing from communication theory the concept of information processing, for which physiological data provide no independent support, does not prove particularly useful, and may actually be misleading.

5. Representing representations

We move now to consider the concepts of *representation* and *intention* that have come to play an increasingly important role in cognitive science and that, unlike information processing, find some plausible epistemologic motivations within the domain of neuroscience. In fact, we do not see how many brain performances could ever be possible unless there are neurological structures that embody the faculty of representing internally the ever-changing interplay between the body and the environment. Nor can we neglect the fact that brains also have the faculty of entertaining desires, beliefs and expectations which, in many cases, provide the sole basis for action. Cognitive psychologists and philosophers are investigating actively the consequences for the structure of the mind of the fact that brains entertain intentional states. In particular, the philosopher and psychologist Fodor has argued that intention and representation are intimately connected because being in an intentional state amounts to being in some computational relation to an internal representation.

If Fodor is right it makes no sense to deal separately with the two faculties. The majority of the neuroscientists, however, do not share his view. There is in fact some confidence that within a reasonable time the computational and algorithmic aspects of the representational faculties will be understood well enough to warrant the final experimental assault to their neural bases. By contrast, no one that I am aware of (not Fodor, at any rate) would at this point express the same confidence concerning desire and belief states. Thus, because of the tutorial nature of this chapter, I will restrict my remark to the problem of what it means to represent something in the brain, and, more specifically, of what that means in the context of sensori-motor coordination.

Surely, the fact noted before that we are unable to figure out how to do away with representations does not guarantee by itself that they are real: experiments disposed of the ether at a time when everyone deemed this concept absolutely indispensable for making sense of electromagnetism. Yet, the case for representations does not rest only on logical ground. There are pathological states in which behavior is altered in a way that suggests a selective impairment of our ability to generate internal models of the world. Moreover, experiments in which the lawful relationship of the organism with the environment is altered in a systematic way, yield results that are congruent with the hypothesis of a *bona fide* representational faculty. Thus, it does not seem a priori inappropriate to investigate this faculty as an identifiable functional module, and even to attempt to capture its salient features in formal models.

When we design a device that must have the ability to represent quantities and relationships, we adopt generally the *algorithmic approach* that is customary in the natural sciences. Specifically, we establish smooth mappings between what has to be represented and the field of real numbers, so as to preserve the structures of proximity implicit in the object of the representation. Under this approach, the act of representing becomes coextensive with the selection of a set of rules for manipulating what has to be represented. Thus, in saying that the paths to be followed by a mobile robot are represented in polar coordinates, we are actually saying that any identifiable aspect of these trajectories - say, the total length, or the number of left-ward turns - can be extracted by following certain recipes that are appropriate to that specific system of representation. More generally, the possible systems of representation for physical laws or other functional relationships are characterized by the rules that one has to apply to move from one system to an other, while preserving invariant the meaning of these laws and relationships.

The point can be exemplified by considering the simplest representational problem, that of locating a point in a plane. Since the plane is a two-dimensional manifold, in any case we have to establish a one-to-one mapping between positions and *pairs* of real numbers. Things become specific when we define the mapping rules. Suppose the rules are the following: take any two straight lines and move along one of them, away from the point of intersection, by an amount corresponding (in some scale) to the first number in the pair; then, move from there along a straight path parallel to the second line by an amount corresponding to the second number. At the end of the procedure one lands in a position of the plane that is uniquely defined by the pair. Of course, the final point is the representation of the pair in the familiar oblique cartesian system of coordinates. Had we opted for a polar representation, the rules would have been different. In all cases, however, the recipes for manipulating or analyzing any geometrical figure or any relationship between two quantities, no matter how complicated, derive unambiguously from the simple rules for locating a point.

The algorithmic approach provides the most general and flexible solution to the representational problem. However, there is another quite different possibility - sometimes referred to as the "*look-up table*" or "*tagging*" solution - that has considerable biological relevance. A trivial example illustrates well the gist of the idea. In old-times hotels, incoming phone calls were dispatched by plugging a jack into the jack receiver with the tag that identified uniquely the desired room. The use of numerical tags, and the fact that jack receivers were generally arranged in rows corresponding to floors, and in columns corresponding to the location of the room along the corridor, were only a matter of practicality; both choices were completely irrelevant to the logic of the system. With respect to the algorithmic solution, the "tagging" method presents the major disadvantage of not providing the representation with a metric structure: there was no way of knowing how far away from each other two rooms were by just looking at their tags. However, this very limitation actually turns out to be an advantage in certain circumstances; for instance, the "tagging" method is robust vis à vis changes in the outlay of the switch-board. Furthermore, tags and look-up tables are ideally suited to solve problems that require fast, stereotyped answers. This is probably the reason why they survived natural selection during the evolution of the nervous system. A well-studied example is the wiping reaction in the frog. A local cause of irritation anywhere on the back of the animal is sensed by the cutaneous receptors and triggers an immediate wiping movement of the hind-limb aimed at the location of the disturbance. The reaction is very accurate, independently of the initial position of the limb. Neurophysiologists have demonstrated that the wiping reaction is subserved by a nervous network that maps sensory representations into motor representations in a manner that is reminiscent of the phone-call dispatching method in the example above. I will mention momentarily another biological implementation of this mode of representation.

To avoid generalities, the ensuing discussion of the systems of representation focusses on just one representative sensori-motor performance for which the issue has been raised and debated in several recent studies. Let us consider the situation, quite common in real life, where one has to grasp an object whose nature and location is provided by vision. The task involves at least two systems of representation. Up to, and including the primary visual cortex, spatial locations are coded topically, that is with the tagging method of representation. At the other end of the sensori-motor chain, moving the hand to a spatial location and pre-shaping the fingers as a function of the size and shape of the object, involve active rotations at the articular joints. Posture and location of the hand throughout the movement are defined geometrically by a set of angles. Therefore, there must be a way for brain to bring into register visual and motor information. One hypothesis is that hand location is represented in an intrinsic geometric system of reference based on angles, and that the correspondence with visual information is established trough the intermediary of a third extrinsic (body-centered) representation, a spherical system of coordinates centered on the shoulder. An alternative hypothesis is that hand position is represented directly in an extrinsic (earth-bound) cartesian system of coordinates, and that articular angles are not relevant variables in the plan for the reaching movement.

Choosing between the two hypotheses is a sensible scientific problem. Yet, if the sensori-motor system were perfect, there would be no way of providing the correct answer: a characteristic consequence of the degree-of-freedom problem mentioned before is that in this ideal case the two hypotheses would yield the same predictions. By outlining one way in which the question can in principle be addressed in real, less-then-perfect systems, I intend to emphasize the equivalence between representations and manipulating rules. For sake of clarity, let us consider a streamlined version of the grasping task in which one is simply requested to rotate in a plane arm and forearm around their respective joints till the extended index finger touches a target point. Imperfections in sensori-motor coordination manifest themselves as small, random deviations of the controlled quantities from the desired values. If the first hypothesis is correct, at least one component of this "noise" affects directly the joint angles at the elbow and shoulder. The deviation of the finger position is then a *derived* random variable whose statistical properties depend on the geometry of the intervening limbs. By contrast, if the correct hypothesis is that end-point position is represented directly in cartesian coordinates, the role of primary and derived variable is interchanged. By adding to this scheme a few technical stipulations on the characteristics of the noise, it is then possible to derive contrasting predictions on the probability distribution of the deviations. Discriminating between the two hypothesis hinges critically on the non-symmetric form of the computations necessary to derive end-point position from articular angles, and viceversa.

6. On representations and modelling

All I said so far about representations was independent of modelling. In this section I will discuss what it takes to represent a representation with a formal model. The electric scheme of a radio is said to be a model of the real thing insofar as a one-to-one correspondence is established between functions of the radio and portions of the scheme. Leaving aside details such as the size and shape of the knobs, the brand of the transformers, and, to some extent, even the specification of how certain functions are implemented, the scheme provides someone *who knows everything about electromagnetism and radio theory* with a sufficient representation for understanding how that particular model of receiver works. We will return in a moment on the significance of the fact that, without previous knowledge of radio theory, no one could possibly make out the functioning of the scheme. Instead, the point that has to be emphasized here is that models of radios are easy to produce because broadcasting and receiving *do not require representing anything*. The whole operation is in fact quite mechanical, and could be implemented (very inefficiently) by replacing hertzian waves with sound waves, tuning circuits with Helmoltz resonators etc.

Much more complex instead are those cases where the device to be modelled, whether artificial or biological, includes a component designed to store and process symbolic messages coming from without. Computers offer a paradigmatic example of the kind of complexity that we have to deal with in these cases. As a piece of hardware, a computer can be modelled every bit as accurately as a radio by an appropriate electric scheme. Moreover, at least in the case of classical von Neumann architectures, Turing has even provided us with the most general and complete functional model for all possible computers. The analogy, however, stops here. Even by assuming complete knowledge of how computers work, there is a fundamental limitation in what one can understand of the function performed by a *running* computer. While the basic function of the radio is inherent to its design (our ignorance concerns just the specific piece of music received when the radio is turned on), the range of functions that can be implemented on a computer is virtually unlimited. Saying that a computer computes does not take us too far; only the knowledge of the software loaded on any one occasion may, under very favorable circumstances, enable us to understand whether the machine is supposed to play chess, solve a differential equation, or edit a text. Of course, we would still be unable to say which chess game will be played, which equation will be solved, or which text will be edited.

Computer programs are internal representations of a plan conceived outside the machine, symbolic messages fed into the machine from without. Thus, in order to capture the function of a given computer at a given time, the hardware scheme must be complemented with an detailed scheme of the program that is about to be run. In other words, a model must represent not only the hardware features, but also the format and content of the internal representations that define the function of the machine. This is actually feasible: a computer can be programmed to simulate how another computer, with a completely different architecture, would execute a given program. However, as anyone who has ever programmed knows only too well, programs are extremely opaque. In all but a few favorable cases, there is just no way of figuring out what a program is supposed to do by simply looking at the sequence of instructions. The reason is twofold. First of all, there is always more than one strategy for mapping a given computational problem into an algorithm. Secondly, any algorithm can be implemented by very many different sequences of instructions that need not to share any identifiable common feature. Thus, modelling a computer that is running entails either copying blindly whatever software is being run, or having access to the plan that originated that software. In neither cases it would seem appropriate to say that such an exercise *per se* helps us to understand the working of the machine.

The implications of this negative conclusion vis à vis the modelling of sensori-motor mechanisms may not seem obvious. After all, nobody seriously believes that these mechanisms - or, for that matter, any other brain function - are subserved by lists of instructions that even remotely resemble computer programs. However, I am here proposing that the line of argument developed above remains valid when the classical concept of program is replaced by the more general notion of representation as it is directly relevant to brain functions. In other words, I suggest that modelling a brain function that is predicated upon the availability of an internal representation incurs the same epistemologic limitations that apply to the computer metaphor, and this quite independently of the fact that computers are poor metaphors of the brain. The argument is based on the algorithmic interpretation of the representational faculty which identifies a representation with the rules that one has to follow for manipulating its content. These rules may be embodied either by *explicit symbolic coding* or *implicitly*, by exploiting the morphogenetic power of complex dynamic systems.

As an example of the first solution, think of the way a sequence of holes on a piece of plastic is used to represent a piece of music on CD's. Surely, the holes as such do not correspond to rules; yet, the *complete system of sound reproduction* with respect to which the sequences of holes are specified, does include an explicit set of rules which are an integral part of the representation. The output of a CD can be simulated accurately by computing the result of applying these prescriptions to the input sequence. This solution has to be contrasted with the implicit method that entails no calculations. Plunge a 3D frame of connected rods into soaped water. When the frame is pulled out, one gets all sorts of beautiful patterns of intersecting water membranes. The patterns represent local solutions to an energy minimization problem that admits many solutions, each uniquely specified by the geometry of the frame. Thus, in order to represent collectively these patterns, it is far easier to prescribe the geometry of the frame than either to describe the equations of the membrane surfaces, or to prescribe the rules for computing the minimum solution (a computation, by the way, that is ferociously complex).

As I said before, it is not known yet which representational faculty in the brain, if any, is supported by explicit symbolic coding, and which, if any, by implicit coding. To the extent that our concern is to make a model of such a faculty, the result is not different. In one case, we must know in advance the underlying symbolic rules. In the other, we must produce a functional replica of the real system, i.e. a hardware, *and* the set of laws that govern nomologically its behavior, such that the faculty under consideration emerges implicitly from their interaction. Either way, the amount of knowledge that is supposed to be available for generating the model is not significantly less than the amount of knowledge that anyone interested in the real system would be happy with. Once again, we reach the same conclusion that we arrived at in the case of computers, namely that modelling fails to provide a synthetic insight into systems that are complex enough to support a representational faculty.

The reasoning above may still seem too general and far removed from the daily worries of the researchers. An example drawn from ongoing research shows that this is not so. Let us consider the task of handwriting a sentence that is being dictated. The acoustic input has to be parsed, analyzed, and ultimately translated into a pointer indicating in real time the entries of graphemic memory that need to be translated into motor commands. A major question in handwriting studies is the format in which lexical items are stored in graphemic memory. The motor plan responsible for the appropriate finger movements depends entirely on such format.

Most tentative answers offered so far to this question fall rather nicely into either one of the two categories of rule-coding strategies outlined above. One line of theorizing, fittingly identified by such labels as "motor drum" or "motor program", holds that graphemic items are coded as explicit sets of strokes complemented by equally explicit sequencing rules. In order to accommodate for the empirical fact that we can produce similar graphic outputs in different sizes, and even with different limbs, the term stroke is generally understood in a symbolic, abstract sense, to denote any set of muscular synergies that, when called upon, result into a specific shape of the pen trajectory.

Another approach that is often associated with the so-called "mass-spring" hypothesis, but has many precursors, takes stock in the fact that to a first approximation muscles are capable of behaving as dumped springs. Moreover, since the bones and flash to which muscles are attached have mass, many body segments react to either external or internal forces as second order dynamic systems, exhibiting a spontaneous propensity to oscillate more or less sinusoidally. This is well in keeping with the fact that both x and y components of handwriting traces can be approximated by periodic oscillations with variable amplitude and phase. The idea then is that handwriting is generated by modulating the parameters of a natural resonant mode of the motor control system. According to the "mass-spring" model, what is specific to each letter and has to be represented in graphemic memory, is a sequence of amplitude and phase settings that bear no explicit connection with the shape of resulting movement.

Finally, a similar lack of isomorphism between internal code and motor output is found in the so-called optimization theories, of which the Minimum-Jerk model is a prototypical example. The model is based on the hypothesis that all planar trajectories comply with a cost-minimization principle involving the mean quadratic value of the derivative of the acceleration. One can then show mathematically that point-to-point trajectories are uniquely defined by their boundary conditions, i.e. by position, velocity and acceleration at the starting and end point. Furthermore, simply by adding one further constraint to the path, in the form of a mandatory via-point, it is possible to generate a large variety of complex trajectories. Just as the principle of minimum energy associates to each frame a set of configurations of water membranes, so does the Minimum-Jerk principle associate to each trajectory a set of boundary conditions that result into that trajectory. In such a framework the internal representation need not to contain all the geometrical and topological details of the motor output (slant, presence of loops etc.). Even a complex movement, such as the writing a word, can be coded in a compact form, simply as a sequence of boundary conditions, and this is all that needs to be represented.

How do these contrasting approaches fare with respect to the epistemologic limitation that I claimed to be inherent to each and every representational faculty ? Well, just as we expected them to fare - that is, not very well -, and precisely for the reasons indicated above. Models based on the motor program metaphor cannot predict much unless one feeds them with *ad hoc* lists of strokes. Worst, the nature of the coded strokes, rather than being derived in a principled manner from some a priori or data-driven intuition, is generally left to the idiosyncratic preferences of the model's proponent. Moreover, most popular choices, such as duration or amplitude of the bursts of the efferent motor commands, do not result into discriminating predictions. Minimum-cost models do slightly better on this score insofar as they are specific about what has to be represented. Yet, when it comes to explain how the list of boundary conditions is set up, we can do no better than adopting the same "just-so" attitude, nor do we have the faintest idea of which aspect of motor control could be intrinsically responsible for minimizing the cost function (notice instead that in the soaped water example we know the principles responsible for the uniqueness of the optimal solution). As far as we can tell, the entire motor system may collectively be responsible for optimal behavior. After all, the only advantage for us of having a motor control system that is many orders of magnitude more complex than that of insects may be that our movements are far less jerky. Finally, mass-spring models account naturally for the oscillatory character of handwriting but do no better than the other competitors in explaining how modulating commands are represented.

7. Where do rules come from ?

I have argued that, in order for a model to have heuristic value, it must take into account not only the hardware of the target module, but also whatever "software" (if any) may be relevant to its function. In this section I would like to dwell upon two related questions: 1) Which part of the software component comes from without and which from within ? 2) What should a model be like to make allowance for the distinction made in 1) ? Having stressed repeatedly that the term "software" is not used here in the strict technical sense, let me argue that both issues can be addressed by restating more formally some of meanings that this term may be given in the context of the brain.

The most intuitive of these meanings is as follows: *any recipe for action supplied by the environment and held in temporary readiness*. Suppose, for instance that a typist is instructed to transcribe a list of spoken words. It must be the case that, before each sequence of key-presses, all the necessary finger movements, in their correct order, be represented in motor memory. Upon execution of a word sequence, this information is likely to be erased and replaced by the information concerning the next word. This is probably the closest approximation to a simple segment of machine code in a computer, and can be dealt with quite naturally by making allowance in the model for some sort of rewritable buffer. We can then proceed to complete the model by adding a permanent matrix of interconnections indicating which finger is associated to which key, and a pointer that feed this matrix by scanning the buffer sequentially.

Next, consider a slightly more complex definition: *any set of rules that dictate how to manipulate recipes of the first kind, should the need arise*. Two novel features characterize this new definition. To begin with, the domain of application of the rule is itself a set of prescriptions; secondly, the rules may still come from without, but, unlike recipes, they may just as well be inherent to the functioning of the module. In the former case a second erasable buffer must be postulated, at a higher hierarchical level; in the latter one has to envisage a more permanent way of fixating the rule, which may or may not be symbolic.

To exemplify the implications for modelling, let us consider again the typing task. Typists can be trained so that words uttered with a high pitch are typed normally, whereas words uttered with a low pitch are typed with just one finger. That people can actually do crazy things like this, switching on command between two very different ways of dealing with the same sequence of letters proves the reality of second-order rules coming from without. The model must be modified accordingly, by adding a switch under the control of a second-order buffer that directs the output of the first buffer either to the standard interconnection matrix, or directly to the control of the one finger to be used.

What about inherent rules ? Experiments show that the total duration of a typing sequence can be modulated by both external and idiosyncratic factors. When this happens, the changes in duration are reflected ratiomorphically on all intervals between keystrokes: a rule is at work that dictates the shortening or the lengthening of individual intervals. The rule is not enforced explicitly during training, and applies only when the typist is a true professional performing at his usual fast rhythm. We have here the example of a prescription that emerges slowing as a result of sensori-motor training. Moreover, since the controlled variable is time, it makes sense to assume that the ratiomorphic rule is in fact the overt manifestation of the manner in which motor recipes are represented in skilled typists. This has far-reaching consequences for modelling, for one can no longer get away from it by simply adding another memory buffer. It is the entire architecture of the model that has to be reformulated, making provision not only for the steady-state behavior of the module, but also for its observed natural proclivity to evolve in time toward that steady state.

A final comment before leaving this point. It may not seem appropriate to stretch the meaning of "software" to include behavioral regularities - such as the ratiomorphic rule - that are likely to result from some intrinsic properties of the representational system. The validity of the definition depends on the extent to which the mode of representation in a given medium is nomologically determined by the properties of the medium. Is there a chain of inferences leading from the known properties of the sensori-motor system to the inescapable conclusion that sequences of keystrokes must be coded so as to exhibit the constancy of interval ratios ? At the moment the answer is negative. In fact it may remain negative even in the future, when neurophysiologists will have charted the sensori-motor system in all its minute details, because that particular mode of coding may be the consequence of the way the system *in toto* is organized. In other words, the mode of coding may be an historical accident rather a necessity. If so, the only way for a model to capture the gist of the coding mechanism - short of making a replica of the entire real system - would be to postulate a set functional primitives, and arrange them according to rules so that they predict the observed phenomena. So long as models adopt such a functionalist-computational approach, it would be justified to refer to second-order rules as a "software" component of the module. We cannot adjudge yet whether or not phenomena like the ratiomorphic rule will ultimately be explained nomologically. There is, however, a third, and even broader construal of the term "software" for which, I shall argue, such an explanation must be definitely ruled out. Consider the following definition: *any set of rules that specifies how recipes and/or second-order rules for one domain interact with recipes and/or second-order rules for another domain.* First, let us go back again to the typing task and show that such third-order rule do in fact exist. Contrary to popular belief, proficiency in typing does not imply isochrony. It can be shown that systematic differences exist among the intervals between keystrokes. The sequence of intervals corresponding to words (i.e. legitimate lexical items in a language) is specific both to the word and to the typist. They represent a highly identifiable motor prosody, just like vocal prosody. Two points need to be stressed. Firstly, even words that share as many as seven consecutive letters out of eight (e.g. *daughter* and *laughter*) have a different motor prosody. Secondly, non-words that conform to the phonological rules of the typist's native language are typed as fast and as fluently as words, and exhibit their own characteristic motor prosody. In contrast, when they have to deal with non-words that are phonologically unacceptable, even professional typists fall back into a letter-by-letter, hesitating performance. Therefore, there is strong evidence that the temporal patterning of the motor recipes for typing is partly under the influence of the phonological rules that normally legislate the processing of speech signals. We do not know yet how and why this cross-domain interaction occurs, but there is convincing evidence that it is itself legislated by rules.

Cross-domain interactions are not a quirk of typing. In particular, there is a growing body of data indicating that visual perception of dynamic events is influenced in a principled manner by implicit procedural knowledge about the working of the motor control system, as well as by factual knowledge about the physical laws. How are we going to deal with these third-order rules? Clearly, the phonological rules adopted by a typist derive from early interactions with his particular linguistic environment; albeit compatible with the brain design, these rules cannot be traced back to any design feature. We have a clear instance of a mechanism that affects profoundly overt motor behavior and cannot be accounted for, even in principle, by hypotheses concerning the brain. Thus, by necessity, a model of the typist must include a "software" component that specifies his linguistic environment, and the *ad hoc* rules that legislate the interaction between his phonological and motor competencies (there is indeed no reason to believe that the same rules apply to all linguistic environments).

We have seen that there are sensori-motor mechanisms whose working can be explained only by taking into account a "software" component resulting from the interaction with the environment. This situation is not unique to sensori-motor coordination. In fact, the same conclusion has been reached independently, in an even stronger form, by considering the process whereby grammatical competence is progressively fixated through the interaction with the linguistic environment. To conclude this section, we consider the question of how this interaction can be simulated in a model. Given that the brute force approach, i.e. plugging by hand the necessary "software" in the model, is unsatisfactory from the epistemologic point of view, can we devise another solution ? Neural network models have been widely advertised as the right answer. Indeed, there are several learning schemes for modifying progressively the strength of the connections between processing units on a trial-and-error basis. An array of units that is completely unstructured before learning can be forced to evolve toward a final steady-state that displays all sorts of production and performance competencies. At this point it should be obvious that if my previous comments are not completely off the mark, neural network model do not in fact deliver what they promise.

A recurring theme in this chapter has been that models are intrinsically limited in their capacity to simulate represented knowledge. The difficulty stems primarily from the fact that brains have neither been designed nor are necessary natural kinds. Therefore, in order to simulate a performance predicated upon the availability of internally represented knowledge, we must concoct with our own intellect appropriate schemes of action and rules that may seem plausible to us and still bear no resemblance whatsoever with the actual schemes and rules. The fact that the learning schemes for neural network models are not based on preconceived schemes and rules gives the illusion of dodging the difficulty. However, consider the fact that to train one such network, say, to form the past tense of english verbs, or to recognize handwritten letters, you have to have a God-like figure that knows in advance the correct solutions. Thus, discriminal information is pumped into the network in a way that has nothing to do with how children learn. The problem is not so much with the difference in time scale, nor even with the fact that trial-and-error is not how we go about learning. The problem is with the very assumption that is presented as the major advantage of the networks, namely that the initial state of the system is *tabula rasa*.

Neural networks can be trained to perform all sorts of useful things, but it is amazing indeed that so many PDP enthusiasts still refuse to accept that both phylogenetic and ontogenetic history is as important for understanding a performing brain as is the analysis of its steady-state performance.

8. What then ?

I have highlighted a few things that models cannot do, and a few things that they can do without advancing our understanding of real brains. Not all limitations, of course, are clear cut, nor are they equally severe. However, there seem to be a common thread that connects them all. Whether the problem is with the meaning of information, with the degrees of freedom, or with the rules and their representation, we are ultimately confronted with a conceptual divide similar to the one evoked by Fodor in his book "Modularity of the Mind" between peripheral and central processing modules. It is a fact that most of the models that really proved useful for understanding brain functions concern the transduction from one form of energy modulation impinging on the sensory system into another form of energy modulation, in other words, functions that according to Fodor are implemented by informationally encapsulated, hardwired, inborn modules. The best examples are to be found in the study of the oculomotor system - particularly in the domain of vestibular and optokinetic compensatory functions - of the early stages of visual processing, and of the articulatory processes. These are all cases in which the underlying neurological modules do not (or at least it is not obvious that they do) have representational faculties, and in which the transduction of energy is negotiated by principles akin to those of the natural sciences. Further progress will certainly be done in this direction. By making a case for the theoretical impossibility of ever being able to understand the logic of the central processor, Fodor went further down the road of pessimism than I wanted to go in this chapter. But, all things considered, to be pessimistic about the possibility of attaining such an understanding through models may boil down to saying the same thing in another way.

Acknowledgements. Partly supported by FNRS, Research Grant 31-32577

SUBJECT INDEX

GPSR Compliance

The European Union's (EU) General Product Safety Regulation (GPSR) is a set of rules that requires consumer products to be safe and our obligations to ensure this.

If you have any concerns about our products, you can contact us on ProductSafety@springernature.com

In case Publisher is established outside the EU, the EU authorized representative is:

Springer Nature Customer Service Center GmbH
Europaplatz 3
69115 Heidelberg, Germany

Batch number: 10809439

Printed by Printforce, the Netherlands